Arduino Microcontroller Processing for Everyone!

Third Edition

Synthesis Lectures on Digital Circuits and Systems

Editor
Mitchell A. Thornton, *Southern Methodist University*

The Synthesis Lectures on Digital Circuits and Systems series is comprised of 50- to 100-page books targeted for audience members with a wide-ranging background. The Lectures include topics that are of interest to students, professionals, and researchers in the area of design and analysis of digital circuits and systems. Each Lecture is self-contained and focuses on the background information required to understand the subject matter and practical case studies that illustrate applications. The format of a Lecture is structured such that each will be devoted to a specific topic in digital circuits and systems rather than a larger overview of several topics such as that found in a comprehensive handbook. The Lectures cover both well-established areas as well as newly developed or emerging material in digital circuits and systems design and analysis.

Arduino Microcontroller Processing for Everyone! Third Edition
Steven F. Barrett
2013

Boolean Differential Equations
Bernd Steinbach and Christian Posthoff
2013

Bad to the Bone: Crafting Electronic Systems with BeagleBone and BeagleBone Black
Steven F. Barrett and Jason Kridner
2013

Introduction to Noise-Resilient Computing
S.N. Yanushkevich, S. Kasai, G. Tangim, A.H. Tran, T. Mohamed, and V.P. Smerko
2013

Atmel AVR Microcontroller Primer: Programming and Interfacing, Second Edition
Steven F. Barrett and Daniel J. Pack
2012

Representation of Multiple-Valued Logic Functions
Radomir S. Stankovic, Jaakko T. Astola, and Claudio Moraga
2012

Progress in Applications of Boolean Functions
Tsutomu Sasao and Jon T. Butler
2009

Embedded Systems Design with the Atmel AVR Microcontroller: Part II
Steven F. Barrett
2009

Embedded Systems Design with the Atmel AVR Microcontroller: Part I
Steven F. Barrett
2009

Embedded Systems Interfacing for Engineers using the Freescale HCS08 Microcontroller II:
Digital and Analog Hardware Interfacing
Douglas H. Summerville
2009

Designing Asynchronous Circuits using NULL Convention Logic (NCL)
Scott C. Smith and JiaDi
2009

Embedded Systems Interfacing for Engineers using the Freescale HCS08 Microcontroller I:
Assembly Language Programming
Douglas H.Summerville
2009

Developing Embedded Software using DaVinci & OMAP Technology
B.I. (Raj) Pawate
2009

Mismatch and Noise in Modern IC Processes
Andrew Marshall
2009

Asynchronous Sequential Machine Design and Analysis: A Comprehensive Development of
the Design and Analysis of Clock-Independent State Machines and Systems
Richard F. Tinder
2009

An Introduction to Logic Circuit Testing
Parag K. Lala
2008

Pragmatic Power
William J. Eccles
2008

Multiple Valued Logic: Concepts and Representations
D. Michael Miller and Mitchell A. Thornton
2007

Finite State Machine Datapath Design, Optimization, and Implementation
Justin Davis and Robert Reese
2007

Atmel AVR Microcontroller Primer: Programming and Interfacing
Steven F. Barrett and Daniel J. Pack
2007

Pragmatic Logic
William J. Eccles
2007

PSpice for Filters and Transmission Lines
Paul Tobin
2007

PSpice for Digital Signal Processing
Paul Tobin
2007

PSpice for Analog Communications Engineering
Paul Tobin
2007

PSpice for Digital Communications Engineering
Paul Tobin
2007

PSpice for Circuit Theory and Electronic Devices
Paul Tobin
2007

Pragmatic Circuits: DC and Time Domain
William J. Eccles
2006

Pragmatic Circuits: Frequency Domain
William J. Eccles
2006

Pragmatic Circuits: Signals and Filters
William J. Eccles
2006

High-Speed Digital System Design
Justin Davis
2006

Introduction to Logic Synthesis using Verilog HDL
Robert B.Reese and Mitchell A.Thornton
2006

Microcontrollers Fundamentals for Engineers and Scientists
Steven F. Barrett and Daniel J. Pack
2006

Arduino Microcontroller Processing for Everyone! Third Edition
Steven F. Barrett

ISBN: 978-3-031-79863-4   paperback
ISBN: 978-3-031-79864-1   ebook

DOI 10.1007/978-3-031-79864-1

A Publication in the Springer series
SYNTHESIS LECTURES ON DIGITAL CIRCUITS AND SYSTEMS

Lecture #43
Series Editor: Mitchell A. Thornton, *Southern Methodist University*
Series ISSN
Synthesis Lectures on Digital Circuits and Systems
Print 1932-3166   Electronic 1932-3174

Arduino Microcontroller Processing for Everyone!

Third Edition

Steven F. Barrett
University of Wyoming, Laramie, WY

SYNTHESIS LECTURES ON DIGITAL CIRCUITS AND SYSTEMS #43

ABSTRACT

This book is about the Arduino microcontroller and the Arduino concept. The visionary Arduino team of Massimo Banzi, David Cuartielles, Tom Igoe, Gianluca Martino, and David Mellis launched a new innovation in microcontroller hardware in 2005, the concept of open source hardware. Their approach was to openly share details of microcontroller-based hardware design platforms to stimulate the sharing of ideas and promote innovation. This concept has been popular in the software world for many years. This book is intended for a wide variety of audiences including students of the fine arts, middle and senior high school students, engineering design students, and practicing scientists and engineers. To meet this wide audience, the book has been divided into sections to satisfy the need of each reader. The book contains many software and hardware examples to assist the reader in developing a wide variety of systems. The book covers two different Arduino products: the Arduino UNO R3 equipped with the Atmel ATmega328 and the Arduino Mega 2560 equipped with the Atmel ATmega2560. The third edition has been updated with the latest on these two processing boards, changes to the Arduino Development Environment and multiple extended examples.

KEYWORDS

Arduino microcontroller, Arduino UNO R3, Atmel microcontroller, Atmel AVR, ATmega328, Arduino Mega 2560, microcontroller interfacing, ATmega2560, embedded systems design

Contents

Preface

This book is about the Arduino microcontroller and the Arduino concept. The visionary Arduino team of Massimo Banzi, David Cuartielles, Tom Igoe, Gianluca Martino, and David Mellis launched a new innovation in microcontroller hardware in 2005, the concept of open source hardware. There approach was to openly share details of microcontroller–based hardware design platforms to stimulate the sharing of ideas and innovation. This concept has been popular in the software world for many years.

This book is written for a number of audiences. First, in keeping with the Arduino concept, the book is written for practitioners of the arts (design students, artists, photographers, etc.) who may need processing power in a project but do not have an in depth engineering background. Second, the book is written for middle school and senior high school students who may need processing power for a school or science fair project. Third, we write for engineering students who require processing power for their senior design project but do not have the background in microcontroller––based applications commonly taught in electrical and computer engineering curricula. Finally, the book provides practicing scientists and engineers an advanced treatment of the Atmel AVR microcontroller. The third edition has been updated with the latest on the Arduino UNO R3 and the Arduino Mega 2560 processors, changes to the Arduino Development Environment and multiple extended examples.

APPROACH OF THE BOOK

To encompass such a wide range of readers, we have divided the book into several portions to address the different readership. Chapters 1 through 2 are intended for novice microcontroller users. Chapter 1 provides a review of the Arduino concept, a description of the Arduino UNO R3 and the Arduino Mega 2560 development boards, and a brief review of the features of the UNO R3's host processor, the Atmel ATmega 328 microcontroller. We also provide a similar treatment of the Arduino Mega 2560 processor board. Chapter 2 provides an introduction to programming for the novice programmer. Chapter 2 also introduces the Arduino Development Environment and how to program sketches. It also serves as a good review for the seasoned developer.

Chapter 3 provides an introduction to embedded system design processes. It provides a systematic, step–by–step approach on how to design complex systems in a stress free manner.

Chapter 4 introduces the extremely important concept of the operating envelope for a microcontroller. The voltage and current electrical parameters for the HC CMOS based Atmel AVR line of microcontrollers is reviewed and applied to properly interface input and output devices to the Arduino UNO R3, the Arduino Mega 2560, the ATmega328 microcontroller, and the ATmega 2560 microcontroller.

Chapters 5 through 8 provide detailed engineering information on the ATmega328 and the AT mega 2560 microcontroller systems including analog-to-digital conversion, interrupts, timing features, and serial communications. These chapters are intended for engineering students and practicing engineers. However, novice microcontroller users will find the information readable and well supported with numerous examples using the Arduino Development Environment and C.

The final chapter, Chapter 9, provides example applications for a wide variety of skill levels.

ACKNOWLEDGMENTS

A number of people have made this book possible. I would like to thank Massimo Banzi of the Arduino design team for his support and encouragement in writing the book. I would also like to thank Joel Claypool of Morgan & Claypool Publishers who has supported a number of writing projects over the last several years. He also provided permission to include portions of background information on the Atmel line of AVR microcontrollers in this book from several of our previous projects. I would also like to thank Sparkfun Electronics of Boulder, Colorado; Atmel Incorporated; the Arduino team; and ImageCraft of Palo Alto, California for use of pictures and figures used within the book.

I would like to dedicate this book to my close friend Dr. Daniel Pack, Ph.D., P.E. Much of the writing is his from earlier Morgan & Claypool projects. In 2000, Daniel suggested that we might write a book together on microcontrollers. I had always wanted to write a book but I thought that's what other people did. With Daniel's encouragement we wrote that first book (and several more since then). Daniel is a good father, good son, good husband, brilliant engineer, a work ethic second to none, and a good friend. To you good friend I dedicate this book. I know that we will do many more together. Finally, I would like to thank my wife and best friend of many years, Cindy.

Laramie, Wyoming, August 2013

Steve Barrett

CHAPTER 1

Getting Started

Objectives: After reading this chapter, the reader should be able to the following:

- Describe the Arduino concept of open source hardware.

- Diagram the layout of the Arduino UNO R3 processor board.

- Name and describe the different features aboard the Arduino UNO R3 processor board.

- Discuss the features and functions of the ATmega328.

- Diagram the layout of the Arduino Mega 2560 processor board.

- Name and describe the different features aboard the Arduino Mega 2560 processor board.

- Discuss the features and functions of the ATmega2560.

- List alternate Arduino processing boards.

- Describe how to extend the hardware features of the Arduino processor.

- Download, configure, and successfully execute a test program using the Arduino Development Environment.

1.1 OVERVIEW

Welcome to the world of Arduino! The Arduino concept of open source hardware was developed by the visionary Arduino team of Massimo Banzi, David Cuartilles, Tom Igoe, Gianluca Martino, and David Mellis in Ivrea, Italy. The team's goal was to develop a line of easy–to–use microcontroller hardware and software such that processing power would be readily available to everyone.

In keeping with the Arduino concept, this book is intended for a wide variety of readers. For those wanting a quick exposure to an Arduino microcontroller board and its easy–to–use software, Chapters 1 and 2 are for you. If you need to tap into some of the other features of the processing power of the ATmega328 or the ATmega2560 host microcontrollers, Chapters 3 through 8 are for you. Chapter 9 contains a series of extended examples at a variety of skill levels.

In keeping with the Arduino open source spirit, you will find a plethora of hardware and software examples throughout the book. I hope you enjoy reading the book, and I also hope you will find it a useful resource in developing Arduino–based projects.

1.2   GETTING STARTED

This chapter is devoted to getting you quickly up and operating with an Arduino–based hardware platform. To get started using an Arduino–based processor, you will need the following hardware and software.

- an Arduino–based hardware processing platform,

- an Arduino compatible power supply, and

- the Arduno software.

Arduino hardware. Throughout the book, we will be using two different Arduino processing boards: the Arduino UNO R3 board and the Arduino Mega 2560 board. A starter kit for the Arduino UNO R3 is available from SparkFun Electronics of Boulder, CO. The starter kit is illustrated in Figure 1.1. The kit is equipped with the processing board, a USB cable to program the board from a host PC, a small breadboard to prototype external hardware, jumper wires, a flex sensor and many external components.

Figure 1.1: Arduino UNO R3 starter kit. (Used with permission from SparkFun Electronics (CC BY–NC–SA).)

Costs for the starter kit and processor boards are:

- Arduino UNO R3–approximately US$30

- Starter Kit for Arduino Flex (includes Arduino UNO R3) –approximately US$60

- Arduino Mega 2560–approximately US$60

Power supply. The Arduino processing boards may be powered from the USB port during project development. However, it is highly recommended that an external power supply be employed. This will allow developing projects beyond the limited current capability of the USB port. Arduino `www.arduino.cc` recommends a power supply from 7–12 VDC with a 2.1 mm center positive plug. A power supply of this type is readily available from a number of electronic parts supply companies. For example, the Jameco #133891 power supply is a 9 VDC model rated at 300 mA and equipped with a 2.1 mm center positive plug. It is available for under US$10.

Arduino software. You will also need the Arduino software called the Arduino Development Environment. It is available as a free download from the Arduino homepage (`www.arduino.cc`). In the next chapter, we will discuss how to program the Arduino processing boards using the Arduino Development Environment. It is essential that you download and get the software operating correctly before proceeding to the next chapter.

The Arduino homepage (`www.arduino.cc`)contains detailed instructions on how to download the software, load the USB drivers, and get a sample program operating on the Arduino processing boards. Due to limited space, these instructions will not be duplicated here. The reader is encouraged to visit the Arduino webpage and get the software up and operating at this time. Following the user–friendly instructions provided at the Arduino home page under the "Getting Started" tab, you will have the software downloaded and communicating with your Arduino processing board within minutes.

This completes a brief overview of the Arduino hardware and software. In the next section, we provide a more detailed overview of the hardware features of the Arduino UNO R3 processor board and the Atmel ATmega328 microcontroller.

1.3   ARDUINO UNO R3 PROCESSING BOARD

The Arduino UNO R3 processing board is illustrated in Figure 1.2. Working clockwise from the left, the board is equipped with a USB connector to allow programming the processor from a host PC. The board may also be programmed using In System Programming (ISP) techniques discussed later in the book. A 6–pin ISP programming connector is on the opposite side of the board from the USB connector.

The board is equipped with a USB–to–serial converter to allow compatibility between the host PC and the serial communications systems aboard the ATmega328 processor. The UNO R3 is also equipped with several small surface mount LEDs to indicate serial transmission (TX) and reception (RX) and an extra LED for project use. The header strip at the top of the board provides access for an analog reference signal, pulse width modulation (PWM) signals,digital input/output (I/O), and serial communications. The header strip at the bottom of the board provides analog inputs for the analog–to–digital (ADC) system and power supply terminals. Finally, the external power supply connector is provided at the bottom left corner of the board. The top and bottom header strips conveniently mate with an Arduino shield (to be discussed shortly) to extend the features of the host processor.

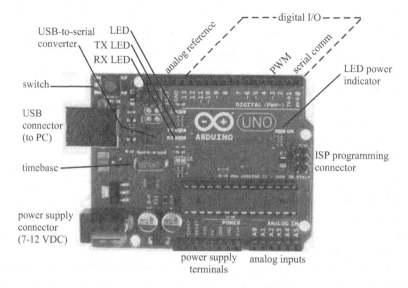

Figure 1.2: Arduino UNO R3 layout. (Figure adapted and used with permission of Arduino Team (CC BY–NC–SA)(www.arduino.cc).)

1.3.1   ARDUINO UNO R3 HOST PROCESSOR–THE ATMEGA328

The host processor for the Arduino UNO R3 is the Atmel Atmega328. The "328" is a 28 pin, 8–bit microcontroller. The architecture is based on the Reduced Instruction Set Computer (RISC) concept which allows the processor to complete 20 million instructions per second (MIPS) when operating at 20 MHz! The "328" is equipped with a wide variety of features as shown in Figure 1.3. The features may be conveniently categorized into the following systems:

- Memory system,

- Port system,

- Timer system,

- Analog–to–digital converter (ADC),

- Interrupt system,

- and Serial communications.

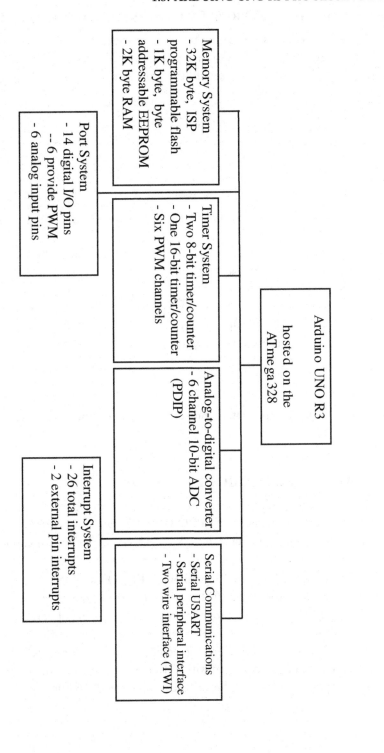

Figure 1.3: Arduino UNO R3 systems.

1.3.2 ARDUINO UNO R3/ATMEGA328 HARDWARE FEATURES

The Arduino UNO R3's processing power is provided by the ATmega328. The pin out diagram and block diagram for this processor are provided in Figures 1.4 and 1.5. In this section, we provide additional detail on the systems aboard the processor.

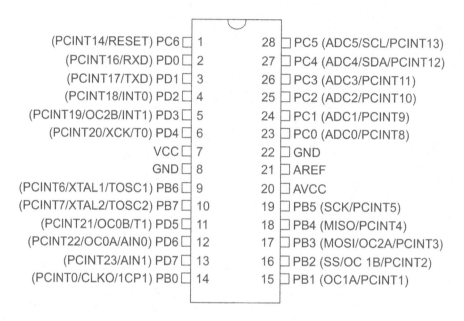

Figure 1.4: ATmega328 pin out. (Figure used with permission of Atmel, Incorporated.)

1.3.3 ATMEGA328 MEMORY

The ATmega328 is equipped with three main memory sections: flash electrically erasable programmable read only memory (EEPROM), static random access memory (SRAM), and byte–addressable EEPROM for data storage. We discuss each memory component in turn.

1.3.3.1 ATmega328 In–System Programmable Flash EEPROM

Bulk programmable flash EEPROM is used to store programs. It can be erased and programmed as a single unit. Also, should a program require a large table of constants, it may be included as a global variable within a program and programmed into flash EEPROM with the rest of the program. Flash EEPROM is nonvolatile meaning memory contents are retained when microcontroller power is lost. The ATmega328 is equipped with 32K bytes of onboard reprogrammable flash memory. This memory component is organized into 16K locations with 16 bits at each location.

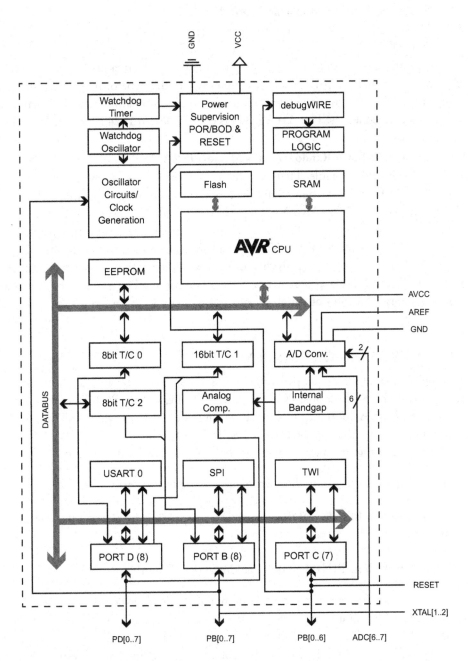

Figure 1.5: ATmega328 block diagram. (Figure used with permission of Atmel, Incorporated.)

1.3.3.2 ATmega328 Byte–Addressable EEPROM

Byte–addressable memory is used to permanently store and recall variables during program execution. It too is nonvolatile. It is especially useful for logging system malfunctions and fault data during program execution. It is also useful for storing data that must be retained during a power failure but might need to be changed periodically. Examples where this type of memory is used are found in applications to store system parameters, electronic lock combinations, and automatic garage door electronic unlock sequences. The ATmega328 is equipped with 1024 bytes of EEPROM.

1.3.3.3 ATmega328 Static Random Access Memory (SRAM)

Static RAM memory is volatile. That is, if the microcontroller loses power, the contents of SRAM memory are lost. It can be written to and read from during program execution. The ATmega328 is equipped with 2K bytes of SRAM. A small portion of the SRAM is set aside for the general purpose registers used by the processor and also for the input/output and peripheral subsystems aboard the microcontroller. A complete ATmega328 register listing and accompanying header file is provided in Appendices A and B, respectively. The header file is included with permission of ImageCraft (`www.imagecraft.com`.)During program execution, RAM is used to store global variables, support dynamic memory allocation of variables, and to provide a location for the stack (to be discussed later).

1.3.4   ATMEGA328 PORT SYSTEM

The Atmel ATmega328 is equipped with four, 8–bit general purpose, digital input/output (I/O) ports designated PORTB (8 bits, PORTB[7:0]), PORTC (7 bits, PORTC[6:0]), and PORTD (8 bits, PORTD[7:0]). All of these ports also have alternate functions which will be described later. In this section, we concentrate on the basic digital I/O port features.

As shown in Figure 1.6, each port has three registers associated with it

- Data Register PORTx –used to write output data to the port.

- Data Direction Register DDRx – used to set a specific port pin to either output (1) or input (0).

- Input Pin Address PINx – used to read input data from the port.

Figure 1.6(b) describes the settings required to configure a specific port pin to either input or output. If selected for input, the pin may be selected for either an input pin or to operate in the high impedance (Hi–Z) mode. In Hi–Z mode, the input appears as high impedance to a particular pin. If selected for output, the pin may be further configured for either logic low or logic high.

Port pins are usually configured at the beginning of a program for either input or output and their initial values are then set. Usually all eight pins for a given port are configured simultaneously. We discuss how to configure specific port pins and how to read/write to them in the next chapter.

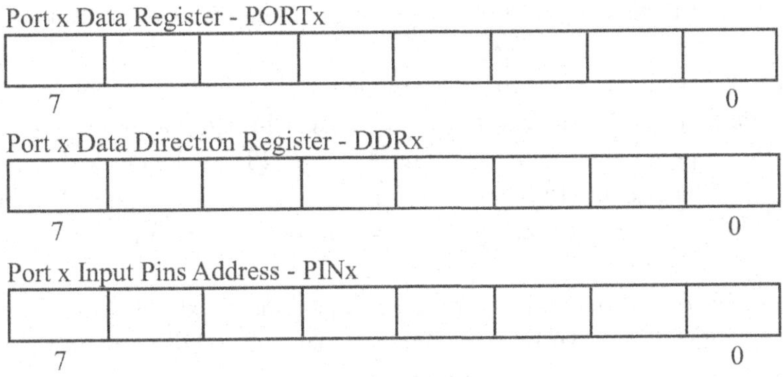

Port x Data Register - PORTx

7   0

Port x Data Direction Register - DDRx

7   0

Port x Input Pins Address - PINx

7   0

a) port associated registers

DDxn	PORTxn	I/O	Comment	Pullup
0	0	input	Tri-state (Hi-Z)	No
0	1	input	source current if externally pulled low	Yes
1	0	output	Output Low (Sink)	No
1	1	output	Output High (Source)	No

x: port designator (B, C, D)
n: pin designator (0 - 7)

b) port pin configuration

Figure 1.6: ATmega328 port configuration registers.

1.3.5   ATMEGA328 INTERNAL SYSTEMS

In this section, we provide a brief overview of the internal features of the ATmega328. It should be emphasized that these features are the internal systems contained within the confines of the microcontroller chip. These built–in features allow complex and sophisticated tasks to be accomplished by the microcontroller.

1.3.5.1 ATmega328 Time Base

The microcontroller is a complex synchronous state machine. It responds to program steps in a sequential manner as dictated by a user–written program. The microcontroller sequences through a predictable fetch–decode–execute sequence. Each unique assembly language program instruction issues a series of signals to control the microcontroller hardware to accomplish instruction related operations.

The speed at which a microcontroller sequences through these actions is controlled by a precise time base called the clock. The clock source is routed throughout the microcontroller to provide a time base for all peripheral subsystems. The ATmega328 may be clocked internally using a user–selectable resistor capacitor (RC) time base or it may be clocked externally. The RC internal time base is selected using programmable fuse bits. You may choose an internal fixed clock operating frequency of 1, 2, 4 or 8 MHz.

To provide for a wider range of frequency selections an external time source may be used. The external time sources, in order of increasing accuracy and stability, are an external RC network, a ceramic resonator, or a crystal oscillator. The system designer chooses the time base frequency and clock source device appropriate for the application at hand.

1.3.5.2 ATmega328 Timing Subsystem

The ATmega328 is equipped with a complement of timers which allows the user to generate a precision output signal, measure the characteristics (period, duty cycle, frequency) of an incoming digital signal, or count external events. Specifically, the ATmega328 is equipped with two 8–bit timer/counters and one 16–bit counter. We discuss the operation, programming, and application of the timing system later in the book.

1.3.5.3 Pulse Width Modulation Channels

A pulse width modulated or PWM signal is characterized by a fixed frequency and a varying duty cycle. Duty cycle is the percentage of time a repetitive signal is logic high during the signal period. It may be formally expressed as:

$$duty\ cycle[\%] = (on\ time/period) \times (100\%)$$

The ATmega328 is equipped with four pulse width modulation (PWM) channels. The PWM channels coupled with the flexibility of dividing the time base down to different PWM subsystem clock source frequencies allows the user to generate a wide variety of PWM signals: from relatively high frequency low duty cycle signals to relatively low frequency high duty cycle signals.

PWM signals are used in a wide variety of applications including controlling the position of a servo motor and controlling the speed of a DC motor. We discuss the operation, programming, and application of the PWM system later in the book.

1.3.5.4 ATmega328 Serial Communications

The ATmega328 is equipped with a host of different serial communication subsystems including the Universal Synchronous and Asynchronous Serial Receiver and Transmitter (USART), the serial peripheral interface (SPI), and the Two–wire Serial Interface. What all of these systems have in common is the serial transmission of data. In a serial communications transmission, serial data is sent a single bit at a time from transmitter to receiver.

ATmega328 Serial USART   The serial USART may be used for full duplex (two way) communication between a receiver and transmitter. This is accomplished by equipping the ATmega328 with independent hardware for the transmitter and receiver. The USART is typically used for asynchronous communication. That is, there is not a common clock between the transmitter and receiver to keep them synchronized with one another. To maintain synchronization between the transmitter and receiver, framing start and stop bits are used at the beginning and end of each data byte in a transmission sequence.

The ATmega328 USART is quite flexible. It has the capability to be set to a variety of data transmission rates known as the Baud (bits per second) rate. The USART may also be set for data bit widths of 5 to 9 bits with one or two stop bits. Furthermore, the ATmega328 is equipped with a hardware generated parity bit (even or odd) and parity check hardware at the receiver. A single parity bit allows for the detection of a single bit error within a byte of data. The USART may also be configured to operate in a synchronous mode. We discuss the operation, programming, and application of the USART later in the book.

ATmega328 Serial Peripheral Interface–SPI   The ATmega328 Serial Peripheral Interface (SPI) can also be used for two–way serial communication between a transmitter and a receiver. In the SPI system, the transmitter and receiver share a common clock source. This requires an additional clock line between the transmitter and receiver but allows for higher data transmission rates as compared to the USART.

The SPI may be viewed as a synchronous 16–bit shift register with an 8–bit half residing in the transmitter and the other 8–bit half residing in the receiver. The transmitter is designated the master since it is providing the synchronizing clock source between the transmitter and the receiver. The receiver is designated as the slave. We discuss the operation, programming, and application of the SPI later in the book.

ATmega328 Two–wire Serial Interface–TWI   The TWI subsystem allows the system designer to network a number of related devices (microcontrollers, transducers, displays, memory storage, etc.) together into a system using a two wire interconnecting scheme. The TWI allows a maximum of 128 devices to be connected together. Each device has its own unique address and may both transmit and receive over the two wire bus at frequencies up to 400 kHz. This allows the device to freely exchange information with other devices in the network within a small area. We discuss the TWI system later in the book.

1.3.5.5   ATmega328 Analog to Digital Converter–ADC

The ATmega328 is equipped with an eight channel analog to digital converter (ADC) subsystem. The ADC converts an analog signal from the outside world into a binary representation suitable for use by the microcontroller. The ATmega328 ADC has 10 bit resolution. This means that an analog voltage between 0 and 5 V will be encoded into one of 1024 binary representations between $(000)_{16}$ and $(3FF)_{16}$. This provides the ATmega328 with a voltage resolution of approximately 4.88 mV. We discuss the operation, programming, and application of the ADC later in the book.

1.3.5.6 ATmega328 Interrupts

The normal execution of a program follows a designated sequence of instructions. However, sometimes this normal sequence of events must be interrupted to respond to high priority faults and status both inside and outside the microcontroller. When these higher priority events occur, the microcontroller must temporarily suspend normal operation and execute event specific actions called an interrupt service routine. Once the higher priority event has been serviced, the microcontroller returns and continues processing the normal program.

The ATmega328 is equipped with a complement of 26 interrupt sources. Two of the interrupts are provided for external interrupt sources while the remaining interrupts support the efficient operation of peripheral subsystems aboard the microcontroller. We discuss the operation, programming, and application of the interrupt system later in the book.

1.3.6   ARDUINO UNO R3 OPEN SOURCE SCHEMATIC

The entire line of Arduino products is based on the visionary concept of open source hardware and software. That is, hardware and software developments are openly shared among users to stimulate new ideas and advance the Arduino concept. In keeping with the Arduino concept, the Arduino team openly shares the schematic of the Arduino UNO R3 processing board. Reference Figure 1.7.

1.4   ARDUINO MEGA 2560 PROCESSING BOARD

The Arduino Mega 2560 processing board is illustrated in Figure 1.8. Working clockwise from the left, the board is equipped with a USB connector to allow programming the processor from a host PC. The board may also be programmed using In System Programming (ISP) techniques discussed later in the book. A 6–pin ISP programming connector is on the opposite side of the board from the USB connector.

The board is equipped with a USB–to–serial converter to allow compatibility between the host PC and the serial communications systems aboard the ATmega2560 processor. The Mega 2560 is also equipped with several small surface mount LEDs to indicate serial transmission (TX) and reception (RX) and an extra LED for project use. The header strip at the top of the board provides access for pulse width modulation (PWM) signals and serial communications. The header strip at the right side of the board provides access to multiple digital input/output pins. The bottom of the board provides analog inputs for the analog–to–digital (ADC) system and power supply terminals. Finally, the external power supply connector is provided at the bottom left corner of the board. The header strips conveniently mate with an Arduino shield (to be discussed shortly) to extend the features of the host processor.

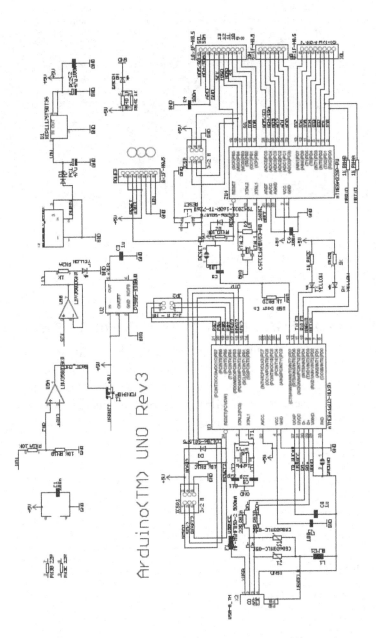

Figure 1.7: Arduino UNO R3 open source schematic. (Figure adapted and used with permission of the Arduino Team (CC BY–NC–SA) (www.arduino.cc).)

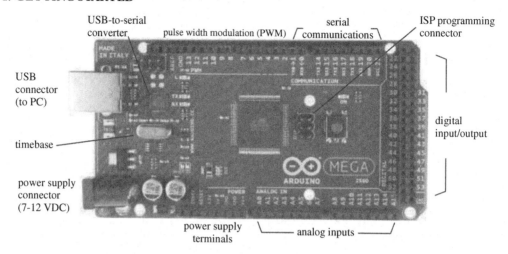

Figure 1.8: Arduino Mega2560 layout. (Figure adapted and used with permission of Arduino Team (CC BY–NC–SA) (www.arduino.cc).)

1.4.1   ARDUINO MEGA 2560 HOST PROCESSOR–THE ATMEGA2560

The host processor for the Arduino Mega 2560 is the Atmel Atmega2560. The "2560" is a 100 pin, surface mount 8–bit microcontroller. The architecture is based on the Reduced Instruction Set Computer (RISC) concept which allows the processor to complete 16 million instructions per second (MIPS) when operating at 16 MHz. The "2560" is equipped with a wide variety of features as shown in Figure 1.9. The features may be conveniently categorized into the following systems:

- Memory system,

- Port system,

- Timer system,

- Analog–to–digital converter (ADC),

- Interrupt system,

- and serial communications.

1.4.2   ARDUINO MEGA 2560 /ATMEGA2560 HARDWARE FEATURES

The Arduino Mega 2560's processing power is provided by the ATmega2560. The pin out diagram and block diagram for this processor are provided in Figures 1.10 and 1.11. In this section, we provide additional detail on the systems aboard the processor.

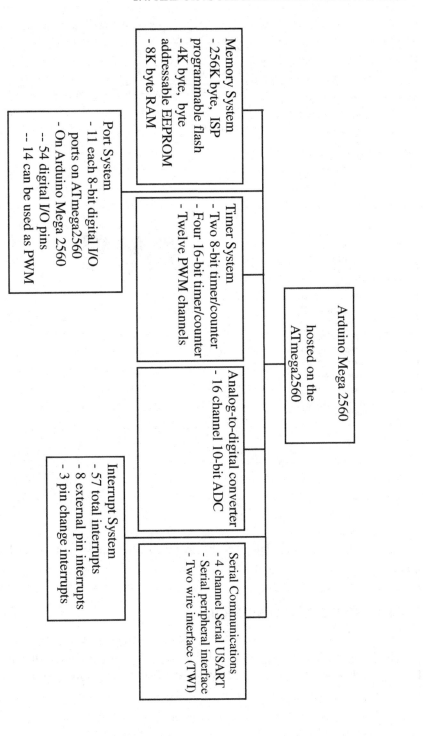

Figure 1.9: Arduino Mega 2560 systems.

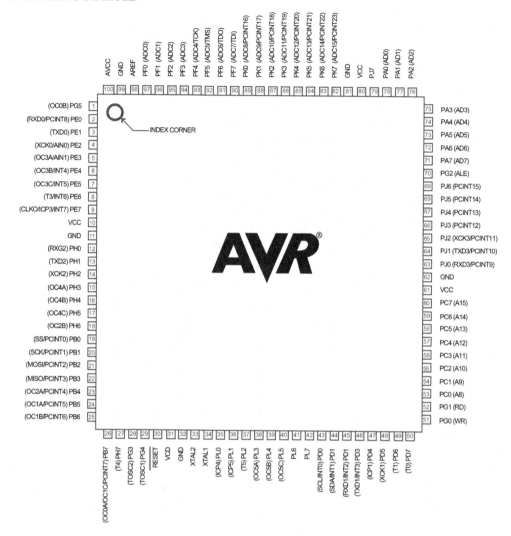

Figure 1.10: ATmega2560 pin out. (Figure used with permission of Atmel, Incorporated.)

1.4.3   ATMEGA2560 MEMORY

The ATmega2560 is equipped with three main memory sections: flash electrically erasable programmable read only memory (EEPROM), static random access memory (SRAM), and byte–addressable EEPROM for data storage. We discuss each memory component in turn.

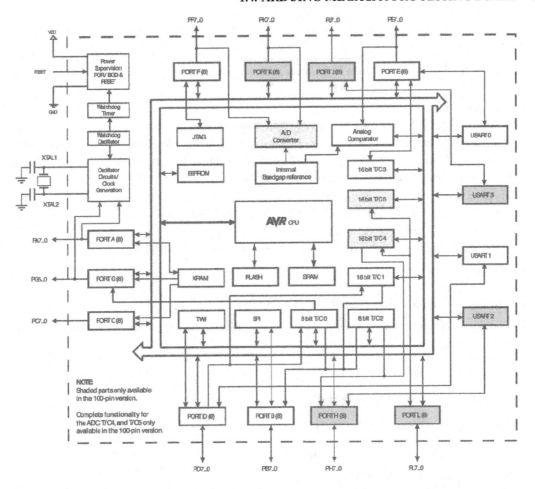

Figure 1.11: ATmega2560 block diagram. (Figure used with permission of Atmel, Incorporated.)

1.4.3.1 ATmega2560 In–System Programmable Flash EEPROM
Bulk programmable flash EEPROM is used to store programs. It can be erased and programmed as a single unit. Also, should a program require a large table of constants, it may be included as a global variable within a program and programmed into flash EEPROM with the rest of the program. Flash EEPROM is nonvolatile meaning memory contents are retained when microcontroller power is lost. The ATmega2560 is equipped with 256K bytes of onboard reprogrammable flash memory.

1.4.3.2 ATmega2560 Byte–Addressable EEPROM
Byte–addressable memory is used to permanently store and recall variables during program execution. It too is nonvolatile. It is especially useful for logging system malfunctions and fault data during

program execution. It is also useful for storing data that must be retained during a power failure but might need to be changed periodically. Examples where this type of memory is used are found in applications to store system parameters, electronic lock combinations, and automatic garage door electronic unlock sequences. The ATmega2560 is equipped with 4096 bytes of byte–addressable EEPROM.

1.4.3.3 ATmega2560 Static Random Access Memory (SRAM)

Static RAM memory is volatile. That is, if the microcontroller loses power, the contents of SRAM memory are lost. It can be written to and read from during program execution. The ATmega2560 is equipped with 8K bytes of SRAM. A small portion of the SRAM is set aside for the general purpose registers used by the processor and also for the input/output and peripheral subsystems aboard the microcontroller. A complete ATmega2560 register listing and accompanying header file is provided in Appendices C and D, respectively. During program execution, RAM is used to store global variables, support dynamic memory allocation of variables, and to provide a location for the stack (to be discussed later).

1.4.4   ATMEGA2560 PORT SYSTEM

The Atmel ATmega2560 is equipped with eleven, 8–bit general purpose, digital input/output (I/O) ports designated:

- PORTA (8 bits, PORTA[7:0])

- PORTB (8 bits, PORTB[7:0])

- PORTC (7 bits, PORTC[7:0])

- PORTD (8 bits, PORTD[7:0])

- PORTE (8 bits, PORTE[7:0])

- PORTF (8 bits, PORTF[7:0])

- PORTG (7 bits, PORTG[7:0])

- PORTH (8 bits, PORTH[7:0])

- PORTJ (8 bits, PORTJ[7:0])

- PORTK (8 bits, PORTK[7:0])

- PORTL (7 bits, PORTL[7:0])

All of these ports also have alternate functions which will be described later. In this section, we concentrate on the basic digital I/O port features.

As shown in Figure 1.6, each port has three registers associated with it

- Data Register PORTx –used to write output data to the port.

- Data Direction Register DDRx –used to set a specific port pin to either output (1) or input (0).

- Input Pin Address PINx –used to read input data from the port.

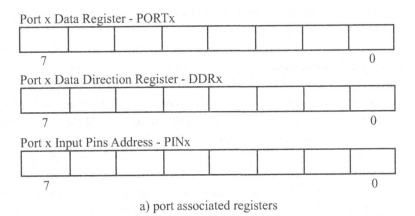

a) port associated registers

DDxn	PORTxn	I/O	Comment	Pullup
0	0	input	Tri-state (Hi-Z)	No
0	1	input	source current if externally pulled low	Yes
1	0	output	Output Low (Sink)	No
1	1	output	Output High (Source)	No

x: port designator (B, C, D)
n: pin designator (0 - 7)

b) port pin configuration

Figure 1.12: ATmega2560 port configuration registers.

Figure 1.6(b) describes the settings required to configure a specific port pin to either input or output. If selected for input, the pin may be selected for either an input pin or to operate in the high impedance (Hi–Z) mode. In Hi–Z mode, the input appears as high impedance to a particular pin. If selected for output, the pin may be further configured for either logic low or logic high.

Port pins are usually configured at the beginning of a program for either input or output and their initial values are then set. Usually all eight pins for a given port are configured simultaneously. We discuss how to configure port pins and how to read/write to them in the next chapter.

1.4.5   ATMEGA2560 INTERNAL SYSTEMS

In this section, we provide a brief overview of the internal features of the ATmega2560. It should be emphasized that these features are the internal systems contained within the confines of the micro-controller chip. These built–in features allow complex and sophisticated tasks to be accomplished by the microcontroller.

1.4.5.1   ATmega2560 Time Base

The microcontroller is a complex synchronous state machine. It responds to program steps in a sequential manner as dictated by a user–written program. The microcontroller sequences through a predictable fetch–decode–execute sequence. Each unique assembly language program instruction issues a series of signals to control the microcontroller hardware to accomplish instruction related operations.

The speed at which a microcontroller sequences through these actions is controlled by a precise time base called the clock. The clock source is routed throughout the microcontroller to provide a time base for all peripheral subsystems. The ATmega2560 may be clocked internally using a user––selectable resistor capacitor (RC) time base or it may be clocked externally. The RC internal time base is selected using programmable fuse bits. You may choose an internal fixed clock operating frequency of 128 kHz or 8 MHz. The clock frequency may be prescaled by a number of different clock division factors (1, 2, 4, etc.).

To provide for a wider range of frequency selections an external time source may be used. The external time sources, in order of increasing accuracy and stability, are an external RC network, a ceramic resonator, or a crystal oscillator. The system designer chooses the time base frequency and clock source device appropriate for the application at hand.

1.4.5.2   ATmega2560 Timing Subsystem

The ATmega2560 is equipped with a complement of timers which allows the user to generate a precision output signal, measure the characteristics (period, duty cycle, frequency) of an incoming digital signal, or count external events. Specifically, the ATmega2560 is equipped with two 8–bit timer/counters and four 16–bit timer/counters. We discuss the operation, programming, and application of the timing system later in the book.

1.4.5.3   Pulse Width Modulation Channels

A pulse width modulated or PWM signal is characterized by a fixed frequency and a varying duty cycle. Duty cycle is the percentage of time a repetitive signal is logic high during the signal period. It may be formally expressed as:

$$duty\ cycle[\%]\ =\ (on\ time/period)\ \times\ (100\%)$$

The ATmega2560 is equipped with four 8–bit pulse width modulation (PWM) channels and 12 PWM channels with programmable resolution. The PWM channels coupled with the flexibility of dividing the time base down to different PWM subsystem clock source frequencies allows the user to generate a wide variety of PWM signals: from relatively high frequency low duty cycle signals to relatively low frequency high duty cycle signals.

PWM signals are used in a wide variety of applications including controlling the position of a servo motor and controlling the speed of a DC motor. We discuss the operation, programming, and application of the PWM system later in the book.

1.4.5.4 ATmega2560 Serial Communications

The ATmega2560 is equipped with a host of different serial communication subsystems including the Universal Synchronous and Asynchronous Serial Receiver and Transmitter (USART), the serial peripheral interface (SPI), and the Two–wire Serial Interface. What all of these systems have in common is the serial transmission of data. In a serial communications transmission, serial data is sent a single bit at a time from transmitter to receiver.

ATmega2560 Serial USART   The serial USART may be used for full duplex (two way) communication between a receiver and transmitter. This is accomplished by equipping the ATmega2560 with independent hardware for the transmitter and receiver. The USART is typically used for asynchronous communication. That is, there is not a common clock between the transmitter and receiver to keep them synchronized with one another. To maintain synchronization between the transmitter and receiver, framing start and stop bits are used at the beginning and end of each data byte in a transmission sequence.

The ATmega2560 USART is quite flexible. It has the capability to be set to a variety of data transmission rates known as the Baud (bits per second) rate. The USART may also be set for data bit widths of 5 to 9 bits with one or two stop bits. Furthermore, the ATmega2560 is equipped with a hardware generated parity bit (even or odd) and parity check hardware at the receiver. A single parity bit allows for the detection of a single bit error within a byte of data. The USART may also be configured to operate in a synchronous mode. We discuss the operation, programming, and application of the USART later in the book.

ATmega2560 Serial Peripheral Interface–SPI   The ATmega2560 Serial Peripheral Interface (SPI) can also be used for two–way serial communication between a transmitter and a receiver. In the SPI system, the transmitter and receiver share a common clock source. This requires an additional clock line between the transmitter and receiver but allows for higher data transmission rates as compared to the USART.

The SPI may be viewed as a synchronous 16–bit shift register with an 8–bit half residing in the transmitter and the other 8–bit half residing in the receiver. The transmitter is designated the master since it is providing the synchronizing clock source between the transmitter and the receiver. The receiver is designated as the slave. We discuss the operation, programming, and application of the SPI later in the book.

ATmega2560 Two–wire Serial Interface–TWI   The TWI subsystem allows the system designer to network a number of related devices (microcontrollers, transducers, displays, memory storage, etc.) together into a system using a two wire interconnecting scheme. The TWI allows a maximum of 128 devices to be connected together. Each device has its own unique address and may both transmit and receive over the two wire bus at frequencies up to 400 kHz. This allows the device to freely exchange information with other devices in the network within a small area. We discuss the TWI system later in the book.

1.4.5.5 ATmega2560 Analog to Digital Converter–ADC

The ATmega2560 is equipped with a 16 channel analog to digital converter (ADC) subsystem. The ADC converts an analog signal from the outside world into a binary representation suitable for use by the microcontroller. The ATmega2560 ADC has 10 bit resolution. This means that an analog voltage between 0 and 5 V will be encoded into one of 1024 binary representations between $(000)_{16}$ and $(3FF)_{16}$. This provides the ATmega2560 with a voltage resolution of approximately 4.88 mV. We discuss the operation, programming, and application of the ADC later in the book.

1.4.5.6 ATmega2560 Interrupts

The normal execution of a program follows a designated sequence of instructions. However, sometimes this normal sequence of events must be interrupted to respond to high priority faults and status both inside and outside the microcontroller. When these higher priority events occur, the microcontroller must temporarily suspend normal operation and execute event specific actions called an interrupt service routine. Once the higher priority event has been serviced, the microcontroller returns and continues processing the normal program.

The ATmega2560 is equipped with a complement of 57 interrupt sources. Eight interrupts are provided for external interrupt sources. Also, the ATmega2560 is equipped with three pin change interrupts. The remaining interrupts support the efficient operation of peripheral subsystems aboard the microcontroller. We discuss the operation, programming, and application of the interrupt system later in the book.

1.5   ARDUINO MEGA 2560 OPEN SOURCE SCHEMATIC

The entire line of Arduino products is based on the visionary concept of open source hardware and software. That is, hardware and software developments are openly shared among users to stimulate new ideas and advance the Arduino concept. In keeping with the Arduino concept, the Arduino team openly shares the schematic of the Arduino Mega 2560 processing board. It is available for download at `www.arduino.cc`.

1.6   EXAMPLE: AUTONOMOUS MAZE NAVIGATING ROBOT

Let's see how the different Arduino processing boards would be used in an application. Graymark (www.graymarkint.com) manufacturers many low–cost, excellent robot platforms. In this example, we will modify the Blinky 602A robot to be controlled by the Arduino UNO R3.

The Blinky 602A kit contains the hardware and mechanical parts to construct a line following robot. The processing electronics for the robot consists of analog circuitry. The robot is controlled by two 3 VDC motors which independently drive a left and right wheel. A third non–powered drag wheel provides tripod stability for the robot.

In this example, we will equip the Blinky 602A robot platform with three Sharp GP12D IR sensors as shown in Figure 1.13. The robot will be placed in a maze with white reflective walls. The goal is for the robot to detect wall placement and navigate through the maze. (Figure 1.14.) The robot will not be provided any information about the maze. The control algorithm for the robot will be hosted on the Arduino UNO R3. The Arduino mega 2560 can also be used for this application.

1.6.1   STRUCTURE CHART

A structure chart is a visual tool used to partition a large project into "doable" smaller parts. It also helps to visualize what systems will be used to control different features of the robot. The arrows within the structure chart indicate the data flow between different portions of the program controlling the robot. The structure chart for the robot project is provided in Figure 1.15. As you can see, the robot has three main systems: the motor control system, the sensor system, and the digital input/output system. These three systems interact with the main control algorithm to allow the robot to autonomously (by itself) navigate through the maze by sensing and avoiding walls.

1.6.2   UML ACTIVITY DIAGRAMS

A Unified Modeling Language (UML) activity diagram, or flow chart, is a tool to help visualize the different steps required for a control algorithm. The UML activity diagram for the robot is provided in Figure 1.16. As you can see, after robot systems are initialized, the robot control system enters a continuous loop to gather data and issue outputs to steer the robot through the maze.

1.6.3   ARDUINO UNO R3 SYSTEMS

The three IR sensors (left, middle, and right) are mounted on the leading edge of the robot to detect maze walls. The output from the sensors is fed to three ADC channels. The robot motors will each be driven by a pulse width modulation (PWM) channel. The Arduino UNO R3 is interfaced to the motors via a transistor with enough drive capability to handle the maximum current requirements of the motor. The robot will be powered by a 9 VDC battery or a power supply (rated at approximately 2A) which is fed to a 5 VDC voltage regulator. We discuss the details of the interface electronics in a later chapter. From this example, you can see how different systems aboard the Arduino processing boards may be used to control different features aboard the Blinky robot.

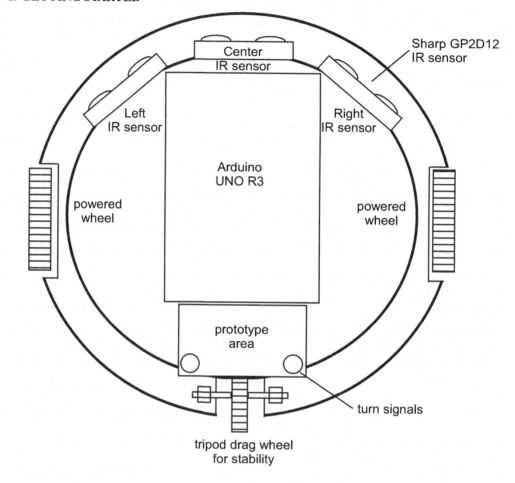

Figure 1.13: Blinky robot layout.

1.7   OTHER ARDUINO–BASED PLATFORMS

There is a wide variety of Arduino–based platforms. The platforms may be purchased from SparkFun Electronics (www.sparkfun.com). Figure 1.17 provides a representative sample. Shown on the left is the Arduino Lily Pad equipped with ATmega328 processor. (The same processor aboard the UNO R3). This processing board can actually be worn and is washable. It was designed to be sewn onto fabric.

In the bottom center figure is the Arduino Mega equipped with ATmega2560 processor. We have already discussed this board in some detail. In the upper right is the Arduino Mini Stamp. This small, but powerful processing board and is equipped with ATmega168 processor.

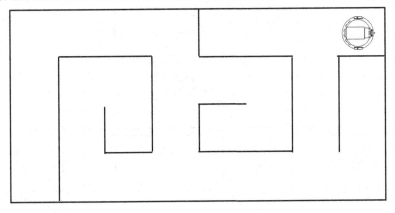

Figure 1.14: Blinky robot navigating maze.

1.8   EXTENDING THE HARDWARE FEATURES OF THE ARDUINO PLATFORMS

Additional features and external hardware may be added to selected Arduino platforms by using a daughter card concept. The daughter card is called an Arduino Shield as shown in Figure 1.18. The shield mates with the header pins on the Arduino board. The shield provides a small fabrication area, a processor reset button, and a general use pushbutton and two light emitting diodes (LEDs).

1.9   APPLICATION: ARDUINO HARDWARE STUDIO

Much like an artist uses a sketch box, we will use an Arduino Hardware Studio throughout the book to develop projects. In keeping with the do–it–yourself (DIY) spirit of Arduino, we have constructed the Studio using readily available off–the–shelf products as shown in Figure 1.19. The Studio includes the following:

- A yellow Pelican Micro Case #1040,

- An Arduino UNO R3 evaluation board,

- Two Jameco JE21 3.3 x 2.1 inch solderless breadboards and

- One piece of black plexiglass.

   We purposely have not provided any construction details. Instead, we encourage you to use your own imagination to develop and construct your own Arduino Hardware Studio.

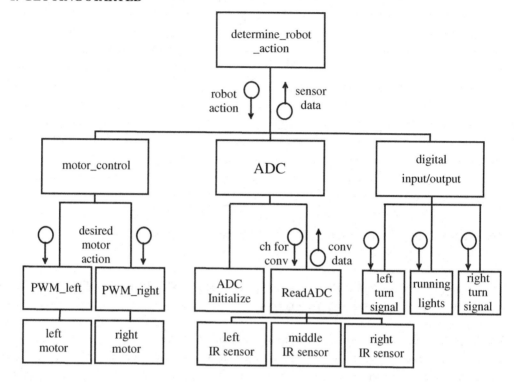

Figure 1.15: Blinky robot structure diagram.

1.10 SUMMARY

In this chapter, we have provided an overview of the Arduino concept of open source hardware. This was followed by a description of the Arduino UNO R3 processor board powered by the ATmega328. An overview of ATmega328 systems followed. This was followed by a description of the Arduino Mega 2560 processor board powered by the ATmega2560. An overview of ATmega2560 systems followed. We then investigated various processing boards in the Arduino line and concluded with brief guidelines on how to download and run the Arduino software environment.

1.11 REFERENCES

- SparkFun Electronics, 6175 Longbow Drive, Suite 200, Boulder, CO 80301 (www.sparkfun.com)

- Arduino homepage (www.arduino.cc)

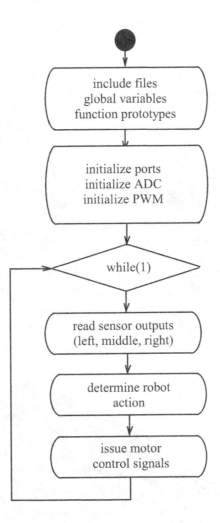

Figure 1.16: Robot UML activity diagram.

- *Atmel 8–bit AVR Microcontroller with 4/8/16/32K Bytes In–System Programmable Flash, AT-mega48PA, 88PA, 168PA, 328P* data sheet: 8171D–AVR–05/11, Atmel Corporation, 2325 Orchard Parkway, San Jose, CA 95131.

- *Atmel 8–bit AVR Microcontroller with 64/128/256K Bytes In-System Programmable Flash, AT-mega640/V, ATmega1280/V, 2560/V* data sheet: 2549P-AVR-10/2012, Atmel Corporation, 2325 Orchard Parkway, San Jose, CA 95131.

Figure 1.17: Arduino variants. (Used with permission from SparkFun Electronics (CC BY–NC–SA).)

Figure 1.18: Arduino shield. (Used with permission from SparkFun Electronics (CC BY–NC–SA).)

1.12 CHAPTER PROBLEMS

1. Describe in your own words the Arduino open source concept.

2. Sketch a block diagram of the ATmega328 or the ATmega2560 and its associated systems. Describe the function of each system.

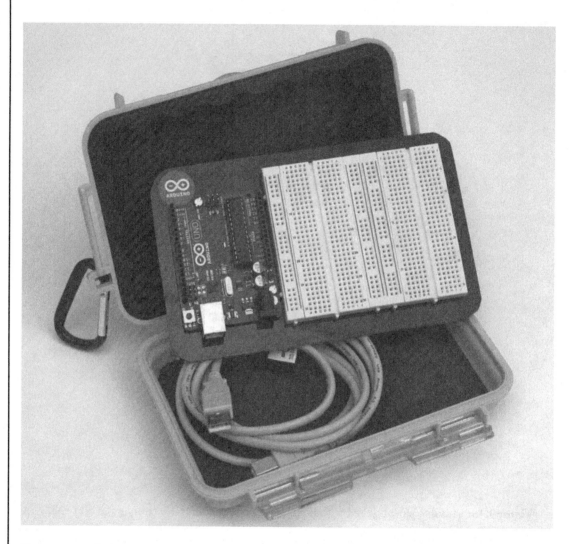

Figure 1.19: Arduino Hardware Studio.

3. What is the purpose of a structure chart?

4. What is the purpose of a UML activity diagram?

5. Describe the different types of memory components within the ATmega328 or the ATmega2560. Describe applications for each memory type.

6. Describe the three different register types associated with each port.

7. How may the features of the Arduino UNO R3 be extended? The Mega 2560?

8. Prepare a table of features for different Arduino products.

9. Discuss different options for the ATmega328 or the ATmega2560 time base. What are the advantages and disadvantages of each type?

10. Summarize the differences between the Arduino UNO R3 and Mega 2560. How would you choose between the two in a given application.

11. Design and fabricate your own Arduino hardware studio.

CHAPTER 2

Programming

Objectives: After reading this chapter, the reader should be able to do the following:

- Successfully download and execute a simple program using the Arduino Development Environment.

- Describe the key features of the Arduino Development Environment.

- Describe what features of the Arduino Development Environment ease the program development process.

- List the programming support information available at the Arduino home page.

- Describe the key components of a program.

- Specify the size of different variables within the C programming language.

- Define the purpose of the main program.

- Explain the importance of using functions within a program.

- Write functions that pass parameters and return variables.

- Describe the function of a header file.

- Discuss different programming constructs used for program control and decision processing.

- Write programs for use on the Arduino UNO R3 and Mega 2560 processing boards.

2.1 OVERVIEW

To the novice, programming a microcontroller may appear mysterious, complicated, overwhelming, and difficult. When faced with a new task, one often does not know where to start. The goal of this chapter is to provide a tutorial on how to begin programming. We will use a top–down design approach. We begin with the "big picture" of the chapter. We then discuss the Ardunio Development Environment and how it may be used to quickly develop a program for the Arduino UNO R3 and the Arduino Mega 2560 processor boards. We then take a closer look at program fundamentals beginning with an overview of the major pieces of a program. We then discuss the basics of the C programming language. Only the most fundamental concepts will be covered. Throughout the chapter, we provide examples and also provide pointers to a number of excellent references.

2.2   THE BIG PICTURE

We begin with the big picture of how to program the Arduino UNO R3 as shown in Figure 2.1. This will help provide an overview of how chapter concepts fit together. It also introduces terms used in writing, editing, compiling, loading and executing a program.

Most microcontrollers are programmed with some variant of the C programming language. The C programming language provides a nice balance between the programmer's control of the microcontroller hardware and time efficiency in program writing.

As you can see in Figure 2.1, the compiler software is hosted on a computer separate from the Arduino UNO R3. The job of the compiler is to transform the program provided by the program writer (filename.c and filename.h) into machine code (filename.hex) suitable for loading into the processor.

Once the source files (filename.c and filename.h) are provided to the compiler, the compiler executes two steps to render the machine code. The first step is the compilation process. Here the program source files are transformed into assembly code (filename.asm). If the program source files contains syntax errors, the compiler reports these to the user. Syntax errors are reported for incorrect use of the C programming language. An assembly language program is not generated until the syntax errors have been corrected.

The assembly language source file (filename.asm) is then passed to the assembler. The assembler transforms the assembly language source file (filename.asm) to machine code (filename.hex) suitable for loading to the Arduino processor.

The Arduino Development Environment provides a user friendly interface to aid in program development, transformation to machine code, and loading into the Arduino processor. The Arduino processor may also be programmed using the In System Programming (ISP) features of the Atmel AVR STK500 (or STK 600) Starter Kit and Development System. We discuss these procedures in a later chapter.

In the next section, we provide an overview of the Arduino Development Environment. You will see how this development tool provides a user–friendly method of quickly developing code applications for the Arduino processing boards.

2.3   ARDUINO DEVELOPMENT ENVIRONMENT

In this section, we provide an overview of the Arduino Development Environment (ADE). We begin with some background information about the ADE and then review its user friendly features. We then introduce the sketchbook concept and provide a brief overview of the built–in software features within the ADE. Our goal is to provide a brief introduction to the features. All Arduino related features are well–documented on the Arduino homepage (www.arduino.cc). We will not duplicate this excellent source of material but merely provide pointers to it. In later chapters, we review the different systems aboard the Arduino processing boards and show how that system may be controlled using the ADE built–in features.

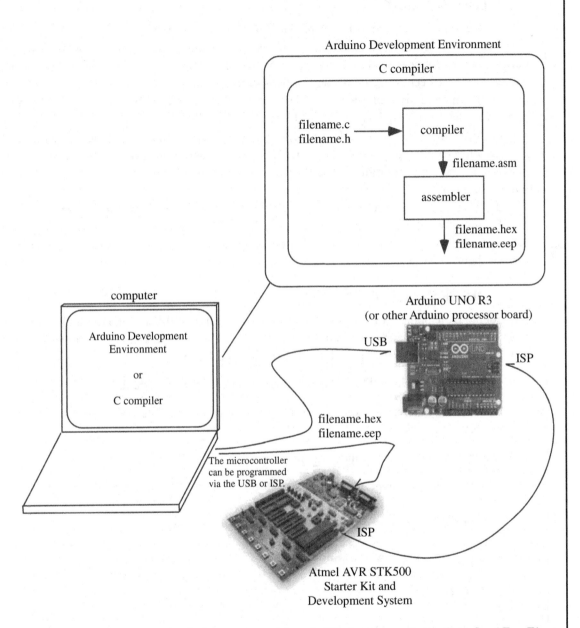

Figure 2.1: Programming the Arduino processor board. (Used with permission from SparkFun Electronics (CC BY–NC–SA), and Atmel, Incorporated.)

2.3.1   BACKGROUND

The first version of the Arduino Development Environment was released in August 2005. It was developed at the Interaction Design Institute in Ivrea, Italy to allow students the ability to quickly put processing power to use in a wide variety of projects. Since that time, updated versions incorporating new features, have been released on a regular basis [www.arduino.cc].

At its most fundamental level, the Arduino Development Environment is a user friendly interface to allow one to quickly write, load, and execute code on a microcontroller. A barebones program need only consist of a setup() and loop() function. The Arduino Development Environment adds the other required pieces such as header files and the main program construct. The ADE is written in Java and has its origins in the Processor programming language and the Wiring Project [www.arduino.cc].

In the next several sections, we introduce the user interface and its large collection of user friendly tools. We also provide an overview of the host of built–in C and C++ software functions that allows the project developer to quickly put the features of the Arduino processing boards to work for them.

2.3.2   QUICK START GUIDE

The Arduino Development Environment may be downloaded from the Arduino website's front page at www.arduino.cc. Versions are available for Windows, Mac OS X, and Linux. Provided below is a Quick start step–by–step approach to blink an onboard LED.

- Download the Arduino Development Environment from www.arduino.cc.

- Connect the Arduino processing board to the host computer via a USB cable (A male to B male).

- Start the Arduino Development Environment.

- Type the following program.

```
//*********************************************

#define LED_PIN 13

void setup()
{
pinMode(LED_PIN, OUTPUT);
}

void loop()
{
```

```
digitalWrite(LED_PIN, HIGH);
delay(500);                    //delay specified in ms
digitalWrite(LED_PIN, LOW);
delay(500);
}


//*********************************************
```

- Upload and execute the program by asserting the "Upload" button.

- The onboard LED should blink at one second intervals.

  With the Arduino ADE downloaded, let's take a closer look at its features.

2.3.3   ARDUINO DEVELOPMENT ENVIRONMENT OVERVIEW

The Arduino Development Environment is illustrated in Figure 2.2. The ADE contains a text editor, a message area for displaying status, a text console, a tool bar of common functions, and an extensive menuing system. The ADE also provides a user friendly interface to the Arduino processor board which allows for a quick upload of code. This is possible because the Arduino processing boards are equipped with a bootloader program.

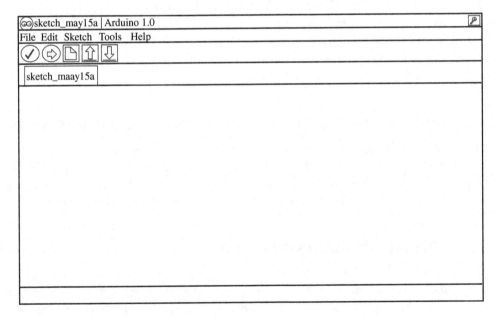

Figure 2.2: Arduino Development Environment [www.arduino.cc].

A close up of the Arduino toolbar is provided in Figure 2.3. The toolbar provides single button access to the more commonly used menu features. Most of the features are self explanatory. As described in the previous section, the "Upload" button compiles your code and uploads it to the Arduino processing board. The "Serial Monitor" button opens the serial monitor feature. The serial monitor feature allows text data to be sent to and received from the Arduino processing board.

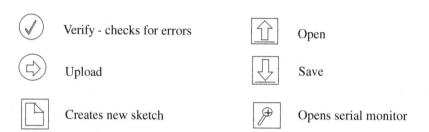

Figure 2.3: Arduino Development Environment buttons.

2.3.4   SKETCHBOOK CONCEPT

In keeping with a hardware and software platform for students of the arts, the Arduino environment employs the concept of a sketchbook. An artist maintains their works in progress in a sketchbook. Similarly, we maintain our programs within a sketchbook in the Arduino environment. Furthermore, we refer to individual programs as sketches. An individual sketch within the sketchbook may be accessed via the Sketchbook entry under the file tab.

2.3.5   ARDUINO SOFTWARE, LIBRARIES, AND LANGUAGE REFERENCES

The Arduino Development Environment has a number of built-in features. Some of the features may be directly accessed via the Arduino Development Environment drop down toolbar illustrated in Figure 2.2. Provided in Figure 2.4 is a handy reference to show all of the available features. The toolbar provides a wide variety of features to compose, compile, load and execute a sketch.

2.3.6   WRITING AN ARDUINO SKETCH

The basic format of the Arduino sketch consists of a "setup" and a "loop" function. The setup function is executed once at the beginning of the program. It is used to configure pins, declare variables and constants, etc. The loop function will execute sequentially step-by-step. When the end of the loop function is reached it will automatically return to the first step of the loop function and execute again. This goes on continuously until the program is stopped.

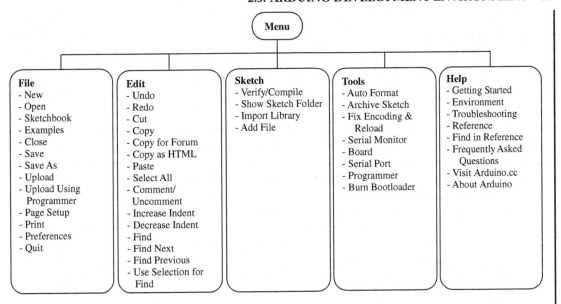

Figure 2.4: Arduino Development Environment menu [www.arduino.cc].

```
//**********************************************************

void setup()
  {
  //place setup code here
  }

void loop()
  {
  //main code steps are provided here
  :
  :

  }

//**********************************************************
```
Example 1: Let's revisit the sketch provided earlier in the chapter.
```
//**********************************************************

#define LED_PIN 13
```

```
void setup()
{
pinMode(LED_PIN, OUTPUT);
}

void loop()
{
digitalWrite(LED_PIN, HIGH);
delay(500);                   //delay specified in ms
digitalWrite(LED_PIN, LOW);
delay(500);
}


//************************************************************
```

In the first line the #define statement links the designator "LED_PIN" to pin 13 on the Arduino processor board. In the setup function, LED_PIN is designated as an output pin. Recall the setup function is only executed once. The program then enters the loop function that is executed sequentially step–by–step and continuously repeated. In this example, the LED_PIN is first set to logic high to illuminate the LED onboard the Arduino processing board. A 500 ms delay then occurs. The LED_PIN is then set low. A 500 ms delay then occurs. The sequence then repeats.

Even the most complicated sketches follow the basic format of the setup function followed by the loop function. To aid in the development of more complicated sketches, the Arduino Development Environment has a number of built–in features that may be divided into the areas of structure, variables and functions. The structure and variable features follow rules similar to the C programming language which is discussed in the next section. The built–in functions consists of a set of pre–defined activities useful to the programmer. These built–in functions are summarized in Figure 2.5.

There are also a number of program examples available to allow the user to quickly construct a sketch. These programs are summarized in Figure 2.6. Complete documentation for these programs are available at the Arduino homepage [www.arduino.cc]. This documentation is easily accessible via the Help tab on the Arduino Development Environment toolbar. This documentation will not be repeated here. Instead, we refer to these features at appropriate places throughout the remainder of the book as we discuss related hardware systems.

Keep in mind the Arduino open source concept. Users throughout the world are constantly adding new built–in features. As new features are added, they will be released in future Arduino Development Environment versions. As an Arduino user, you too may add to this collection of useful tools. It is also important to keep in mind the Arduino boards have an onboard microcontroller

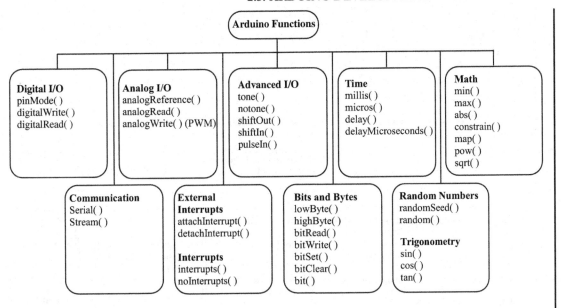

Figure 2.5: Arduino Development Environment functions [www.arduino.cc].

that can be programmed using the C programming language In the next section, we discuss the fundamental components of a C program.

Example 2: In this example we connect an external LED to Arduino UNO R3 pin 12. The onboard LED will blink alternately with the external LED. The external LED is connected to the Arduino UNO R3 as shown in Figure 2.7.

```
//*********************************************************

#define int_LED 13
#define ext_LED 12

void setup()
{
pinMode(int_LED, OUTPUT);
pinMode(ext_LED, OUTPUT);
}

void loop()
{
digitalWrite(int_LED, HIGH);
digitalWrite(ext_LED, LOW);
```

```
delay(500);                    //delay specified in ms
digitalWrite(int_LED, LOW);
digitalWrite(ext_LED, HIGH);
delay(500);
}


//*********************************************************
```

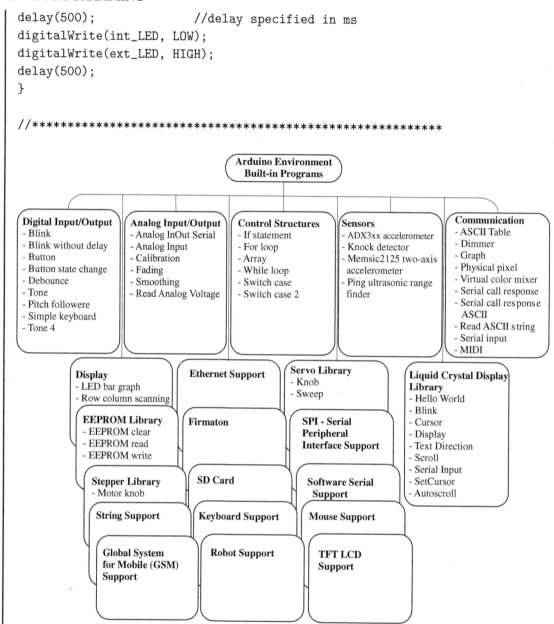

Figure 2.6: Arduino Development Environment built–in features [www.arduino.cc].

Example 3: In this example we connect an external LED to Arduino UNO R3 pin 12 and an external switch attached to pin 11. The onboard LED will blink alternately with the external LED when the switch is depressed. The external LED and switch is connected to the Arduino UNO R3 as shown in Figure 2.8.

```
//******************************************************

#define int_LED 13
#define ext_LED 12
#define ext_sw  11

int switch_value;

void setup()
{
pinMode(int_LED, OUTPUT);
pinMode(ext_LED, OUTPUT);
pinMode(ext_sw,  INPUT);
}

void loop()
{
switch_value = digitalRead(ext_sw);
  if(switch_value == LOW)
    {
    digitalWrite(int_LED, HIGH);
    digitalWrite(ext_LED, LOW);
    delay(50);
    digitalWrite(int_LED, LOW);
    digitalWrite(ext_LED, HIGH);
    delay(50);
    }
  else
    {
    digitalWrite(int_LED, LOW);
    digitalWrite(ext_LED, LOW);
    }
}
//******************************************************
```

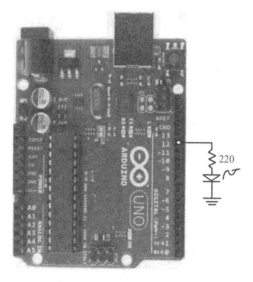

a) schematic

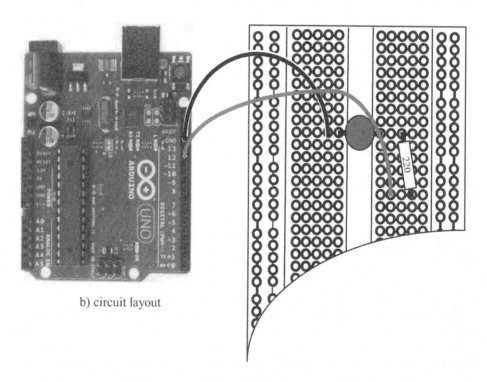

b) circuit layout

Figure 2.7: Arduino UNO R3 with an external LED. (UNO R3 illustration used with permission of the Arduino Team (CC BY–NC–SA) www.arduino.cc).

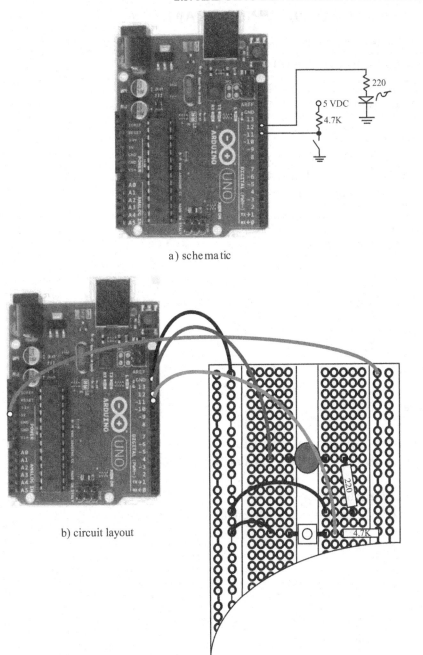

a) schematic

b) circuit layout

Figure 2.8: Arduino UNO R3 with an external LED. (UNO R3 illustration used with permission of the Arduino Team (CC BY–NC–SA) www.arduino.cc).

2.4   ANATOMY OF A PROGRAM

Programs written for a microcontroller have a fairly repeatable format. Slight variations exist but many follow the format provided.

```
//Comments containing program information
// - file name:
// - author:
// - revision history:
// - compiler setting information:
// - hardware connection description to microcontroller pins
// - program description


//include files
#include<file_name.h>


//function prototypes
A list of functions and their format used within the program


//program constants
#define     TRUE    1
#define     FALSE   0
#define     ON      1
#define     OFF     0


//interrupt handler definitions
Used to link the software to hardware interrupt features


//global variables
Listing of variables used throughout the program


//main program

void main(void)
{

body of the main program

}

//function definitions
```

```
A detailed function body and definition
for each function used within the program
```

Let's take a closer look at each piece.

2.4.1   COMMENTS

Comments are used throughout the program to document what and how things were accomplished within a program. The comments help you reconstruct your work at a later time. Imagine that you wrote a program a year ago for a project. You now want to modify that program for a new project. The comments will help you remember the key details of the program.

Comments are not compiled into machine code for loading into the microcontroller. Therefore, the comments will not fill up the memory of your microcontroller. Comments are indicated using double slashes (//). Anything from the double slashes to the end of a line is then considered a comment. A multi–line comment can be constructed using a /* at the beginning of the comment and a */ at the end of the comment. These are handy to block out portions of code during troubleshooting or providing multi–line comments.

At the beginning of the program, comments may be extensive. Comments may include some of the following information:

- file name

- program author

- revision history or a listing of the key changes made to the program

- compiler setting information

- hardware connection description to microcontroller pins

- program description

2.4.2   INCLUDE FILES

Often you need to add extra files to your project besides the main program. For example, most compilers require a "personality file" on the specific microcontroller that you are using. This file is provided with the compiler and provides the name of each register used within the microcontroller. It also provides the link between a specific register's name within software and the actual register location within hardware. These files are typically called header files and their name ends with a ".h". Within the C compiler there will also be other header files to include in your program such as the "math.h" file when programming with advanced math functions.

To include header files within a program, the following syntax is used:

```
//include files
```

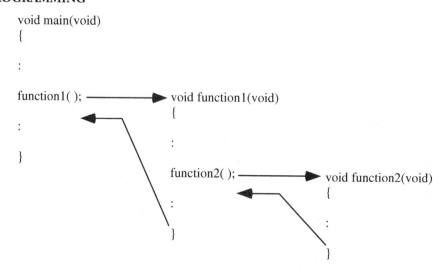

Figure 2.9: Function calling.

```
#include<file_name1.h>
#include<file_name2.h>
```

In an upcoming section, we see how the Arduino Development Environment makes it quite easy to include a header file within a program.

2.4.3 FUNCTIONS

In the next chapter, we discuss in detail the top down design, bottom up implementation approach to designing microcontroller based systems. In this approach, a microcontroller based project including both hardware and software is partitioned into systems, subsystems, etc. The idea is to take a complex project and break it into doable pieces with a defined action.

We use the same approach when writing computer programs. At the highest level is the main program which calls functions that have a defined action. When a function is called, program control is released from the main program to the function. Once the function is complete, program control reverts back to the main program.

Functions may in turn call other functions as shown in Figure 2.9. This approach results in a collection of functions that may be reused over and over again in various projects. Most importantly, the program is now subdivided into doable pieces, each with a defined action. This makes writing the program easier but also makes it much easier to modify the program since every action is in a known location.

There are three different pieces of code required to properly configure and call the function:

- the function prototype,

- the function call, and

- the function body.

Function prototypes are provided early in the program as previously shown in the program template. The function prototype provides the name of the function and any variables required by the function and any variable returned by the function.

The function prototype follows this format:

```
return_variable  function_name(required_variable1, required_variable2);
```

If the function does not require variables or sends back a variable the word "void" is placed in the variable's position.

The **function call** is the code statement used within a program to execute the function. The function call consists of the function name and the actual arguments required by the function. If the function does not require arguments to be delivered to it for processing, the parenthesis containing the variable list is left empty.

The function call follows this format:

```
function_name(required_variable1, required_variable2);
```

A function that requires no variables follows this format:

```
function_name( );
```

When the function call is executed by the program, program control is transferred to the function, the function is executed, and program control is then returned to the portion of the program that called it.

The **function body** is a self–contained "mini–program." The first line of the function body contains the same information as the function prototype: the name of the function, any variables required by the function, and any variable returned by the function. The last line of the function contains a "return" statement. Here a variable may be sent back to the portion of the program that called the function. The processing action of the function is contained within the open ({) and close brackets (}). If the function requires any variables within the confines of the function, they are declared next. These variable are referred to as local variables. A local variable is known only within the confines of a specific function. The actions required by the function follow.

The function prototype follows this format:

```
return_variable  function_name(required_variable1, required_variable2)
{
//local variables required by the function
unsigned int  variable1;
unsigned char variable2;

//program statements required by the function
```

```
//return variable
return return_variable;
}
```

Example: In this example, we describe how to configure the ports of the microcontroller to act as input or output ports. Briefly, associated with each port is a register called the data direction register (DDR). Each bit in the DDR corresponds to a bit in the associated PORT. For example, PORTB has an associated data direction register DDRB. If DDRB[7] is set to a logic 1, the corresponding port pin PORTB[7] is configured as an output pin. Similarly, if DDRB[7] is set to logic 0, the corresponding port pin is configured as an input pin.

During some of the early steps of a program, a function is called to initialize the ports as input, output, or some combination of both. This is illustrated in Figure 2.10.

```
//function prototypes
void initialize_ports(void);


//main function
void main(void)
{

  :

  initialize_ports( );  ──────────▶    //function body
                                       void initialize_ports(void)
  :              ◀──────────            {
                                        DDRB  = 0x00;      //initialize PORTB as input
  }                                     PORTB = 0x00;

                                        DDRC = 0xFF;       //initialize PORTC as output
                                        PORTC = 0x00;      //set pins to logic 0

                                        DDRD = 0xFF;       //initialize PORTD as output
                                        PORTD = 0x00;      //set pins to logic 0
                                        }
```

Figure 2.10: Configuring ports.

2.4.4   PROGRAM CONSTANTS

The #define statement is used to associate a constant name with a numerical value in a program. It can be used to define common constants such as pi. It may also be used to give terms used within a program a numerical value. This makes the code easier to read. For example, the following constants may be defined within a program:

```
//program constants
#define    TRUE   1
#define    FALSE  0
#define    ON     1
#define    OFF    0
```

2.4.5   INTERRUPT HANDLER DEFINITIONS

Interrupts are functions that are written by the programmer but usually called by a specific hardware event during system operation. We discuss interrupts and how to properly configure them in an upcoming chapter.

2.4.6   VARIABLES

There are two types of variables used within a program: global variables and local variables. A global variable is available and accessible to all portions of the program. Whereas, a local variable is only known and accessible within the function where it is declared.

When declaring a variable in C, the number of bits used to store the operator is also specified. In Figure 2.11, we provide a list of common C variable sizes used with the ImageCraft ICC AVR compiler. The size of other variables such as pointers, shorts, longs, etc. are contained in the compiler documentation [ImageCraft].

When programming microcontrollers, it is important to know the number of bits used to store the variable and also where the variable will be assigned. For example, assigning the contents of an unsigned char variable, which is stored in 8–bits, to an 8–bit output port will have a predictable result. However, assigning an unsigned int variable, which is stored in 16–bits, to an 8–bit output port does not provide predictable results. It is wise to insure your assignment statements are balanced for accurate and predictable results. The modifier "unsigned" indicates all bits will be used to specify the magnitude of the argument. Signed variables will use the left most bit to indicate the polarity (±) of the argument.

A global variable is declared using the following format provided below. The type of the variable is specified, followed by its name, and an initial value if desired.

```
//global variables
unsigned int  loop_iterations = 6;
```

Type	Size	Range
unsigned char	1	0..255
signed char	1	-128..127
unsigned int	2	0..65535
signed int	2	-32768..32767
float	4	+/-1.175e-38.. +/-3.40e+38
double	4	+/-1.175e-38.. +/-3.40e+38

Figure 2.11: C variable sizes used with the ImageCraft ICC AVR compiler [ImageCraft].

2.4.7   MAIN PROGRAM

The main program is the hub of activity for the entire program. The main program typically consists of program steps and function calls to initialize the processor followed by program steps to collect data from the environment external to the microcontroller, process the data and make decisions, and provide external control signals back to the environment based on the data collected.

2.5   FUNDAMENTAL PROGRAMMING CONCEPTS

In the previous section, we covered many fundamental concepts. In this section we discuss operators, programming constructs, and decision processing constructs to complete our fundamental overview of programming concepts.

2.5.1   OPERATORS

There are a wide variety of operators provided in the C language. An abbreviated list of common operators are provided in Figures 2.12 and 2.13. The operators have been grouped by general category. The symbol, precedence, and brief description of each operator are provided. The precedence column indicates the priority of the operator in a program statement containing multiple operators. Only the fundamental operators are provided. For more information on this topic, see Barrett and Pack in the Reference section at the end of the chapter.

General		
Symbol	Precedence	Description
{ }	1	Brackets, used to group program statements
( )	1	Parenthesis, used to establish precedence
=	12	Assignment

Arithmetic Operations		
Symbol	Precedence	Description
*	3	Multiplication
/	3	Division
+	4	Addition
-	4	Substraction

Logical Operations		
Symbol	Precedence	Description
<	6	Less than
<=	6	Less than or equal to
>	6	Greater
>=	6	Greater than or equal to
==	7	Equal to
!=	7	Not equal to
&&	9	Logical AND
‖	10	Logical OR

Figure 2.12: C operators. (Adapted from [Barrett and Pack]).

Bit Manipulation Operations		
Symbol	Precedence	Description
<<	5	Shift left
>>	5	Shift right
&	8	Bitwise AND
^	8	Bitwise exclusive OR
		8	Bitwise OR

Unary Operations		
Symbol	Precedence	Description
!	2	Unary negative
~	2	One's complement (bit-by-bit inversion)
++	2	Increment
--	2	Decrement
type(argument)	2	Casting operator (data type conversion)

Figure 2.13: C operators (continued). (Adapted from [Barrett and Pack]).

2.5.1.1 General operations

Within the general operations category are brackets, parenthesis, and the assignment operator. We have seen in an earlier example how bracket pairs are used to indicate the beginning and end of the main program or a function. They are also used to group statements in programming constructs and decision processing constructs. This is discussed in the next several sections.

The parenthesis is used to boost the priority of an operator. For example, in the mathematical expression $7 \times 3 + 10$, the multiplication operation is performed before the addition since it has a higher precedence. Parenthesis may be used to boost the precedence of the addition operation. If we contain the addition operation within parenthesis $7 \times (3 + 10)$, the addition will be performed before the multiplication operation and yield a different result from the earlier expression.

The assignment operator (=) is used to assign the argument(s) on the right–hand side of an equation to the left–hand side variable. It is important to insure that the left and the right–hand side of the equation have the same type of arguments. If not, unpredictable results may occur.

2.5.1.2 Arithmetic operations

The arithmetic operations provide for basic math operations using the various variables described in the previous section. As described in the previous section, the assignment operator (=) is used to assign the argument(s) on the right–hand side of an equation to the left–hand side variable.

Example: In this example, a function returns the sum of two unsigned int variables passed to the function.

```
unsigned int  sum_two(unsigned int variable1, unsigned int variable2)
{
unsigned int  sum;

sum = variable1 + variable2;

return sum;
}
```

2.5.1.3 Logical operations

The logical operators provide Boolean logic operations. They can be viewed as comparison operators. One argument is compared against another using the logical operator provided. The result is returned as a logic value of one (1, true, high) or zero (0 false, low). The logical operators are used extensively in program constructs and decision processing operations to be discussed in the next several sections.

2.5.1.4 Bit manipulation operations

There are two general types of operations in the bit manipulation category: shifting operations and bitwise operations. Let's examine several examples:

Example: Given the following code segment, what will the value of variable2 be after execution?

```
unsigned char  variable1 = 0x73;
unsigned char  variable2;

variable2 = variable1 << 2;
```

Answer: Variable "variable1" is declared as an eight bit unsigned char and assigned the hexadecimal value of $(73)_{16}$. In binary this is $(0111\_0011)_2$. The << 2 operator provides a left shift of the argument by two places. After two left shifts of $(73)_{16}$, the result is $(cc)_{16}$ and will be assigned to the variable "variable2."

Note that the left and right shift operation is equivalent to multiplying and dividing the variable by a power of two.

The bitwise operators perform the desired operation on a bit–by–bit basis. That is, the least significant bit of the first argument is bit–wise operated with the least significant bit of the second argument and so on.

Example: Given the following code segment, what will the value of variable3 be after execution?

```
unsigned char    variable1 = 0x73;
unsigned char    variable2 = 0xfa;
unsigned char    variable3;

variable3 = variable1 & variable2;
```

Answer: Variable "variable1" is declared as an eight bit unsigned char and assigned the hexadecimal value of $(73)_{16}$. In binary, this is $(0111\_0011)_2$. Variable "variable2" is declared as an eight bit unsigned char and assigned the hexadecimal value of $(fa)_{16}$. In binary, this is $(1111\_1010)_2$. The bitwise AND operator is specified. After execution variable "variable3," declared as an eight bit unsigned char, contains the hexadecimal value of $(72)_{16}$.

2.5.1.5   Unary operations

The unary operators, as their name implies, require only a single argument.

For example, in the following code segment, the value of the variable "i" is incremented. This is a shorthand method of executing the operation "$i = i + 1;$"

```
unsigned int    i;

i++;
```

Example: It is not uncommon in embedded system design projects to have every pin on a microcontroller employed. Furthermore, it is not uncommon to have multiple inputs and outputs assigned to the same port but on different port input/output pins. Some compilers support specific pin reference. Another technique that is not compiler specific is **bit twiddling**. Figure 2.14 provides bit twiddling examples on how individual bits may be manipulated without affecting other bits using bitwise and unary operators. The information provided here was extracted from the ImageCraft ICC AVR compiler documentation [ImageCraft].

2.5.2   PROGRAMMING CONSTRUCTS

In this section, we discuss several methods of looping through a piece of code. We will examine the "for" and the "while" looping constructs.

The **for** loop provides a mechanism for looping through the same portion of code a fixed number of times. The for loop consists of three main parts:

- loop initiation,

- loop termination testing, and

- the loop increment.

In the following code fragment the for loop is executed ten times.

Syntax	Description	Example
a | b	bitwise or	PORTA |=0x80; // turn on bit 7 (msb)
a & b	bitwise and	if ((PINA & 0x81)==0) // check bit 7 and bit 0
a ^ b	bitwise exclusive or	PORTA ^=0x80; // flip bit 7
~a	bitwise complement	PORTA &=~0x80; // turn off bit 7

Figure 2.14: Bit twiddling [ImageCraft].

```
unsigned int  loop_ctr;

for(loop_ctr = 0; loop_ctr < 10; loop_ctr++)
  {
                    //loop body


  }
```

The for loop begins with the variable "loop_ctr" equal to 0. During the first pass through the loop, the variable retains this value. During the next pass through the loop, the variable "loop_ctr" is incremented by one. This action continues until the "loop_ctr" variable reaches the value of ten. Since the argument to continue the loop is no longer true, program execution continues after the close bracket for the for loop.

In the previous example, the for loop counter was incremented at the beginning of each loop pass. The "loop_ctr" variable can be updated by any amount. For example, in the following code fragment the "loop_ctr" variable is increased by three for every pass of the loop.

```
unsigned int  loop_ctr;

for(loop_ctr = 0; loop_ctr < 10; loop_ctr=loop_ctr+3)
  {
                    //loop body


  }
```

The "loop_ctr" variable may also be initialized at a high value and then decremented at the beginning of each pass of the loop.

```
unsigned int  loop_ctr;

for(loop_ctr = 10; loop_ctr > 0; loop_ctr-)
  {
                        //loop body

  }
```

As before, the "loop_ctr" variable may be decreased by any numerical value as appropriate for the application at hand.

The **while** loop is another programming construct that allows multiple passes through a portion of code. The while loop will continue to execute the statements within the open and close brackets while the condition at the beginning of the loop remains logically true. The code snapshot below will implement a ten iteration loop. Note how the "loop_ctr" variable is initialized outside of the loop and incremented within the body of the loop. As before, the variable may be initialized to a greater value and then decremented within the loop body.

```
unsigned int  loop_ctr;

loop_ctr = 0;
while(loop_ctr < 10)
  {
                        //loop body
  loop_ctr++;
  }
```

Frequently, within a microcontroller application, the program begins with system initialization actions. Once initialization activities are complete, the processor enters a continuous loop. This may be accomplished using the following code fragment.

```
while(1)
  {


  }
```

2.5.3 DECISION PROCESSING

There are a variety of constructs that allow decision making. These include the following:

- the **if** statement,

- the **if–else** construct,

- the **if–else if–else** construct, and the

- **switch** statement.

The **if** statement will execute the code between an open and close bracket set should the condition within the if statement be logically true.

Example: To help develop the algorithm for steering the Blinky 602A robot through a maze, a light emitting diode (LED) is connected to PORTB pin 1 on the ATmega328. The robot's center IR sensor is connected to an analog–to–digital converter at PORTC, pin 1. The IR sensor provides a voltage output that is inversely proportional to distance of the sensor from the maze wall. It is desired to illuminate the LED if the robot is within 10 cm of the maze wall. The sensor provides an output voltage of 2.5 VDC at the 10 cm range. The following **if** statement construct will implement this LED indicator. We provide the actual code to do this later in the chapter.

```
if (PORTC[1] > 2.5)              //Center
IR sensor voltage greater than 2.5 VDC
   {
   PORTB = 0x02;                 //illuminate LED on PORTB[1]
   }
```

In the example provided, there is no method to turn off the LED once it is turned on. This will require the **else** portion of the construct as shown in the next code fragment.

```
if (PORTC[1] > 2.5)              //Center
IR sensor voltage greater than 2.5 VDC
   {
   PORTB = 0x02;                 //illuminate LED on PORTB[1]
   }
else
   {
   PORTB = 0x00;                 //extinguish the LED on PORTB[1]
   }
```

The **if–else if–else** construct may be used to implement a three LED system. In this example, the left, center, and right IR sensors are connected to analog–to–digital converter channels on PORTC pins 2, 1, and 0, respectively. The LED indicators are connected to PORTB pins 2, 1, and 0. The following code fragment implements this LED system.

```
if (PORTC[2] > 2.5)              //Left IR
sensor voltage greater than 2.5 VDC
   {
   PORTB = 0x04;                 //illuminate LED on PORTB[2]
   }
else if (PORTC[1] > 2.5)         //Center
IR sensor voltage greater than 2.5 VDC
   {
```

```
    PORTB = 0x02;                    //illuminate the LED on PORTB[1]
    }
else if (PORTC[0] > 2.5)        //Right
IR sensor voltage greater than 2.5 VDC
    {
    PORTB = 0x01;                    //illuminate the LED on PORTB[0]
    }
else                             //no walls sensed within 10 cm
    {
    PORTB = 0x00;                    //extinguish LEDs
    }
```

The **switch** statement is used when multiple if–else conditions exist. Each possible condition is specified by a case statement. When a match is found between the switch variable and a specific case entry, the statements associated with the case are executed until a **break** statement is encountered.

Example: Suppose eight pushbutton switches are connected to PORTD. Each switch will implement a different action. A switch statement may be used to process the multiple possible decisions as shown in the following code fragment.

```
void read_new_input(void)
{
new_PORTD = PIND;

if(new_PORTD != old_PORTD)               //check for status change PORTD

switch(new_PORTD)
    {                                    //process change in PORTD input
    case 0x01:                           //PD0
                                         //PD0 related actions
            break;

    case 0x02:                           //PD1
                                         //PD1 related actions
            break;

    case 0x04:                           //PD2
                                         //PD2 related actions
            break;

    case 0x08:                           //PD3
                                         //PD3 related actions
```

```
            break;

case 0x10:                              //PD4
                                        //PD4 related actions
            break;

case 0x20:                              //PD5
                                        //PD5 related actions
            break;

case 0x40:                              //PD6
                                        //PD6 related actions
            break;

case 0x80:                              //PD7
                                        //PD7 related actions
            break;

            default:;                   //all other cases
    }                                   //end switch(new_PORTD)
 }                                      //end if new_PORTD
old_PORTD=new_PORTD;                    //update PORTD
}
```

That completes our brief overview of the C programming language. In the next section, we illustrate how to use the Arduino processing board in several applications. We use the Arduino UNO R3. The examples can be adapted for use with other boards in the Arduino family.

2.6   APPLICATION 1: ROBOT IR SENSOR

To demonstrate how to construct a sketch in the Arduino Development Environment, we revisit the robot IR sensor application provided earlier in the chapter. We also investigate the sketches's interaction with the Arduino UNO R3 processing board and external sensors and indicators. We will use the robot project as an ongoing example throughout the remainder of the book.

Recall from Chapter 1, the Blinky 602A kit contains the hardware and mechanical parts to construct a line following robot. In this example, we modify the robot platform by equipping it with three Sharp GP12D IR sensors as shown in Figure 2.15. The sensors are available from SparkFun Electronics (www.sparkfun.com). The sensors are mounted to a bracket constructed from thin aluminum. Dimensions for the bracket are provided in the figure. Alternatively, the IR sensors may be mounted to the robot platform using "L" brackets available from a local hardware store. In later

Application sections, we equip the robot with all three IR sensors. In this example, we equip the robot with a single sensor and test its function as a proof of concept.

The IR sensor provides a voltage output that is inversely proportional to the sensor distance from the maze wall. It is desired to illuminate the LED if the robot is within 10 cm of the maze wall. The sensor provides an output voltage of 2.5 VDC at the 10 cm range. The interface between the IR sensor and the Arduino UNO R3 board is provided in Figure 2.16.

The IR sensor's power (red wire) and ground (black wire) connections are connected to the 5V and Gnd pins on the Arduino UNO R3 board, respectively. The IR sensor's output connection (yellow wire) is connected to the ANALOG IN 5 pin on the Arduino UNO R3 board. The LED circuit shown in the top right corner of the diagram is connected to the DIGITAL 0 pin on the Arduino UNO R3 board. We discuss the operation of this circuit in the Interfacing chapter later in the book.

Earlier in the chapter, we provided a framework for writing the if–else statement to turn the LED on and off. Here is the actual sketch to accomplish this.

```
//************************************************************************
#define LED_PIN    0              //digital pin - LED connection
#define IR_sensor_pin 5           //analog pin - IR sensor

int IR_sensor_value;              //declare variable for IR sensor value

void setup()
  {
  pinMode(LED_PIN, OUTPUT);       //configure pin 0 for digital output
  }

void loop()
  {
                                  //read analog output from IR sensor
    IR_sensor_value = analogRead(IR_sensor_pin);

    if(IR_sensor_value > 512)     //0 to 1023 maps to 0 to 5 VDC
      {
      digitalWrite(LED_PIN, HIGH); //turn LED on
      }
    else
      {
      digitalWrite(LED_PIN, LOW);  //turn LED off
      }
  }
//************************************************************************
```

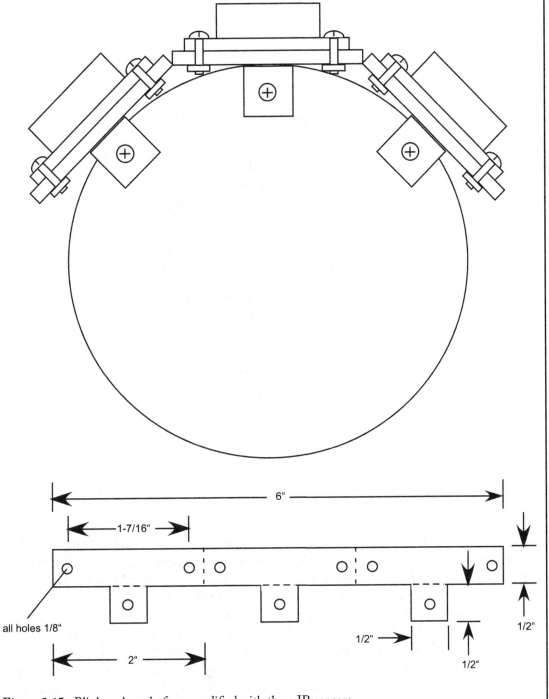

Figure 2.15: Blinky robot platform modified with three IR sensors.

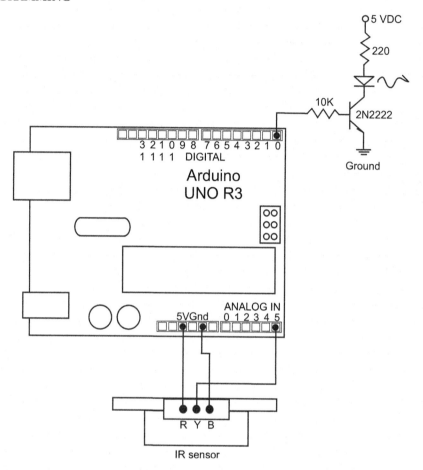

Figure 2.16: IR sensor interface.

The sketch begins by providing names for the two Arduino UNO R3 board pins that will be used in the sketch. This is not required but it makes the code easier to read. We define the pin for the LED as "LED_PIN." Any descriptive name may be used here. Whenever the name is used within the sketch, the number "0" will be substituted for the name by the compiler.

After providing the names for pins, the next step is to declare any variables required by the sketch. In this example, the output from the IR sensor will be converted from an analog to a digital value using the built–in Arduino "analogRead" function. A detailed description of the function may be accessed via the Help menu. It is essential to carefully review the support documentation for a built–in Arduino function the first time it is used. The documentation provides details on variables required by the function, variables returned by the function, and an explanation on function operation.

The "analogRead" function requires the pin for analog conversion variable passed to it and returns the analog signal read as an integer value (int) from 0 to 1023. So, for this example, we need to declare an integer value to receive the returned value. We have called this integer variable "IR_sensor_value."

Following the declaration of required variables are the two required functions for an Arduino UNO R3 program: setup and loop. The setup function calls an Arduino built–in function, pinMode, to set the "LED_PIN" as an output pin. The loop function calls several functions to read the current analog value on pin 5 (the IR sensor output) and then determine if the reading is above 512 (2.5 VDC). If the reading is above 2.5 VDC, the LED on DIGITAL pin 0 is illuminated, else it is turned off.

After completing writing the sketch with the Arduino Development Environment, it must be compiled and then uploaded to the Arduino UNO R3 board. These two steps are accomplished using the "Sketch – Verify/Compile" and the "File – Upload to I/O Board" pull down menu selections.

In the next example, we adapt the IR sensor project to provide custom lighting for an art piece.

2.7   APPLICATION 2: ART PIECE ILLUMINATION SYSTEM

My oldest son Jonathan Barrett is a gifted artist (www.closertothesuninternational.com). Although I own several of his pieces, my favorite one is a painting he did during his early student days. The assignment was to paint your favorite place. Jonny painted Lac Laronge, Saskatchewan as viewed through the window of a pontoon plane. Jonny, his younger brother Graham, and I have gone on several fishing trips with friends to this very special location. An image of the painting is provided in Figure 2.17.

The circuit and sketch provided in the earlier example may be slightly modified to provide custom lighting for an art piece. The IR sensor could be used to detect the presence and position of those viewing the piece. Custom lighting could then be activated by the Arduino UNO R3 board via the DIGITAL output pins. In the Lac Laronge piece, lighting could be provided from behind the painting using high intensity white LEDs. The intensity of the LEDs could be adjusted by changing the value of the resistor is series with the LED. The lighting intensity could also be varied based on the distance the viewer is from the painting. We cover the specifics in an upcoming chapter.

2.8   APPLICATION 3: FRIEND OR FOE SIGNAL

In aerial combat a "friend or foe" signal is used to identify aircraft on the same side. The signal is a distinctive pattern only known by aircraft on the same side. In this example, we generate a friend signal on the internal LED on pin 13 that consists of 10 pulses of 40 ms each followed by a 500 ms pulse and then 1000 ms when the LED is off.

Figure 2.17: Lac Laronge, Saskatchewan. Image used with permission, Jonny Barrett, Closer to the Sun Fine Art and Design, Steamboat Springs, CO [www.closertothesuninternational.com].

```
//**********************************************

#define LED_PIN 13

void setup()
{
pinMode(LED_PIN, OUTPUT);
}

void loop()
{
int i;
```

```
for(i=0; i<=9; i++)
  {
  digitalWrite(LED_PIN, HIGH);
  delay(20);                 //delay specified in ms
  digitalWrite(LED_PIN, LOW);
  delay(20);
  }

digitalWrite(LED_PIN, HIGH);
delay(500);                  //delay specified in ms
digitalWrite(LED_PIN, LOW);
delay(1000);
}


//*************************************************
```

2.9 SUMMARY

The goal of this chapter was to provide a tutorial on how to begin programming. We used a top-
-down design approach. We began with the "big picture" of the chapter followed by an overview
of the Arduino Development Environment. We then discussed the basics of the C programming
language. Only the most fundamental concepts were covered. We concluded with several extended
examples. Throughout the chapter, we provided examples and also provided references to a number
of excellent references.

2.10 REFERENCES

- *Atmel 8–bit AVR Microcontroller with 4/8/16/32K Bytes In–System Programmable Flash, AT-mega48PA, 88PA, 168PA, 328P* data sheet: 8171D–AVR–05/11, Atmel Corporation, 2325 Orchard Parkway, San Jose, CA 95131.

- *Atmel 8–bit AVR Microcontroller with 64/128/256K Bytes In–System Programmable Flash, AT-mega640/V, ATmega1280/V, 2560/V* data sheet: 2549P-AVR-10/2012, Atmel Corporation, 2325 Orchard Parkway, San Jose, CA 95131.

- ImageCraft Embedded Systems C Development Tools, 706 Colorado Avenue, #10–88, Palo Alto, CA, 94303, www.imagecraft.com

- S. F. Barrett and D.J. Pack, Embedded Systems Design and Applications with the 68HC12 and HCS12, Pearson Prentice Hall, 2005.

- Arduino homepage, www.arduino.cc

- Jonathan Barrett, Closer to the Sun International.
  `www.closertothesuninternational.com`

- Barrett S, Pack D (2006) Microcontrollers Fundamentals for Engineers and Scientists. Morgan and Claypool Publishers. DOI: 10.2200/S00025ED1V01Y200605DCS001

- Barrett S and Pack D (2008) Atmel AVR Microcontroller Primer Programming and Interfacing. Morgan and Claypool Publishers. DOI: 10.2200/S00100ED1V01Y200712DCS015

- Barrett S (2010) Embedded Systems Design with the Atmel AVR Microcontroller. Morgan and Claypool Publishers. DOI: 10.2200/S00225ED1V01Y200910DCS025

2.11   CHAPTER PROBLEMS

1. Describe the steps in writing a sketch and executing it on an Arduino UNO R3 processing board.

2. Describe the key portions of a C program.

3. Describe two different methods to program an Arduino processing board.

4. What is an include file?

5. What are the three pieces of code required for a program function?

6. Describe how to define a program constant.

7. Provide the C program statement to set PORTB pins 1 and 7 to logic one. Use bit–twiddling techniques.

8. Provide the C program statement to reset PORTB pins 1 and 7 to logic zero. Use bit–twiddling techniques.

9. What is the difference between a for and while loop?

10. When should a switch statement be used versus the if–then statement construct?

11. What is the serial monitor feature used for in the Arduino Development Environment?

12. Describe what variables are required and returned and the basic function of the following built–in Arduino functions: Blink, Analog Input.

13. Adapt Application 1: Robot IR sensor for the Arduino Mega 2560 processor board.

14. Adapt Application 2: Art Piece Illumination System for the Arduino Mega 2560 processor board.

CHAPTER 3

Embedded Systems Design

Objectives: After reading this chapter, the reader should be able to do the following:

- Define an embedded system.

- List all aspects related to the design of an embedded system.

- Provide a step–by–step approach to embedded system design.

- Discuss design tools and practices related to embedded systems design.

- Apply embedded system design practices in the design of an Arduino–based microcontroller system employing several interacting subsystems.

- Provide a detailed design for an autonomous maze navigating robot controlled by the Arduino processing board including hardware layout and interface, structure and UML activity diagrams, and coded algorithm.

In Chapters 1 and 2, we provided an overview of the Arduino hardware and the Arduino Development Environment to get you quickly up and operating with this user friendly processor. In the remainder of the book, we will take a second and detailed pass through this information. This chapter provides a step–by–step methodical approach to designing advanced embedded system. Chapters 4 through 7 take an advanced look at the serial communications systems, the analog–to–digital converter system, the interrupt system, and the timing system. Chapter 8 provides detailed information on how to interface a wide variety of peripheral devices to Arduino processors. Throughout these chapters, we show how the built–in features of the Arduino Development Environment may be employed in different applications.

In this chapter, we begin with a definition of just what is an embedded system. We then explore the process of how to successfully (and with low stress) develop an embedded system prototype that meets established requirements. We conclude the chapter with an extended example. The example illustrates the embedded system design process in the development and prototype of the autonomous maze navigating robot based on the Blinky 602A (www.graymarkint.com) controlled by the Arduino UNO R3 processing board.

3.1 WHAT IS AN EMBEDDED SYSTEM?

An embedded system contains a microcontroller to accomplish its job of processing system inputs and generating system outputs. The link between system inputs and outputs is provided by a coded algorithm stored within the processor's resident memory. What makes embedded systems design so interesting and challenging is the design must also take into account the proper electrical interface for the input and output devices, limited on–chip resources, human interface concepts, the operating environment of the system, cost analysis, related standards, and manufacturing aspects [Anderson]. Through careful application of this material you will be able to design and prototype embedded systems based on the Arduino microcontroller.

3.2   EMBEDDED SYSTEM DESIGN PROCESS

In this section, we provide a step–by–step approach to develop the first prototype of an embedded system that will meet established requirements. There are many formal design processes that we could study. We concentrate on the steps that are common to most. We purposefully avoid formal terminology of a specific approach and instead concentrate on the activities that are accomplished as a system prototype is developed. The design process we describe is illustrated in Figure 3.1 using a Unified Modeling Language (UML) activity diagram. We discuss the UML activity diagrams later in the chapter.

3.2.1   PROJECT DESCRIPTION

The goal of the project description step is to determine what the system is ultimately supposed to do. To achieve this step you must thoroughly investigate what the system is supposed to do. Questions to raise and answer during this step include but are not limited to the following:

- What is the system supposed to do?

- Where will it be operating and under what conditions?

- Are there any restrictions placed on the system design?

To answer these questions, the designer interacts with the client to ensure clear agreement on what is to be done. If you are completing this project for yourself, you must still carefully and thoughtfully complete this step. The establishment of clear, definable system requirements may require considerable interaction between the designer and the client. It is essential that both parties agree on system requirements before proceeding further in the design process. The final result of this step is a detailed listing of system requirements and related specifications.

3.2.2   BACKGROUND RESEARCH

Once a detailed list of requirements has been established, the next step is to perform background research related to the design. In this step, the designer will ensure they understand all requirements and features required by the project. This will again involve interaction between the designer and the client. The designer will also investigate applicable codes, guidelines, protocols, and standards related to the project. This is also a good time to start thinking about the interface between different portions

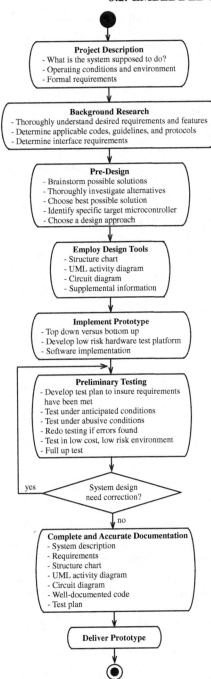

Figure 3.1: Embedded system design process.

of the project particularly the input and output devices peripherally connected to the microcontroller. The ultimate objective of this step is to have a thorough understanding of the project requirements, related project aspects, and any interface challenges within the project.

3.2.3   PRE–DESIGN

The goal of the pre–design step is to convert a thorough understanding of the project into possible design alternatives. Brainstorming is an effective tool in this step. Here, a list of alternatives is developed. Since an embedded system typically involves both hardware and/or software, the designer can investigate whether requirements could be met with a hardware only solution or some combination of hardware and software. Generally, speaking a hardware only solution executes faster; however, the design is fixed once fielded. On the other hand, a software implementation provides flexibility and a typically slower execution speed. Most embedded design solutions will use a combination of both hardware and software to capitalize on the inherent advantages of each.

Once a design alternative has been selected, the general partition between hardware and software can be determined. It is also an appropriate time to select a specific hardware device to implement the prototype design. If a microcontroller technology has been chosen, it is now time to select a specific controller. This is accomplished by answering the following questions:

- What microcontroller systems or features i.e., ADC, PWM, timer, etc.) are required by the design?

- How many input and output pins are required by the design?

- What is the maximum anticipated operating speed of the microcontroller expected to be?

Recall from Chapter 1 there are a wide variety of Arduino–based microcontrollers available to the designer.

3.2.4   DESIGN

With a clear view of system requirements and features, a general partition determined between hardware and software, and a specific microcontroller chosen, it is now time to tackle the actual design. It is important to follow a systematic and disciplined approach to design. This will allow for low stress development of a documented design solution that meets requirements. In the design step, several tools are employed to ease the design process. They include the following:

- Employing a top–down design, bottom up implementation approach,

- Using a structure chart to assist in partitioning the system,

- Using a Unified Modeling Language (UML) activity diagram to work out program flow, and

- Developing a detailed circuit diagram of the entire system.

Let's take a closer look at each of these. The information provided here is an abbreviated version of the one provided in "Microcontrollers Fundamentals for Engineers and Scientists." The interested reader is referred there for additional details and an in
—depth example [Barrett and Pack].

Top down design, bottom up implementation. An effective tool to start partitioning the design is based on the techniques of top–down design, bottom–up implementation. In this approach, you start with the overall system and begin to partition it into subsystems. At this point of the design, you are not concerned with how the design will be accomplished but how the different pieces of the project will fit together. A handy tool to use at this design stage is the structure chart. The structure chart shows the hierarchy of how system hardware and software components will interact and interface with one another. You should continue partitioning system activity until each subsystem in the structure chart has a single definable function.

UML Activity Diagram. Once the system has been partitioned into pieces, the next step in the design process is to start working out the details of the operation of each subsystem we previously identified. Rather than beginning to code each subsystem as a function, we will work out the information and control flow of each subsystem using another design tool: the Unified Modeling Language (UML) activity diagram. The activity diagram is simply a UML compliant flow chart. UML is a standardized method of documenting systems. The activity diagram is one of the many tools available from UML to document system design and operation. The basic symbols used in a UML activity diagram for a microcontroller based system are provided in Figure 3.2[Fowler].

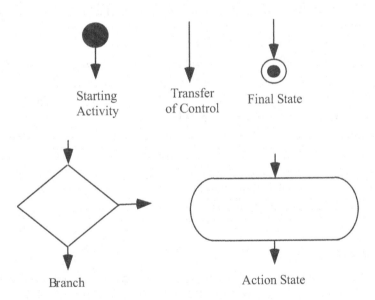

Figure 3.2: UML activity diagram symbols. Adapted from [source].

To develop the UML activity diagram for the system, we can use a top–down, bottom–up, or a hybrid approach. In the top–down approach, we begin by modeling the overall flow of the algorithm from a high level. If we choose to use the bottom–up approach, we would begin at the bottom of the structure chart and choose a subsystem for flow modeling. The specific course of action chosen depends on project specifics. Often, a combination of both techniques, a hybrid approach, is used. You should work out all algorithm details at the UML activity diagram level prior to coding any software. If you can not explain system operation at this higher level, first, you have no business being down in the detail of developing the code. Therefore, the UML activity diagram should be of sufficient detail so you can code the algorithm directly from it [Dale].

In the design step, a detailed circuit diagram of the entire system is developed. It will serve as a roadmap to implement the system. It is also a good idea at this point to investigate available design information relative to the project. This would include hardware design examples, software code examples, and application notes available from manufacturers.

At the completion of this step, the prototype design is ready for implementation and testing.

3.2.5   IMPLEMENT PROTOTYPE

To successfully implement a prototype, an incremental approach should be followed. Again, the top-–down design, bottom–up implementation provides a solid guide for system implementation. In an embedded system design involving both hardware and software, the hardware system including the microcontroller should be assembled first. This provides the software the required signals to interact with. As the hardware prototype is assembled on a prototype board, each component is tested for proper operation as it is brought online. This allows the designer to pinpoint malfunctions as they occur.

Once the hardware prototype is assembled, coding may commence. As before, software should be incrementally brought online. You may use a top down, bottom up, or hybrid approach depending on the nature of the software. The important point is to bring the software online incrementally such that issues can be identified and corrected early on.

It is highly recommended that low cost stand–in components be used when testing the software with the hardware components. For example, push buttons, potentiometers, and LEDs may be used as low cost stand–in component simulators for expensive input instrumentation devices and expensive output devices such as motors. This allows you to insure the software is properly operating before using it to control the actual components.

3.2.6   PRELIMINARY TESTING

To test the system, a detailed test plan must be developed. Tests should be developed to verify that the system meets all of its requirements and also intended system performance in an operational environment. The test plan should also include scenarios in which the system is used in an unintended manner. As before a top–down, bottom–up, or hybrid approach can be used to test the system.

Once the test plan is completed, actual testing may commence. The results of each test should be carefully documented. As you go through the test plan, you will probably uncover a number of run time errors in your algorithm. After you correct a run time error, the entire test plan must be performed again. This ensures that the new fix does not have an unintended affect on another part of the system. Also, as you process through the test plan, you will probably think of other tests that were not included in the original test document. These tests should be added to the test plan. As you go through testing, realize your final system is only as good as the test plan that supports it!

Once testing is complete, you might try another level of testing where you intentionally try to "jam up" the system. In another words, try to get your system to fail by trying combinations of inputs that were not part of the original design. A robust system should continue to operate correctly in this type of an abusive environment. It is imperative that you design robustness into your system. When testing on a low cost simulator is complete, the entire test plan should be performed again with the actual system hardware. Once this is completed you should have a system that meets its requirements!

3.2.7   COMPLETE AND ACCURATE DOCUMENTATION

With testing complete, the system design should be thoroughly documented. Much of the documentation will have already been accomplished during system development. Documentation will include the system description, system requirements, the structure chart, the UML activity diagrams documenting program flow, the test plan, results of the test plan, system schematics, and properly documented code. To properly document code, you should carefully comment all functions describing their operation, inputs, and outputs. Also, comments should be included within the body of the function describing key portions of the code. Enough detail should be provided such that code operation is obvious. It is also extremely helpful to provide variables and functions within your code names that describe their intended use.

You might think that a comprehensive system documentation is not worth the time or effort to complete it. Complete documentation pays rich dividends when it is time to modify, repair, or update an existing system. Also, well–documented code may be often reused in other projects: a method for efficient and timely development of new systems.

3.3   EXAMPLE: BLINKY 602A AUTONOMOUS MAZE NAVIGATING ROBOT SYSTEM DESIGN

To illustrate the design process, we provide an in depth example using the Blinky 602A robot manufactured by Graymark (www.graymarkint.com). In the upcoming paragraphs, we progress through the design process a step at a time.

Problem description and background research. Graymark manufacturers many low–cost, excellent robot platforms. In this project, we will modify the Blinky 602A robot to be controlled by an Arduino UNO R3 processing board. The Blinky 602A kit contains the hardware and mechanical parts to construct a line following robot. The processing electronics for the robot consists of analog circuitry. The robot is controlled by two 3 VDC motors, which independently drive a left and right wheel. A third non–powered drag wheel provides tripod stability for the robot.

In this project, we replace the analog control circuitry with the Arduino UNO R3 as the processing element. We also modify the robot's mission from being a line follower to autonomous maze navigator. To detect maze walls, we equip the Blinky 602A robot platform with three Sharp GP12D IR sensors as shown in Figure 3.3. The robot will be placed in a maze with reflective white walls. The goal of the project is for the robot to detect maze walls and navigate through the maze without touching the walls. The robot will not be provided any information about the maze. It must gather maze information on its own as it progresses through the maze. The control algorithm for the robot will be hosted on the Arduino UNO R3 processing board.

Requirements: The requirements for this project are straight forward; the robot will autonomously (on its own) navigate through the maze without touching maze walls. It is important to note that a map of the maze will not be programmed into the robot. The robot will sense the presence of the maze walls using the Sharp IR sensors and then make decisions to avoid the walls and process through the maze. From this description, the following requirements result. The robot will be:

- Equipped with three Sharp IR sensors to sense maze walls.

- Propelled through the maze using the two powered wheels provided in the Blinky 602A kit and a third drag wheel for stability.

- Controlled by the Arduino UNO R3 processing board.

- Equipped with turn signals (LEDs) to indicate a turn.

- Equipped with LEDs, one for each IR sensor, to indicate a wall has been detected by a specific sensor.

Pre–design. With requirements clearly understood, the next step is normally to brainstorm possible solutions. In this example, we have already decided to use the Arduino UNO R3 processing board. Other alternatives include using analog or digital hardware to determine robot action or another microcontroller.

Circuit diagram. The circuit diagram for the robot is provided in Figure 3.4. The three IR sensors (left, middle, and right) will be mounted on the leading edge of the robot to detect maze walls. The output from the sensors is fed to three Arduino UNO R3 ADC channels (ANALOG IN 0–2). The robot motors will be driven by PWM channels (PWM: DIGITAL 11 and PWM: DIGITAL 10). The Arduino UNO R3 is interfaced to the motors via a Darlington NPN transistor (TIP120) with enough drive capability to handle the maximum current requirements of the motor. Since the microcontroller is powered at 5 VDC and the motors are rated at 3 VDC, two 1N4001 diodes are placed in series with the motor. This reduces the supply voltage to the motor to be approximately 3

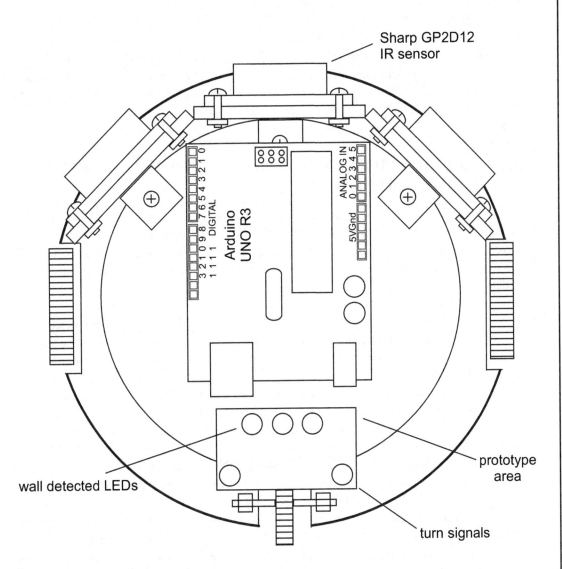

Figure 3.3: Robot layout with the Arduino UNO R3 processing board.

VDC. The robot will be powered by a 9 VDC battery which is fed to a 5 VDC voltage regulator. The details of the interface electronics are provided in a later chapter. To save on battery expense, it is recommended to use a 9 VDC, 2A rated inexpensive, wall–mount power supply to provide power to the 5 VDC voltage regulator. A power umbilical of braided wire may be used to provide power to the robot while navigating about the maze.

Structure chart: The structure chart for the robot project is provided in Figure 3.5.

UML activity diagrams: The UML activity diagram for the robot is provided in Figure 3.6.

Arduino UNO R3 Program: We will develop the entire control algorithm for the Arduino UNO R3 board in the Application sections in the remainder of the book. We get started on the control algorithm in the next section.

3.4   APPLICATION: CONTROL ALGORITHM FOR THE BLINKY 602A ROBOT

, we provide the basic framework for the robot control algorithm. The control algorithm will read the IR sensors attached to the Arduino UNO R3 ANALOG IN (pins 0–2). In response to the wall placement detected, it will render signals to turn the robot to avoid the maze walls. Provided in Figure 3.7 is a truth table that shows all possibilities of maze placement that the robot might encounter. A detected wall is represented with a logic one. An asserted motor action is also represented with a logic one.

The robot motors may only be moved in the forward direction. We review techniques to provide bi–directional motor control in an upcoming chapter. To render a left turn, the left motor is stopped and the right motor is asserted until the robot completes the turn. To render a right turn, the opposite action is required.

The task in writing the control algorithm is to take the UML activity diagram provided in Figure 3.6 and the actions specified in the robot action truth table (Figure 3.7 and transform both into an Arduino sketch. This may seem formidable but we take it a step at a time. The sketch written in the Applications section of the previous chapter will serve as our starting point.

The control algorithm begins with Arduino UNO R3 pin definitions. Variables are then declared for the readings from the three IR sensors. The two required Arduino functions follow: setup() and loop(). In the setup() function, Arduino UNO R3 pins are declared as output. The loop() begins by reading the current value of the three IR sensors. Recall from the Application section in the previous chapter, the 512 value corresponds to a particular IR sensor range. This value may be adjusted to change the range at which the maze wall is detected. The read of the IR sensors is followed by an eight part if–else if statement. The statement contains a part for each row of the truth table provided in Figure 3.7. For a given configuration of sensed walls, the appropriate wall detection LEDs are illuminated followed by commands to activate the motors (analogWrite) and illuminate the appropriate turn signals. The analogWrite command issues a signal from 0 to 5 VDC by sending a constant from 0 to 255 using pulse width modulation (PWM) techniques. PWM techniques will be discussed in an upcoming chapter. The turn signal commands provide to actions: the appropriate

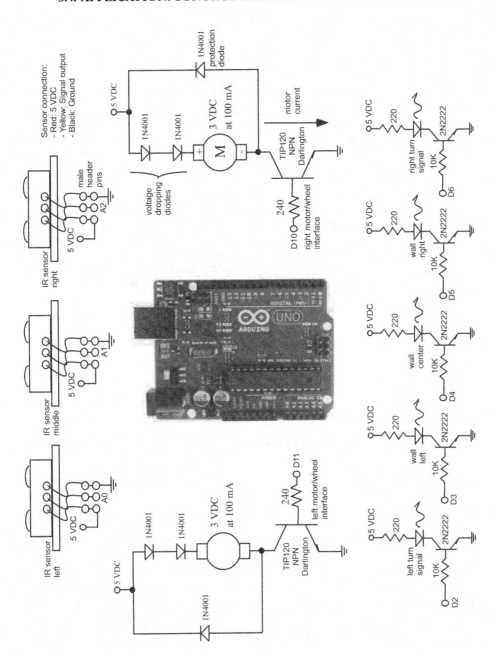

Figure 3.4: Robot circuit diagram. (UNO R3 illustration used with permission of the Arduino Team (CC BY–NC–SA) www.arduino.cc).

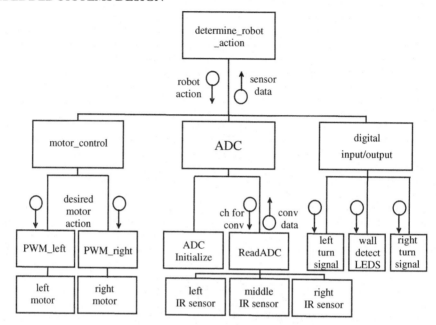

Figure 3.5: Robot structure diagram.

turns signals are flashed and a 1.5 s total delay is provided. This provides the robot 1.5 s to render a turn. This delay may need to be adjusted during the testing phase.

```
//****************************************************************************
                                    //analog input pins
#define left_IR_sensor    A0        //analog pin - left IR sensor

#define center_IR_sensor  A1        //analog pin - center IR sensor
#define right_IR_sensor   A2        //analog pin - right IR sensor

                                    //digital output pins
                                    //LED indicators - wall detectors
#define wall_left         3         //digital pin - wall_left
#define wall_center       4         //digital pin - wall_center
#define wall_right        5         //digital pin - wall_right

                                    //LED indicators - turn signals
#define left_turn_signal  2         //digital pin - left_turn_signal
#define right_turn_signal 6         //digital pin - right_turn_signal
```

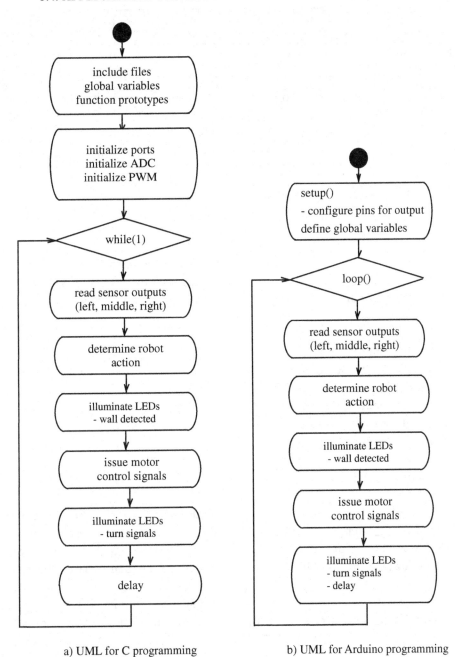

a) UML for C programming

b) UML for Arduino programming

Figure 3.6: Robot UML activity diagram.

	Left Sensor	Middle Sensor	Right Sensor	Wall Left	Wall Middle	Wall Right	Left Motor	Right Motor	Left Signal	Right Signal	Comments
0	0	0	0	0	0	0	1	1	0	0	Forward
1	0	0	1	0	0	1	1	1	0	0	Forward
2	0	1	0	0	1	0	1	0	0	1	Right
3	0	1	1	0	1	1	0	1	1	0	Left
4	1	0	0	1	0	0	1	1	0	0	Forward
5	1	0	1	1	0	1	1	1	0	0	Forward
6	1	1	0	1	1	0	1	0	0	1	Right
7	1	1	1	1	1	1	1	0	0	1	Right

Figure 3.7: Truth table for robot action.

```
                                       //motor outputs
#define left_motor       11            //digital pin - left_motor
#define right_motor      10            //digital pin - right_motor


int left_IR_sensor_value;              //declare
variable for left IR sensor
int center_IR_sensor_value;            //declare
variable for center IR sensor
int right_IR_sensor_value;             //declare
variable for right IR sensor

void setup()
  {
                                       //LED indicators - wall detectors
  pinMode(wall_left,   OUTPUT);        //configure pin 1 for digital output
  pinMode(wall_center, OUTPUT);        //configure pin 2 for digital output
  pinMode(wall_right,  OUTPUT);        //configure pin 3 for digital output

                                       //LED indicators - turn signals
  pinMode(left_turn_signal,OUTPUT);    //configure pin 0 for digital output
  pinMode(right_turn_signal,OUTPUT);   //configure pin 4 for digital output

                                       //motor outputs - PWM
  pinMode(left_motor,  OUTPUT);        //configure pin 11 for digital output
  pinMode(right_motor, OUTPUT);        //configure pin 10 for digital output
```

```
  }

void loop()
  {
                                        //read analog output from IR sensors
    left_IR_sensor_value   = analogRead(left_IR_sensor);
    center_IR_sensor_value = analogRead(center_IR_sensor);
    right_IR_sensor_value  = analogRead(right_IR_sensor);

    //robot action table row 0
    if((left_IR_sensor_value < 512)&&(center_IR_sensor_value < 512)&&
      (right_IR_sensor_value < 512))
      {
                                              //wall detection LEDs
      digitalWrite(wall_left,   LOW);         //turn LED off
      digitalWrite(wall_center, LOW);         //turn LED off
      digitalWrite(wall_right,  LOW);         //turn LED off
                                              //motor control
      analogWrite(left_motor,  128);
        //0 (off) to 255 (full speed)
      analogWrite(right_motor, 128);
        //0 (off) to 255 (full speed)
                                              //turn signals
      digitalWrite(left_turn_signal,  LOW);   //turn LED off
      digitalWrite(right_turn_signal, LOW);   //turn LED off
      delay(500);                             //delay 500 ms
      digitalWrite(left_turn_signal,  LOW);   //turn LED off
      digitalWrite(right_turn_signal, LOW);   //turn LED off
      delay(500);                             //delay 500 ms
      digitalWrite(left_turn_signal,  LOW);   //turn LED off
      digitalWrite(right_turn_signal, LOW);   //turn LED off
      delay(500);                             //delay 500 ms
      digitalWrite(left_turn_signal,  LOW);   //turn LED off
      digitalWrite(right_turn_signal, LOW);   //turn LED off
      analogWrite(left_motor, 0);             //turn motor off
      analogWrite(right_motor,0);             //turn motor off
      }

    //robot action table row 1
```

```
     else if((left_IR_sensor_value < 512)&&(center_IR_sensor_value < 512)&&
             (right_IR_sensor_value > 512))
       {
                                              //wall detection LEDs
       digitalWrite(wall_left,   LOW);        //turn LED off
       digitalWrite(wall_center, LOW);        //turn LED off
       digitalWrite(wall_right,  HIGH);       //turn LED on
                                              //motor control
       analogWrite(left_motor,  128);
          //0 (off) to 255 (full speed)
       analogWrite(right_motor, 128);
          //0 (off) to 255 (full speed)

                                              //turn signals
       digitalWrite(left_turn_signal,  LOW);  //turn LED off
       digitalWrite(right_turn_signal, LOW);  //turn LED off
       delay(500);                            //delay 500 ms
       digitalWrite(left_turn_signal,  LOW);  //turn LED off
       digitalWrite(right_turn_signal, LOW);  //turn LED off
       delay(500);                            //delay 500 ms
       digitalWrite(left_turn_signal,  LOW);  //turn LED off
       digitalWrite(right_turn_signal, LOW);  //turn LED off
       delay(500);                            //delay 500 ms
       digitalWrite(left_turn_signal,  LOW);  //turn LED off
       digitalWrite(right_turn_signal, LOW);  //turn LED off
       analogWrite(left_motor, 0);            //turn motor off
       analogWrite(right_motor,0);            //turn motor off
       }

   //robot action table row 2
   else if((left_IR_sensor_value < 512)&&(center_IR_sensor_value > 512)&&
           (right_IR_sensor_value < 512))
     {
                                            //wall detection LEDs
     digitalWrite(wall_left,   LOW);        //turn LED off
     digitalWrite(wall_center, HIGH);       //turn LED on
     digitalWrite(wall_right,  LOW);        //turn LED off
                                            //motor control
     analogWrite(left_motor,  128);
        //0 (off) to 255 (full speed)
```

```
analogWrite(right_motor, 0);
  //0 (off) to 255 (full speed)
                                          //turn signals
digitalWrite(left_turn_signal,  LOW);     //turn LED off
digitalWrite(right_turn_signal, HIGH);    //turn LED on
delay(500);                               //delay 500 ms
digitalWrite(left_turn_signal,  LOW);     //turn LED off
digitalWrite(right_turn_signal, LOW);     //turn LED off
delay(500);                               //delay 500 ms
digitalWrite(left_turn_signal,  LOW);     //turn LED off
digitalWrite(right_turn_signal, HIGH);    //turn LED on
delay(500);                               //delay 500 ms
digitalWrite(left_turn_signal,  LOW);     //turn LED off
digitalWrite(right_turn_signal, LOW);     //turn LED off
analogWrite(left_motor, 0);               //turn motor off
analogWrite(right_motor,0);               //turn motor off
}


//robot action table row 3
else if((left_IR_sensor_value < 512)&&(center_IR_sensor_value > 512)&&
      (right_IR_sensor_value > 512))
{
                                          //wall detection LEDs
digitalWrite(wall_left,   LOW);           //turn LED off
digitalWrite(wall_center, HIGH);          //turn LED on
digitalWrite(wall_right,  HIGH);          //turn LED on
                                          //motor control
analogWrite(left_motor,  0);
  //0 (off) to 255 (full speed)
analogWrite(right_motor, 128);
  //0 (off) to 255 (full speed)
                                          //turn signals
digitalWrite(left_turn_signal,  HIGH);    //turn LED on
digitalWrite(right_turn_signal, LOW);     //turn LED off
delay(500);                               //delay 500 ms
digitalWrite(left_turn_signal,  LOW);     //turn LED off
digitalWrite(right_turn_signal, LOW);     //turn LED off
delay(500);                               //delay 500 ms
digitalWrite(left_turn_signal,  HIGH);    //turn LED on
```

```
                digitalWrite(right_turn_signal, LOW);    //turn LED off
                delay(500);                              //delay 500 ms
                digitalWrite(left_turn_signal,  LOW);    //turn LED off
                digitalWrite(right_turn_signal, LOW);    //turn LED off
                analogWrite(left_motor, 0);              //turn motor off
                analogWrite(right_motor,0);              //turn motor off
                }

        //robot action table row 4
        else if((left_IR_sensor_value > 512)&&(center_IR_sensor_value < 512)&&
                (right_IR_sensor_value < 512))
                {
                                                         //wall detection LEDs
                digitalWrite(wall_left,   HIGH);         //turn LED on
                digitalWrite(wall_center, LOW);          //turn LED off
                digitalWrite(wall_right,  LOW);          //turn LED off
                                                         //motor control
                analogWrite(left_motor,  128);
                  //0 (off) to 255 (full speed)
                analogWrite(right_motor, 128);
                  //0 (off) to 255 (full speed)

                                                         //turn signals
                digitalWrite(left_turn_signal,  LOW);    //turn LED off
                digitalWrite(right_turn_signal, LOW);    //turn LED off
                delay(500);                              //delay 500 ms
                digitalWrite(left_turn_signal,  LOW);    //turn LED off
                digitalWrite(right_turn_signal, LOW);    //turn LED off
                delay(500);                              //delay 500 ms
                digitalWrite(left_turn_signal,  LOW);    //turn LED off
                digitalWrite(right_turn_signal, LOW);    //turn LED off
                delay(500);                              //delay 500 ms
                digitalWrite(left_turn_signal,  LOW);    //turn LED off
                digitalWrite(right_turn_signal, LOW);    //turn LED off
                analogWrite(left_motor, 0);              //turn motor off
                analogWrite(right_motor,0);              //turn motor off
                }

        //robot action table row 5
        else if((left_IR_sensor_value > 512)&&(center_IR_sensor_value < 512)&&
```

```
                      (right_IR_sensor_value > 512))
{
                                              //wall detection LEDs
digitalWrite(wall_left,   HIGH);              //turn LED on
digitalWrite(wall_center, LOW);               //turn LED off
digitalWrite(wall_right,  HIGH);              //turn LED on
                                              //motor control
analogWrite(left_motor,  128);
  //0 (off) to 255 (full speed)
analogWrite(right_motor, 128);
  //0 (off) to 255 (full speed)

                                              //turn signals
digitalWrite(left_turn_signal,  LOW);         //turn LED off
digitalWrite(right_turn_signal, LOW);         //turn LED off
delay(500);                                   //delay 500 ms
digitalWrite(left_turn_signal,  LOW);         //turn LED off
digitalWrite(right_turn_signal, LOW);         //turn LED off
delay(500);                                   //delay 500 ms
digitalWrite(left_turn_signal,  LOW);         //turn LED off
digitalWrite(right_turn_signal, LOW);         //turn LED off
delay(500);                                   //delay 500 ms
digitalWrite(left_turn_signal,  LOW);         //turn LED off
digitalWrite(right_turn_signal, LOW);         //turn LED off
analogWrite(left_motor, 0);                   //turn motor off
analogWrite(right_motor,0);                   //turn motor off
}

//robot action table row 6
else if((left_IR_sensor_value > 512)&&(center_IR_sensor_value > 512)&&
        (right_IR_sensor_value < 512))
{
                                              //wall detection LEDs
digitalWrite(wall_left,   HIGH);              //turn LED on
digitalWrite(wall_center, HIGH);              //turn LED on
digitalWrite(wall_right,  LOW);               //turn LED off
                                              //motor control
analogWrite(left_motor,  128);
  //0 (off) to 255 (full speed)
analogWrite(right_motor, 0);
```

```
      //0 (off) to 255 (full speed)

                                          //turn signals
  digitalWrite(left_turn_signal,  LOW);   //turn LED off
  digitalWrite(right_turn_signal, HIGH);  //turn LED on
  delay(500);                             //delay 500 ms
  digitalWrite(left_turn_signal,  LOW);   //turn LED off
  digitalWrite(right_turn_signal, LOW);   //turn LED off
  delay(500);                             //delay 500 ms
  digitalWrite(left_turn_signal,  LOW);   //turn LED off
  digitalWrite(right_turn_signal, HIGH);  //turn LED off
  delay(500);                             //delay 500 ms
  digitalWrite(left_turn_signal,  LOW);   //turn LED OFF
  digitalWrite(right_turn_signal, LOW);   //turn LED OFF
  analogWrite(left_motor, 0);             //turn motor off
  analogWrite(right_motor,0);             //turn motor off
  }


//robot action table row 7
else if((left_IR_sensor_value > 512)&&(center_IR_sensor_value > 512)&&
        (right_IR_sensor_value > 512))
  {
                                          //wall detection LEDs
  digitalWrite(wall_left,   HIGH);        //turn LED on
  digitalWrite(wall_center, HIGH);        //turn LED on
  digitalWrite(wall_right,  HIGH);        //turn LED on
                                          //motor control
  analogWrite(left_motor,  128);
    //0 (off) to 255 (full speed)
  analogWrite(right_motor, 0);
    //0 (off) to 255 (full speed)

                                          //turn signals
  digitalWrite(left_turn_signal,  LOW);   //turn LED off
  digitalWrite(right_turn_signal, HIGH);  //turn LED on
  delay(500);                             //delay 500 ms
  digitalWrite(left_turn_signal,  LOW);   //turn LED off
  digitalWrite(right_turn_signal, LOW);   //turn LED off
  delay(500);                             //delay 500 ms
  digitalWrite(left_turn_signal,  LOW);   //turn LED off
  digitalWrite(right_turn_signal, HIGH);  //turn LED on
```

```
    delay(500);                             //delay 500 ms
    digitalWrite(left_turn_signal,  LOW);   //turn LED off
    digitalWrite(right_turn_signal, LOW);   //turn LED off
    analogWrite(left_motor, 0);             //turn motor off
    analogWrite(right_motor,0);             //turn motor off
    }
}
//*****************************************************************************
```

Testing the control algorithm: It is recommended that the algorithm be first tested without the entire robot platform. This may be accomplished by connecting the three IR sensors and LEDS to the appropriate pins on the Arduino UNO R3 as specified in Figure 3.4. In place of the two motors and their interface circuits, two LEDs with the required interface circuitry may be used. The LEDs will illuminate to indicate the motors would be on during different test scenarios. Once this algorithm is fully tested in this fashion, the Arduino UNO R3 may be mounted to the robot platform and connected to the motors. Full up testing in the maze may commence. Enjoy!

3.5 SUMMARY

In this chapter, we discussed the design process, related tools, and applied the process to a real world design. As previously mentioned, this design example will be periodically revisited throughout the text. It is essential to follow a systematic, disciplined approach to embedded systems design to successfully develop a prototype that meets established requirements.

3.6 REFERENCES

- M. Anderson, Help Wanted: Embedded Engineers Why the United States is losing its edge in embedded systems, *IEEE–USA Today's Engineer*, Feb 2008.

- Barrett S, Pack D (2006) Microcontrollers Fundamentals for Engineers and Scientists. Morgan and Claypool Publishers. DOI: 10.2200/S00025ED1V01Y200605DCS001

- Barrett S and Pack D (2008) Atmel AVR Microcontroller Primer Programming and Interfacing. Morgan and Claypool Publishers. DOI: 10.2200/S00100ED1V01Y200712DCS015

- Barrett S (2010) Embedded Systems Design with the Atmel AVR Microcontroller. Morgan and Claypool Publishers. DOI: 10.2200/S00225ED1V01Y200910DCS025

- M. Fowler with K. Scott "UML Distilled – A Brief Guide to the Standradr Object Modeling Language," 2nd edition. Boston:Addison–Wesley, 2000.

- N. Dale and S.C. Lilly "Pascal Plus Data Structures," 4th edition. Englewood Cliffs, NJ: Jones and Bartlett, 1995.

- *Atmel 8–bit AVR Microcontroller with 4/8/16/32K Bytes In–System Programmable Flash, AT-mega48PA, 88PA, 168PA, 328P* data sheet: 8171D–AVR–05/11, Atmel Corporation, 2325 Orchard Parkway, San Jose, CA 95131.

- *Atmel 8–bit AVR Microcontroller with 64/128/256K Bytes In–System Programmable Flash, AT-mega640/V, ATmega1280/V, 2560/V* data sheet: 2549P-AVR-10/2012, Atmel Corporation, 2325 Orchard Parkway, San Jose, CA 95131.

3.7   CHAPTER PROBLEMS

1. What is an embedded system?

2. What aspects must be considered in the design of an embedded system?

3. What is the purpose of the structure chart, UML activity diagram, and circuit diagram?

4. Why is a system design only as good as the test plan that supports it?

5. During the testing process, when an error is found and corrected, what should now be accomplished?

6. Discuss the top–down design, bottom–up implementation concept.

7. Describe the value of accurate documentation.

8. What is required to fully document an embedded systems design?

9. Update the robot action truth table if the robot was equipped with four IR sensors.

10. Adapt the Blinky 602 control algorithm for the Arduino Mega 2560 processor board.

CHAPTER 4

Atmel AVR Operating
Parameters and Interfacing

Objectives: After reading this chapter, the reader should be able to

- Describe the voltage and current parameters for the Arduino UNO R3, the Arduino Mega 2560, and the Atmel AVR HC CMOS type microcontroller.

- Specify a battery system to power an Arduino processor board and the Atmel AVR based system.

- Apply the voltage and current parameters toward properly interfacing input and output devices to an Arduino processing board and the Atmel AVR microcontroller.

- Interface a wide variety of input and output devices to an Arduino processing board and the Atmel AVR microcontroller.

- Describe the special concerns that must be followed when an Arduino processing board and the Atmel AVR microcontroller is used to interface to a high power DC or AC device.

- Discuss the requirement for an optical based interface.

- Describe how to control the speed and direction of a DC motor.

- Describe how to control several types of AC loads.

4.1 OVERVIEW

The textbook from Morgan & Claypool Publishers (M&C) titled, "Microcontrollers Fundamentals for Engineers and Scientists," contains a chapter entitled "Operating Parameters and Interfacing." With M&C permission, portions of the chapter have been repeated here for your convenience. However, we have customized the information provided to the Arduino UNO R3, the Arduino Mega 2560 and the Atmel AVR line of microcontrollers and have also expanded the coverage of the chapter to include interface techniques for a number of additional input and output devices.

In this chapter, we introduce you to the extremely important concepts of the operating envelope for a microcontroller. We begin by reviewing the voltage and current electrical parameters for the HC CMOS based Atmel AVR line of microcontrollers. We then show how to apply this information to properly interface input and output devices to the Arduino UNO R3, the Arduino Mega 2560, and the ATmega328 microcontroller. We then discuss the special considerations for controlling a high power DC or AC load such as a motor and introduce the concept of an optical interface. Throughout the chapter, we provide a number of detailed examples.

The importance of this chapter can not be emphasized enough. Any time an input or an output device is connected to a microcontroller, the interface between the device and the microcontroller must be carefully analyzed and designed. This will ensure the microcontroller will continue to operate within specified parameters. Should the microcontroller be operated outside its operational envelope, erratic, unpredictable, and an unreliable system may result.

4.2   OPERATING PARAMETERS

Any time a device is connected to a microcontroller, careful interface analysis must be performed. Most microcontrollers are members of the "HC," or high–speed CMOS, family of chips. As long as all components in a system are also of the "HC" family, as is the case for the Arduino UNO R3, the Arduino Mega 2560, and the Atmel AVR line of microcontrollers, electrical interface issues are minimal. If the microcontroller is connected to some component not in the "HC" family, electrical interface analysis must be completed. Manufacturers readily provide the electrical characteristic data necessary to complete this analysis in their support documentation.

To perform the interface analysis, there are eight different electrical specifications required for electrical interface analysis. The electrical parameters are:

- V_{OH}: the lowest guaranteed output voltage for a logic high,

- V_{OL}: the highest guaranteed output voltage for a logic low,

- I_{OH}: the output current for a V_{OH} logic high,

- I_{OL}: the output current for a V_{OL} logic low,

- V_{IH}: the lowest input voltage guaranteed to be recognized as a logic high,

- V_{IL}: the highest input voltage guaranteed to be recognized as a logic low,

- I_{IH}: the input current for a V_{IH} logic high, and

- I_{IL}: the input current for a V_{IL} logic low.

These electrical characteristics are required for both the microcontroller and the external components. Typical values for a microcontroller in the HC CMOS family assuming $V_{DD} = 5.0$ volts and $V_{SS} = 0$ volts are provided below. The minus sign on several of the currents indicates a current flow out of the device. A positive current indicates current flow into the device.

- V_{OH} = 4.2 volts,

- V_{OL} = 0.4 volts,

- I_{OH} = −0.8 milliamps,

- I_{OL} = 1.6 milliamps,

- V_{IH} = 3.5 volts,

- V_{IL} = 1.0 volt,

- I_{IH} = 10 microamps, and

- I_{IL} = −10 microamps.

It is important to realize that these are static values taken under very specific operating conditions. If external circuitry is connected such that the microcontroller acts as a current source (current leaving the microcontroller) or current sink (current entering the microcontroller), the voltage parameters listed above will also be affected.

In the current source case, an output voltage V_{OH} is provided at the output pin of the microcontroller when the load connected to this pin draws a current of I_{OH}. If a load draws more current from the output pin than the I_{OH} specification, the value of V_{OH} is reduced. If the load current becomes too high, the value of V_{OH} falls below the value of V_{IH} for the subsequent logic circuit stage and not be recognized as an acceptable logic high signal. When this situation occurs, erratic and unpredictable circuit behavior results.

In the sink case, an output voltage V_{OL} is provided at the output pin of the microcontroller when the load connected to this pin delivers a current of I_{OL} to this logic pin. If a load delivers more current to the output pin of the microcontroller than the I_{OL} specification, the value of V_{OL} increases. If the load current becomes too high, the value of V_{OL} rises above the value of V_{IL} for the subsequent logic circuit stage and not be recognized as an acceptable logic low signal. As before, when this situation occurs, erratic and unpredictable circuit behavior results.

For convenience this information is illustrated in Figure 4.1. In (a), we provided an illustration of the direction of current flow from the HC device and also a comparison of voltage levels. As a reminder current flowing out of a device is considered a negative current (source case) while current flowing into the device is considered positive current(sink case). The magnitude of the voltage and current for HC CMOS devices are shown in (b). As more current is sinked or sourced from a microcontroller pin, the voltage will be pulled up or pulled down, respectively, as shown in (c). If input and output devices are improperly interfaced to the microcontroller, these loading conditions may become excessive, and voltages will not be properly interpreted as the correct logic levels.

You must also ensure that total current limits for an entire microcontroller port and the overall bulk port specifications are met. For planning purposes the sum of current sourced or sinked from

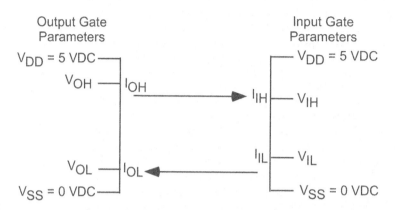

a) Voltage and current electrical parameters

Output Parameters	Input Parameters
V_{OH} = 4.2 V	V_{IH} = 3.5 V
V_{OL} = 0.4 V	V_{IL} = 1.0 V
I_{OH} = - 0.8 mA	I_{IH} = 10 μA
I_{OL} = 1.6 mA	I_{IL} = - 10 μA

b) HC CMOS voltage and current parameters

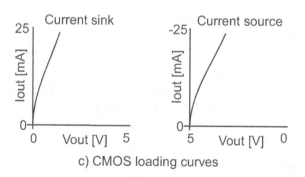

c) CMOS loading curves

Figure 4.1: Electrical voltage and current parameters.

a port should not exceed 100 mA. Furthermore, the sum of currents for all ports should not exceed 200 mA. As before, if these guidelines are not followed, erratic microcontroller behavior may result.

The procedures presented in the following sections, when followed carefully, will ensure the microcontroller will operate within its designed envelope. The remainder of the chapter is divided into input device interface analysis followed by output device interface analysis. Since many embedded systems operate from a DC battery source, we begin by examining several basic battery supply circuits.

4.3   BATTERY OPERATION

Many embedded systems are remote, portable systems operating from a battery supply. To properly design a battery source for an embedded system, the operating characteristics of the embedded system must be matched to the characteristics of the battery supply.

4.3.1   EMBEDDED SYSTEM VOLTAGE AND CURRENT DRAIN SPECIFICATIONS

An embedded system has a required supply voltage and an overall current requirement. For the purposes of illustration, we will assume our microcontroller based embedded system operates from 5 VDC. The overall current requirements of the system are determined by the worst case current requirements when all embedded system components are operational.

4.3.2   BATTERY CHARACTERISTICS

To properly match a battery to an embedded system, the battery voltage and capacity must be specified. Battery capacity is typically specified as a mAH rating. For example, a typical 9 VDC non--rechargeable alkaline battery has a capacity of 550 mAH. If the embedded system has a maximum operating current of 50 mA, it will operate for approximately eleven hours before battery replacement is required.

A battery is typically used with a voltage regulator to maintain the voltage at a prescribed level. Figure 4.2 provides sample circuits to provide a +5 VDC and a ±5 VDC portable battery source. Additional information on battery capacity and characteristics may be found in Barrett and Pack [Prentice–Hall, 2005].

4.4   INPUT DEVICES

In this section, we discuss how to properly interface input devices to a microcontroller. We will start with the most basic input component, a simple on/off switch.

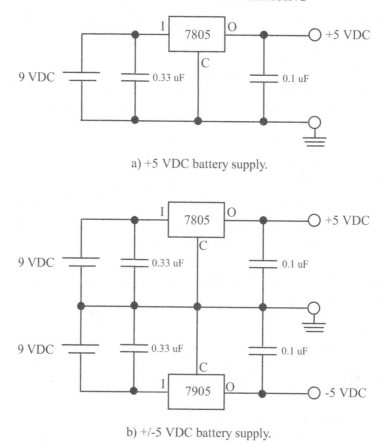

a) +5 VDC battery supply.

b) +/-5 VDC battery supply.

Figure 4.2: Battery supply circuits employing a 9 VDC battery with a 5 VDC regulators.

4.4.1   SWITCHES

Switches come in a variety of types. As a system designer it is up to you to choose the appropriate switch for a specific application. Switch varieties commonly used in microcontroller applications are illustrated in Figure 4.3(a). Here is a brief summary of the different types:

- **Slide switch:** A slide switch has two different positions: on and off. The switch is manually moved to one position or the other. For microcontroller applications, slide switches are available that fit in the profile of a common integrated circuit size dual inline package (DIP). A bank of four or eight DIP switches in a single package is commonly available.

- **Momentary contact pushbutton switch:** A momentary contact pushbutton switch comes in two varieties: normally closed (NC) and normally open (NO). A normally open switch, as its name implies, does not normally provide an electrical connection between its contacts. When the pushbutton portion of the switch is depressed, the connection between the two switch contacts is made. The connection is held as long as the switch is depressed. When the switch is released the connection is opened. The converse is true for a normally closed switch. For microcontroller applications, pushbutton switches are available in a small tact type switch configuration.

- **Push on/push off switches:** These type of switches are also available in a normally open or normally closed configuration. For the normally open configuration, the switch is depressed to make connection between the two switch contacts. The pushbutton must be depressed again to release the connection.

- **Hexadecimal rotary switches:** Small profile rotary switches are available for microcontroller applications. These switches commonly have sixteen rotary switch positions. As the switch is rotated to each position, a unique four bit binary code is provided at the switch contacts.

A common switch interface is shown in Figure 4.3(b). This interface allows a logic one or zero to be properly introduced to a microcontroller input port pin. The basic interface consists of the switch in series with a current limiting resistor. The node between the switch and the resistor is provided to the microcontroller input pin. In the configuration shown, the resistor pulls the microcontroller input up to the supply voltage V_{DD}. When the switch is closed, the node is grounded and a logic zero is provided to the microcontroller input pin. To reverse the logic of the switch configuration, the position of the resistor and the switch is simply reversed.

4.4.2   PULLUP RESISTORS IN SWITCH INTERFACE CIRCUITRY

Many microcontrollers are equipped with pullup resistors at the input pins. The pullup resistors are asserted with the appropriate register setting. The pullup resistor replaces the external resistor in the switch configuration as shown in Figure 4.3b) right.

4.4.3   SWITCH DEBOUNCING

Mechanical switches do not make a clean transition from one position (on) to another (off). When a switch is moved from one position to another, it makes and breaks contact multiple times. This activity may go on for tens of milliseconds. A microcontroller is relatively fast as compared to the action of the switch. Therefore, the microcontroller is able to recognize each switch bounce as a separate and erroneous transition.

To correct the switch bounce phenomena additional external hardware components may be used or software techniques may be employed. A hardware debounce circuit is illustrated in Figure 4.3(c). The node between the switch and the limiting resistor of the basic switch circuit is fed to a low pass filter (LPF) formed by the 470 k ohm resistor and the capacitor. The LPF prevents abrupt changes (bounces) in the input signal from the microcontroller. The LPF is followed by a 74HC14

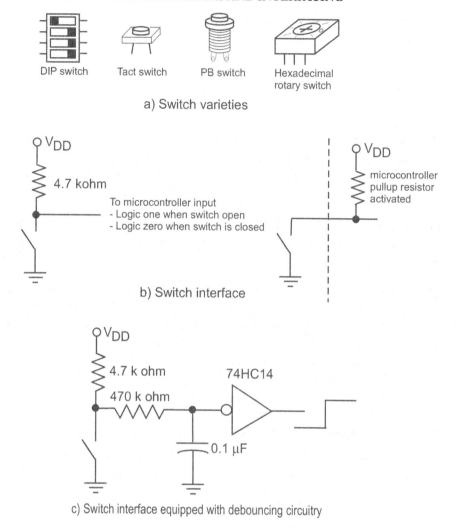

a) Switch varieties

b) Switch interface

c) Switch interface equipped with debouncing circuitry

Figure 4.3: Switch interface.

Schmitt Trigger, which is simply an inverter equipped with hysteresis. This further limits the switch bouncing.

Switches may also be debounced using software techniques. This is accomplished by inserting a 30 to 50 ms lockout delay in the function responding to port pin changes. The delay prevents the microcontroller from responding to the multiple switch transitions related to bouncing.

You must carefully analyze a given design to determine if hardware or software switch debouncing techniques will be used. It is important to remember that all switches exhibit bounce phenomena and, therefore, must be debounced.

4.4.4 KEYPADS

A keypad is simply an extension of the simple switch configuration. A typical keypad configuration and interface are shown in Figure 4.4. As you can see the keypad is simply multiple switches in the same package. A hexadecimal keypad is provided in the figure. A single row of keypad switches are asserted by the microcontroller and then the host keypad port is immediately read. If a switch has been depressed, the keypad pin corresponding to the column the switch is in will also be asserted. The combination of a row and a column assertion can be decoded to determine which key has been pressed as illustrated in the table. Keypad rows are continually asserted one after the other in sequence. Since the keypad is a collection of switches, debounce techniques must also be employed.

The keypad may be used to introduce user requests to a microcontroller. A standard keypad with alphanumeric characters may be used to provide alphanumeric values to the microcontroller such as providing your personal identification number (PIN) for a financial transaction. However, some keypads are equipped with removable switch covers such that any activity can be associated with a key press.

In Figure 4.5, we have connected the ATmega328 to a hexadecimal keypad via PORTB. PORTB[3:0] is configured as output to selectively assert each row. PORTB[7:4] is configured as input. Each row is sequentially asserted low. Each column is then read via PORTB[7:4] to see if any switch in that row has been depressed. If no switches have been depressed in the row, an "F" will be read from PORTB[7:4]. If a switch has been depressed, some other value than "F" will be read. The read value is then passed into a switch statement to determine the ASCII equivalent of the depressed switch. The function is not exited until the switch is released. This prevents a switch "double hit."

```
//*************************************************************************


unsigned char get_keypad_value(void)
{
unsigned char  PORTB_value, PORTB_value_masked;
unsigned char  ascii_value;

DDRC = 0x0F;
       //set PORTB[7:4] to input,
                                          //PORTB[3:0] to output

                                          //switch depressed in row 0?
PORTB = 0xFE;                             //assert row 0 via PORTB[0]
PORTB_value = PINB;                       //read PORTB
PORTB_value_masked = (PORTB_value & 0xf0); //mask PORTB[3:0]
```

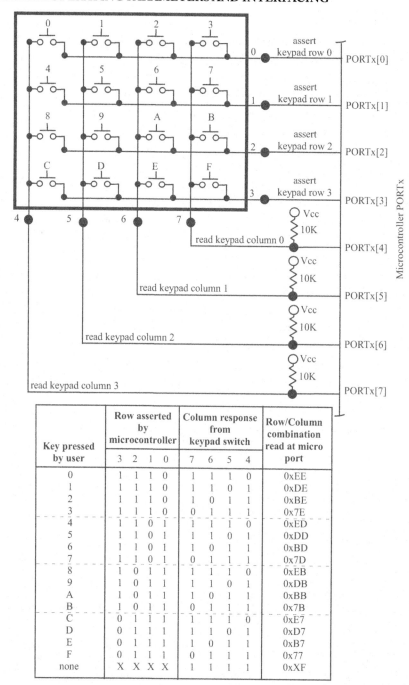

Figure 4.4: Keypad interface.

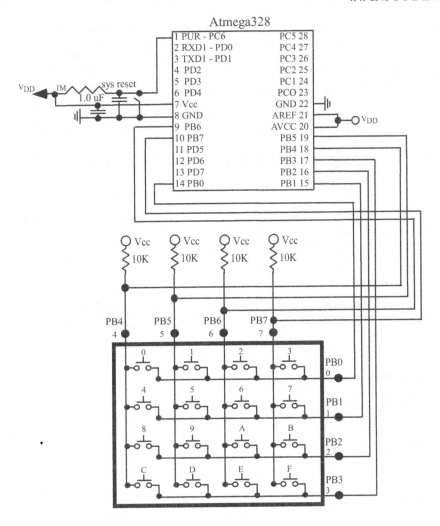

Figure 4.5: Hexadecimal keypad interface to microcontroller.

```
                                                //switch depressed in row 1?
if(PORTB_value_masked == 0xf0)
      //no switches depressed in row 0
   {
   PORTB = 0xFD;                                //assert Row 1 via PORTB[1]
   PORTB_value = PINB;                          //read PORTB
   PORTB_value_masked = (PORTB_value & 0xf0);//mask PORTB[3:0]
```

```
    }


                                                  //switch depressed in row 2?
if(PORTB_value_masked == 0xf0)
        //no switches depressed in row 0
    {
    PORTB = 0xFB;                          //assert Row 2 via PORTC[2]
    PORTB_value = PINB;                    //read PORTB
    PORTB_value_masked = (PORTB_value & 0xf0);//mask PORTB[3:0]
    }


                                                  //switch depressed in row 3?
if(PORTB_value_masked == 0xf0)
        //no switches depressed in row 0
    {
    PORTB = 0xF7;                          //assert Row 3 via PORTB[3]
    PORTB_value = PINB;                    //read PORTB
    PORTB_value_masked = (PORTB_value & 0xf0);//mask PORTB[3:0]
    }

if(PORTB_value_masked != 0xf0)
  {
  switch(PORTB_value_masked)
    {
    case 0xEE: ascii_value = '0';
              break;

    case 0xDE: ascii_value = '1';
              break;

    case 0xBE: ascii_value = '2';
              break;

    case 0x7E: ascii_value = '3';
              break;

    case 0xED: ascii_value = '4';
              break;
```

```
    case 0xDD: ascii_value = '5';
               break;

    case 0xBD: ascii_value = '6';
               break;

    case 0x7D: ascii_value = '7';
               break;

    case 0xEB: ascii_value = '8';
               break;

    case 0xDB: ascii_value = '9';
               break;

    case 0xBB: ascii_value = 'a';
               break;

    case 0x&B: ascii_value = 'b';
               break;

    case 0xE7: ascii_value = 'c';
               break;

    case 0xD7: ascii_value = 'd';
               break;

    case 0xB7: ascii_value = 'e';
               break;

    case 0x77: ascii_value = 'f';
               break;

    default:;
    }

while(PORTB_value_masked != 0xf0);
       //wait for key to be released
```

```
return ascii_value;
}


//****************************************************************************
```

4.4.5   SENSORS

A microcontroller is typically used in control applications where data is collected, the data is assimilated and processed by the host algorithm, and a control decision and accompanying signals are provided by the microcontroller. Input data for the microcontroller is collected by a complement of input sensors. These sensors may be digital or analog in nature.

4.4.5.1   Digital Sensors

Digital sensors provide a series of digital logic pulses with sensor data encoded. The sensor data may be encoded in any of the parameters associated with the digital pulse train such as duty cycle, frequency, period, or pulse rate. The input portion of the timing system may be configured to measure these parameters.

An example of a digital sensor is the optical encoder. An optical encoder consists of a small plastic transparent disk with opaque lines etched into the disk surface. A stationary optical emitter and detector pair is placed on either side of the disk. As the disk rotates, the opaque lines break the continuity between the optical source and detector. The signal from the optical detector is monitored to determine disk rotation as shown in Figure 4.6.

Optical encoders are available in a variety of types depending on the information desired. There are two major types of optical encoders: incremental encoders and absolute encoders. An absolute encoder is used when it is required to retain position information when power is lost. For example, if you were using an optical encoder in a security gate control system, an absolute encoder would be used to monitor the gate position. An incremental encoder is used in applications where a velocity or a velocity and direction information is required.

The incremental encoder types may be further subdivided into tachometers and quadrature encoders. An incremental tachometer encoder consists of a single track of etched opaque lines as shown in Figure 4.6(a). It is used when the velocity of a rotating device is required. To calculate velocity, the number of detector pulses are counted in a fixed amount of time. Since the number of pulses per encoder revolution is known, velocity may be calculated.

The quadrature encoder contains two tracks shifted in relationship to one another by 90 degrees. This allows the calculation of both velocity and direction. To determine direction, one would monitor the phase relationship between Channel A and Channel B as shown in Figure 4.6(b). The absolute encoder is equipped with multiple data tracks to determine the precise location of the encoder disk [Sick Stegmann].

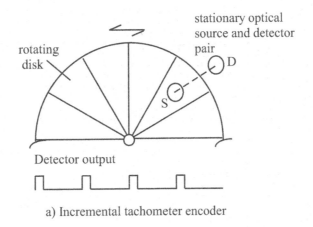

a) Incremental tachometer encoder

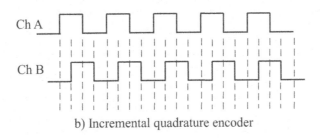

b) Incremental quadrature encoder

Figure 4.6: Optical encoder.

4.4.5.2 Analog Sensors

Analog sensors provide a DC voltage that is proportional to the physical parameter being measured. As discussed in the analog to digital conversion chapter, the analog signal may be first preprocessed by external analog hardware such that it falls within the voltage references of the conversion subsystem. The analog voltage is then converted to a corresponding binary representation.

Example 1: Flex Sensor An example of an analog sensor is the flex sensor shown in Figure 4.7(a). The flex sensor provides a change in resistance for a change in sensor flexure. At 0 degrees flex, the sensor provides 10k Ohms of resistance. For 90 degrees flex, the sensor provides 30–40k Ohms of resistance. Since the processor can not measure resistance directly, the change in flex sensor resistance must be converted to a change in a DC voltage. This is accomplished using the voltage divider network shown in Figure 4.7(c). For increased flex, the DC voltage will increase. The voltage can be measured using the analog–to–digital converter subsystem.

The flex sensor may be used in applications such as virtual reality data gloves, robotic sensors, biometric sensors, and in science and engineering experiments [Images Company]. The author used

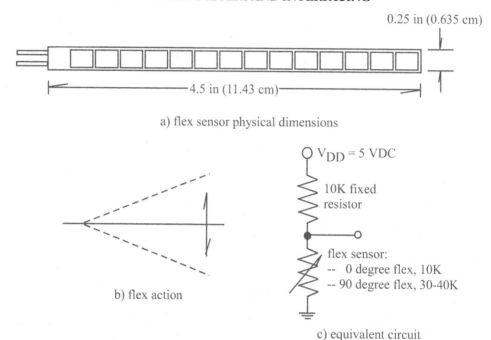

a) flex sensor physical dimensions

b) flex action

c) equivalent circuit

Figure 4.7: Flex sensor.

the circuit provided in Figure 4.7 to help a colleague in Zoology monitor the movement of a newt salamander during a scientific experiment.

Example 2: Ultrasonic sensor The ultrasonic sensor pictured in Figure 4.8 is an example of an analog based sensor. The sensor is based on the concept of ultrasound or sound waves that are at a frequency above the human range of hearing (20 Hz to 20 kHz). The ultrasonic sensor pictured in Figure 4.8c) emits a sound wave at 42 kHz. The sound wave reflects from a solid surface and returns back to the sensor. The amount of time for the sound wave to transit from the surface and back to the sensor may be used to determine the range from the sensor to the wall. Pictured in Figure 4.8(c) and (d) is an ultrasonic sensor manufactured by Maxbotix (LV–EZ3). The sensor provides an output that is linearly related to range in three different formats: a) a serial RS–232 compatible output at 9600 bits per second, b) a pulse width modulated (PWM) output at a 147 us/inch duty cycle, and c) an analog output at a resolution of 10 mV/inch. The sensor is powered from a 2.5 to 5.5 VDC source [www.sparkfun.com].

Example 3: Inertial Measurement Unit Pictured in Figure 4.9 is an inertial measurement unit (IMU) combination which consists of an IDG5000 dual–axis gyroscope and an ADXL335 triple axis accelerometer. This sensor may be used in unmanned aerial vehicles (UAVs), autonomous

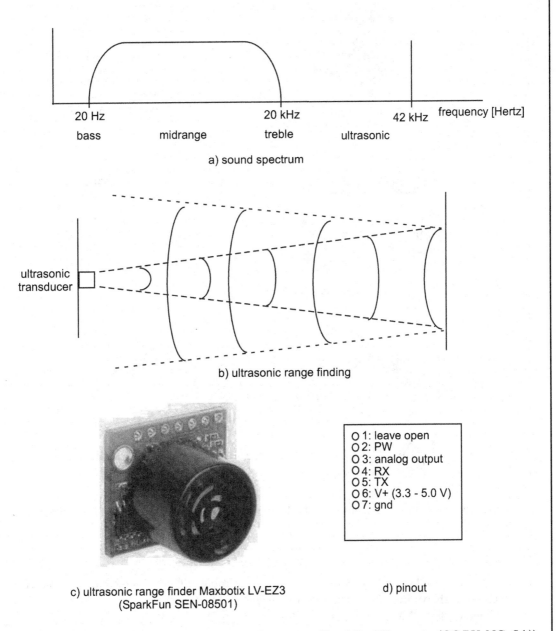

a) sound spectrum

b) ultrasonic range finding

c) ultrasonic range finder Maxbotix LV-EZ3
(SparkFun SEN-08501)

d) pinout

Figure 4.8: Ultrasonic sensor. (Sensor image used courtesy of SparkFun, Electronics (CC BY–NC–SA)).

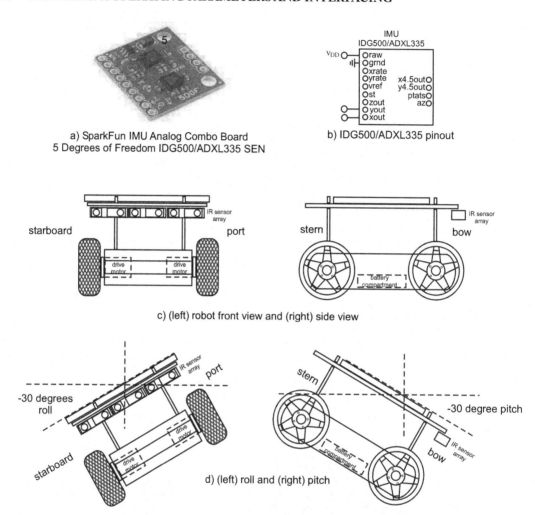

Figure 4.9: Inertial measurement unit. (IMU image used courtesy of SparkFun, Electronics (CC BY–NC–SA)).

helicopters and robots. For robotic applications the robot tilt may be measured in the X and Y directions as shown in Figure 4.9(c) and (d) [www.sparkfun.com].

Example 4: LM34 Temperature Sensor Example Temperature may be sensed using an LM34 (Fahrenheit) or LM35 (Centigrade) temperature transducer. The LM34 provides an output voltage that is linearly related to temperature. For example, the LM34D operates from 32 degrees F to 212 degrees F providing +10mV/degree Fahrenheit resolution with a typical accuracy of ±0.5 degrees Fahrenheit [National]. This sensor is used in the automated cooling fan example at the end of the chapter. The output from the sensor is typically connected to the ADC input of the microcontroller.

4.5   OUTPUT DEVICES

As previously mentioned, an external device should not be connected to a microcontroller without first performing careful interface analysis to ensure the voltage, current, and timing requirements of the microcontroller and the external device. In this section, we describe interface considerations for a wide variety of external devices. We begin with the interface for a single light emitting diode.

4.5.1   LIGHT EMITTING DIODES (LEDS)

An LED is typically used as a logic indicator to inform the presence of a logic one or a logic zero at a specific pin of a microcontroller. An LED has two leads: the anode or positive lead and the cathode or negative lead. To properly bias an LED, the anode lead must be biased at a level approximately 1.7 to 2.2 volts higher than the cathode lead. This specification is known as the forward voltage (V_f) of the LED. The LED current must also be limited to a safe level known as the forward current (I_f). The diode voltage and current specifications are usually provided by the manufacturer.

An example of an LED biasing circuit is provided in Figure 4.10. A logic one is provided by the microcontroller to the input of the inverter. The inverter provides a logic zero at its output which provides a virtual ground at the cathode of the LED. Therefore, the proper voltage biasing for the LED is provided. The resistor (R) limits the current through the LED. A proper resistor value can be calculated using $R = (V_{DD} - V_{DIODE})/I_{DIODE}$. It is important to note that a 7404 inverter must be used due to its capability to safely sink 16 mA of current. Alternately, an NPN transistor such as a 2N2222 (PN2222 or MPQ2222) may be used in place of the inverter as shown in the figure. In Chapter 1, we used large (10 mm) red LEDs in the KNH instrumentation project. These LEDs have V_f of 6 to 12 VDC and I_f of 20 mA at 1.85 VDC. This requires the interface circuit shown in Figure 4.10c) right.

4.5.2   SEVEN SEGMENT LED DISPLAYS

To display numeric data, seven segment LED displays are available as shown in Figure 4.11(a). Different numerals can be displayed by asserting the proper LED segments. For example, to display the number five, segments a, c, d, f, and g would be illuminated. Seven segment displays are available in common cathode (CC) and common anode (CA) configurations. As the CC designation implies, all seven individual LED cathodes on the display are tied together.

The microcontroller is not capable of driving the LED segments directly. As shown in Figure 4.11(a), an interface circuit is required. We use a 74LS244 octal buffer/driver circuit to boost the current available for the LED. The LS244 is capable of providing 15 mA per segment (I_{OH}) at 2.0 VDC (V_{OH}). A limiting resistor is required for each segment to limit the current to a safe value for the LED. Conveniently, resistors are available in DIP packages of eight for this type of application.

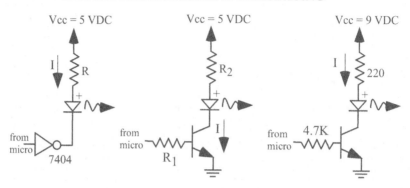

Figure 4.10: LED display devices.

Seven segment displays are available in multi–character panels. In this case, separate micro-controller ports are not used to provide data to each seven segment character. Instead, a single port is used to provide character data. A portion of another port is used to sequence through each of the characters as shown in Figure 4.11(b). An NPN (for a CC display) transistor is connected to the common cathode connection of each individual character. As the base contact of each transistor is sequentially asserted, the specific character is illuminated. If the microcontroller sequences through the display characters at a rate greater than 30 Hz, the display will have steady illumination.

4.5.3 CODE EXAMPLE

Provided below is a function used to illuminate the correct segments on a multi–numeral seven display. The numeral is passed in as an argument to the function along with the numerals position on the display and also an argument specifying whether or not the decimal point (dp) should be displayed at that position. The information to illuminate specific segments are provided in Figure 4.11c).

```
//******************************************************************
void LED_character_display(unsigned int numeral, unsigned int position,
                           unsigned int decimal_point)
{
unsigned char output_value;

                                //illuminate numerical segments
switch(numeral)
  {
  case 0: output_value = 0x7E;
    break;
```

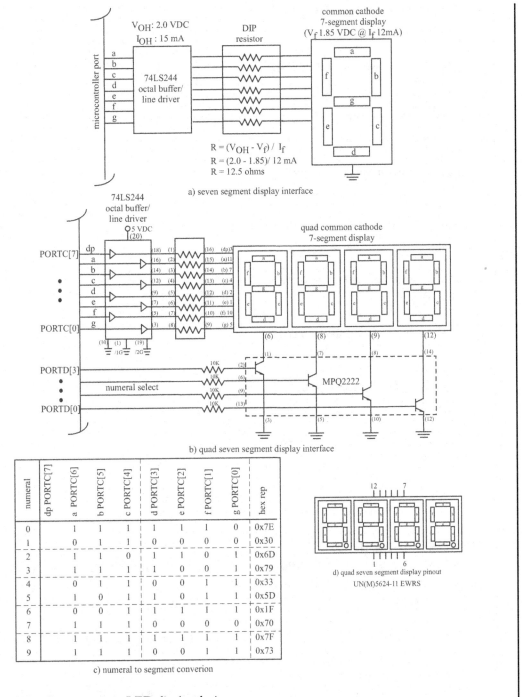

Figure 4.11: Seven segment LED display devices.

```
   case 1: output_value = 0x30;
     break;

   case 2: output_value = 0x6D;
     break;

   case 3: output_value = 0x79;
     break;

   case 4: output_value = 0x33;
     break;

   case 5: output_value = 0x5D;
     break;

   case 6: output_value = 0x1F;
     break;

   case 7: output_value = 0x70;
     break;

   case 8: output_value = 0x7F;
     break;

   case 9: output_value = 0x73;
     break;

   default:;
   }

if(decimal_point != 0)
  PORTB = output_value | 0x80;        //illuminate decimal point

switch(position)                      //assert position
  {
  case 0: PORTD = 0x01;               //least significant bit
          break;

  case 1: PORTD = 0x02;               //least significant bit + 1
```

```
                break;

    case 2: PORTD = 0x04;                   //least significant bit + 2
                break;

    case 3: PORTD = 0x08;                   //most significant bit
                break;

    default:;
        }
}
//*****************************************************************************
```

4.5.4   TRI–STATE LED INDICATOR

The tri–state LED indicator is shown in Figure 4.12. It is used to provide the status of an entire microcontroller port. The indicator bank consists of eight green and eight red LEDs. When an individual port pin is logic high, the green LED is illuminated. When logic low, the red LED is illuminated. If the port pin is at a tri–state high impedance state, no LED is illuminated.

The NPN/PNP transistor pair at the bottom of the figure provides a 2.5 VDC voltage reference for the LEDs. When a specific port pin is logic high (5.0 VDC), the green LED will be forward biased since its anode will be at a higher potential than its cathode. The 47 ohm resistor limits current to a safe value for the LED. Conversely, when a specific port pin is at a logic low (0 VDC) the red LED will be forward biased and illuminate. For clarity, the red and green LEDs are shown as being separate devices. LEDs are available that have both LEDs in the same device.

4.5.5   DOT MATRIX DISPLAY

The dot matrix display consists of a large number of LEDs configured in a single package. A typical 5 x 7 LED arrangement is a matrix of five columns of LEDs with seven LEDs per row as shown in Figure 4.13. Display data for a single matrix column [R6–R0] is provided by the microcontroller. That specific row is then asserted by the microcontroller using the column select lines [C2–C0]. The entire display is sequentially built up a column at a time. If the microcontroller sequences through each column fast enough (greater than 30 Hz), the matrix display appears to be stationary to a human viewer.

In Figure 4.13(a), we have provided the basic configuration for the dot matrix display for a single display device. However, this basic idea can be expanded in both dimensions to provide a multi–character, multi–line display. A larger display does not require a significant number of microcontroller pins for the interface. The dot matrix display may be used to display alphanumeric data as well as graphics data. In Figure 4.13(b), we have provided additional detail of the interface circuit.

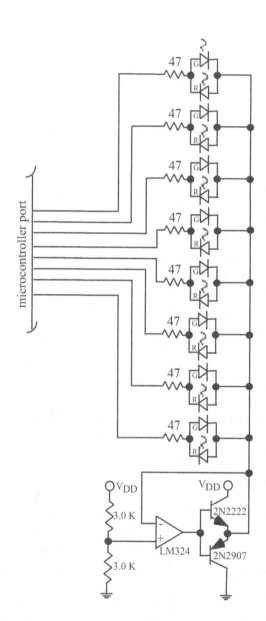

Figure 4.12: Tri-state LED display.

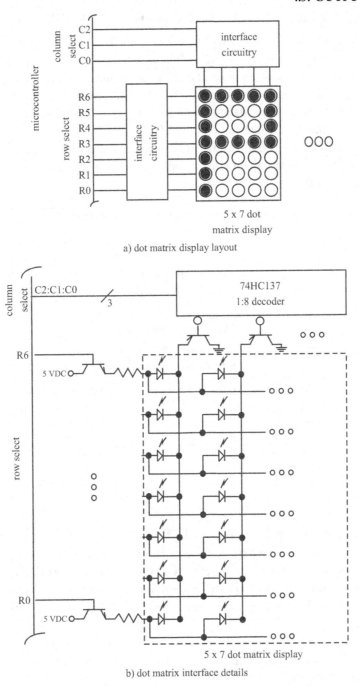

a) dot matrix display layout

b) dot matrix interface details

Figure 4.13: Dot matrix display.

4.5.6   LIQUID CRYSTAL CHARACTER DISPLAY (LCD) IN C

An LCD is an output device to display text information as shown in Figure 4.14. LCDs come in a wide variety of configurations, including multi–character, multi–line format. A 16 x 2 LCD format is common. That is, it has the capability of displaying two lines of 16 characters each. The characters are sent to the LCD via American Standard Code for Information Interchange (ASCII) format a single character at a time. For a parallel configured LCD, an eight bit data path and two lines are required between the microcontroller and the LCD. A small microcontroller mounted to the back panel of the LCD translates the ASCII data characters and control signals to properly display the characters. LCDs are configured for either parallel or serial data transmission format. In the example provided, we use a parallel configured display.

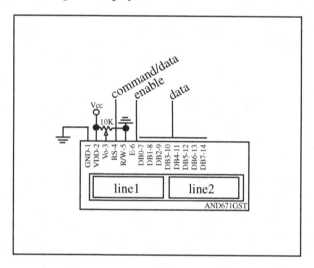

Figure 4.14: LCD display.

4.5.6.1   Programming a parallel configured LCD in C

Some sample C code is provided below to send data and control signals to an LCD. In this specific example, an AND671GST 1 x 16 character LCD was connected to the Atmel ATmega328 micro-controller. One 8–bit port and two extra control lines are required to connect the microcontroller to the LCD. Note: The initialization sequence for the LCD is specified within the manufacturer's technical data.

```
//********************************************************************
//Internal Oscillator: 1 MHz
//ATMEL AVR ATmega328
//Chip Port Function I/O Source/Dest Asserted Notes
//********************************************************************
```

```
//Pin 1: /Reset
//Pin 2: PD0 to DB0 LCD
//Pin 3: PD1 to DB1 LCD
//Pin 4: PD2 to DB2 LCD
//Pin 5: PD3 to DB3 LCD
//Pin 6: PD4 to DB4 LCD
//Pin 7: Vcc
//Pin 8: Gnd
//Pin 11: PD5 to DB6 LCD
//Pin 12: PD6 to DB6 LCD
//Pin 13: PD7 to DB7 LCD
//Pin 20: AVCC to Vcc
//Pin 21: AREF to Vcc
//Pin 22 Gnd
//Pin 27 PC4 to LCD Enable (E)
//Pin 28 PC5 to LCD RS


//include files********************************************************

//ATMEL register definitions for ATmega328
#include<iom328v.h>



//function prototypes*************************************************
void delay(unsigned int number_of_65_5ms_interrupts);
void init_timer0_ovf_interrupt(void);
void initialize_ports(void);            //initializes ports
void power_on_reset(void);              //returns system to startup state
void clear_LCD(void);                   //clears LCD display
void LCD_Init(void);                    //initialize AND671GST LCD
void putchar(unsigned char c);          //send character to LCD
void putcommand(unsigned char c);       //send command to LCD
void timer0_interrupt_isr(void);
void perform_countdown(void);
void clear_LCD(void);
void systems_A_OK(void);
void print_mission_complete(void);
void convert_int_to_string_display_LCD(unsigned int total_integer_value);
```

```
//program constants
#define TRUE    1
#define FALSE   0
#define OPEN    1
#define CLOSE   0
#define YES     1
#define NO      0
#define SAFE    1
#define UNSAFE  0
#define ON      1
#define OFF     0


//interrupt handler definition
#pragma interrupt_handler timer0_interrupt_isr:17


//main program*******************************************************

//global variables

void main(void)
{
init_timer0_ovf_interrupt();
//initialize Timer0 to serve as elapsed
initialize_ports();                     //initialize ports

perform_countdown();
delay(46);
:
:
:
systems_A_OK();

}//end main


//function definitions************************************************

//*******************************************************************
//initialize_ports: provides initial configuration for I/O ports
//*******************************************************************
```

```
void initialize_ports(void)
{
DDRB = 0xff;                    //PORTB[7:0] as output
PORTB= 0x00;                    //initialize low
DDRC = 0xff;                    //PORTC[7:0] as output
PORTC= 0x00;                    //initialize low

DDRD = 0xff;                    //PORTB[7:0] as output
PORTD= 0x00;                    //initialize low
}


//******************************************************************
//delay(unsigned int num_of_65_5ms_interrupts): this generic delay function
//provides the specified delay as the number of 65.5 ms "clock ticks" from
//the Timer0 interrupt.
//Note: this function is only valid when using a 1 MHz crystal or ceramic
//      resonator
//******************************************************************

void delay(unsigned int number_of_65_5ms_interrupts)
{
TCNT0 = 0x00;                           //reset timer0
delay_timer = 0;
while(delay_timer <= number_of_65_5ms_interrupts)
  {
  ;
  }
}


//******************************************************************
//int_timer0_ovf_interrupt(): The Timer0 overflow interrupt is being
//employed as a time base for a master timer for this project.
//The internal time base is set to operate at 1 MHz and then
//is divided by 256.  The 8-bit Timer0
//register (TCNT0) overflows every 256 counts or every 65.5 ms.
//******************************************************************

void init_timer0_ovf_interrupt(void)
```

```
{
TCCR0 = 0x04; //divide timer0 timebase by 256, overflow occurs every 65.5ms
TIMSK = 0x01; //enable timer0 overflow interrupt
asm("SEI");    //enable global interrupt
}


//******************************************************************************
//LCD_Init: initialization for an LCD connected in the following manner:
//LCD: AND671GST 1x16 character display
//LCD configured as two 8 character lines in a 1x16 array
//LCD data bus (pin 14-pin7) ATMEL 8: PORTD
//LCD RS (pin 28)   ATMEL 8: PORTC[5]
//LCD E  (pin 27)   ATMEL 8: PORTC[4]
//******************************************************************************

void LCD_Init(void)
{
delay(1);
delay(1);
delay(1);
                    // output command string to initialize LCD
putcommand(0x38);   //function set 8-bit
delay(1);
putcommand(0x38);   //function set 8-bit
putcommand(0x38);   //function set 8-bit
putcommand(0x38);   //one line, 5x7 char
putcommand(0x0C);   //display on
putcommand(0x01);   //display clear-1.64 ms
putcommand(0x06);   //entry mode set
putcommand(0x00);   //clear display, cursor at home
putcommand(0x00);   //clear display, cursor at home
}


//******************************************************************************
//putchar:prints specified ASCII character to LCD
//******************************************************************************

void putchar(unsigned char c)
{
```

```
DDRD  = 0xff;                                  //set PORTD as output
DDRC  = DDRC|0x30;                             //make PORTC[5:4] output
PORTD = c;
PORTC = (PORTC|0x20)|PORTC_pullup_mask;   //RS=1
PORTC = (PORTC|0x10)|PORTC_pullup_mask;;  //E=1
PORTC = (PORTC&0xef)|PORTC_pullup_mask;;  //E=0
delay(1);
}


//*******************************************************************
//putcommand: performs specified LCD related command
//*******************************************************************

void putcommand(unsigned char d)
{
DDRD  = 0xff;                                  //set PORTD as output
DDRC  = DDRC|0xC0;                             //make PORTA[5:4] output
PORTC = (PORTC&0xdf)|PORTC_pullup_mask;   //RS=0
PORTD = d;
PORTC = (PORTC|0x10)|PORTC_pullup_mask;   //E=1
PORTC = (PORTC&0xef)|PORTC_pullup_mask;   //E=0
delay(1);
}


//*******************************************************************
//clear_LCD: clears LCD
//*******************************************************************

void clear_LCD(void)
{
putcommand(0x01);
}


//*******************************************************************
//void timer0_interrupt_isr(void)
//*******************************************************************

void timer0_interrupt_isr(void)
{
```

```
delay_timer++;
}
//*****************************************************************************
//void perform_countdown(void)
//*****************************************************************************

void perform_countdown(void)
{
clear_LCD();
putcommand(0x01);                    //cursor home
putcommand(0x80);                    //DD RAM location 1 - line 1
putchar('1'); putchar ('0'); //print 10
delay(15);                           //delay 1s

putcommand(0x01);                    //cursor home
putcommand(0x80);                    //DD RAM location 1 - line 1
putchar('9');                        //print 9
delay(15);                           //delay 1s

putcommand(0x01);                    //cursor home
putcommand(0x80);                    //DD RAM location 1 - line 1
putchar('8');                        //print 8
delay(15);                           //delay 1s

putcommand(0x01);                    //cursor home
putcommand(0x80);                    //DD RAM location 1 - line 1
putchar('7');                        //print 7
delay(15);                           //delay 1s

putcommand(0x01);                    //cursor home
putcommand(0x80);                    //DD RAM location 1 - line 1
putchar('6');                        //print 6
delay(15);                           //delay 1s

putcommand(0x01);                    //cursor home
putcommand(0x80);                    //DD RAM location 1 - line 1
putchar('5');                        //print 5
delay(15);                           //delay 1s
```

```
putcommand(0x01);              //cursor home
putcommand(0x80);              //DD RAM location 1 - line 1
putchar('4');                  //print 4
delay(15);                     //delay 1s

putcommand(0x01);              //cursor home
putcommand(0x80);              //DD RAM location 1 - line 1
putchar('3');                  //print 3
delay(15);                     //delay 1s

putcommand(0x01);              //cursor home
putcommand(0x80);              //DD RAM location 1 - line 1
putchar('2');                  //print 2
delay(15);                     //delay 1s

putcommand(0x01);              //cursor home
putcommand(0x80);              //DD RAM location 1 - line 1
putchar('1');                  //print 1
delay(15);                     //delay 1s


putcommand(0x01);              //cursor home
putcommand(0x80);              //DD RAM location 1 - line 1
putchar('0');                  //print 0
delay(15);                     //delay 1s

//BLASTOFF!
putcommand(0x01);              //cursor home

putcommand(0x80);              //DD RAM location 1 - line 1
putchar('B'); putchar('L'); putchar('A'); putchar('S'); putchar('T');
putchar('O'); putchar('F'); putchar('F'); putchar('!');

}


//****************************************************************************
//void systems_A_OK(void)
//****************************************************************************
```

```
void systems_A_OK(void)
{
clear_LCD();
putcommand(0x01);              //cursor home
putcommand(0x80);                //DD RAM location 1 - line 1
putchar('S'); putchar('Y'); putchar('S'); putchar('T'); putchar('E');
putchar('M'); putchar('S'); putchar(' '); putchar('A'); putchar('-');
putchar('O'); putchar('K'); putchar('!'); putchar('!'); putchar('!');
}


//**********************************************************************
//void print_mission_complete(void)
//**********************************************************************

void print_mission_complete(void)
{
clear_LCD();
putcommand(0x01);              //cursor home
putcommand(0x80);                //DD RAM location 1 - line 1
putchar('M'); putchar('I'); putchar('S'); putchar('S'); putchar('I');
putchar('O'); putchar('N');

putcommand(0xC0);//DD RAM location 1 - line 2
putchar('C'); putchar('O'); putchar('M'); putchar('P'); putchar('L');
putchar('E'); putchar('T'); putchar('E'); putchar('!');
}


//**********************************************************************
//end of file

//**********************************************************************
```

4.5.7   PROGRAMMING A SERIAL CONFIGURED LCD

We demonstrate how to connect a serial configured LCD in Section 8.10.

4.5.8   LIQUID CRYSTAL CHARACTER DISPLAY (LCD) USING THE ARDUINO DEVELOPMENT ENVIRONMENT

The Arduino Development Environment provides full support to interface an Arduino UNO R3 to either a 4–bit or 8–bit configured LCD. In the 4–bit mode precious output pins are preserved for other uses over the 8–bit configuration [www.arduino.cc].

The Arduino Development Environment hosts the following LCD related functions:

- **Hello World:** This function displays the greeting "hello world" and also displays elapsed time since the last reset.

- **Blink:** The Blink function provides cursor control.

- **Cursor:** This function provides control over the underscore style cursor.

- **Display:** This function blanks the display without the loss of displayed information.

- **Text Direction:** This function controls which direction text flows from the cursor.

- **Autoscroll:** This function provides automatic scrolling to the new text.

- **Serial Input:** This function accepts serial input and displays it on the LCD.

- **SetCursor:** This function sets the cursor position on the LCD.

- **Scroll:** This function scrolls text left and right.

Rather than include the excellent Arduino resource here, the interested reader is referred to the Arduino website for sample code and full documentation [www.arduino.cc].

4.5.9   HIGH POWER DC DEVICES

A number of direct current devices may be controlled with an electronic switching device such as a MOSFET. Specifically, an N–channel enhancement MOSFET (metal oxide semiconductor field effect transistor) may be used to switch a high current load on and off (such as a motor) using a low current control signal from a microcontroller as shown in Figure 4.15(a). The low current control signal from the microcontroller is connected to the gate of the MOSFET. The MOSFET switches the high current load on and off consistent with the control signal. The high current load is connected between the load supply and the MOSFET drain. It is important to note that the load supply voltage and the microcontroller supply voltage do not have to be at the same value. When the control signal on the MOSFET gate is logic high, the load current flows from drain to source. When the control signal applied to the gate is logic low, no load current flows. Thus, the high power load is turned on and off by the low power control signal from the microcontroller.

Often the MOSFET is used to control a high power motor load. A motor is a notorious source of noise. To isolate the microcontroller from the motor noise an optical isolator may be used as an

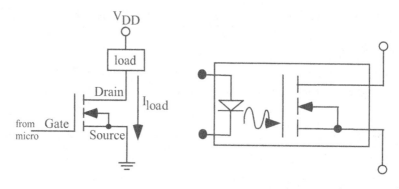

a) N-channel enhance MOSFET   b) solid state relay with optical interface

Figure 4.15: MOSFET circuits.

interface as shown in Figure 4.15(b). The link between the control signal from the microcontroller to the high power load is via an optical link contained within a Solid State Relay (SSR). The SSR is properly biased using techniques previously discussed.

4.6   DC SOLENOID CONTROL

The interface circuit for a DC solenoid is provided in Figure 4.16. A solenoid provides a mechanical insertion (or extraction) when asserted. In the interface, an optical isolator is used between the microcontroller and the MOSFET used to activate the solenoid. A reverse biased diode is placed across the solenoid. Both the solenoid power supply and the MOSFET must have the appropriate voltage and current rating to support the solenoid requirements.

4.7   DC MOTOR SPEED AND DIRECTION CONTROL

Often, a microcontroller is used to control a high power motor load. To properly interface the motor to the microcontroller, we must be familiar with the different types of motor technologies. Motor types are illustrated in Figure 4.17.

- **DC motor:** A DC motor has a positive and negative terminal. When a DC power supply of suitable current rating is applied to the motor it will rotate. If the polarity of the supply is switched with reference to the motor terminals, the motor will rotate in the opposite direction. The speed of the motor is roughly proportional to the applied voltage up to the rated voltage of the motor.

- **Servo motor:** A servo motor provides a precision angular rotation for an applied pulse width modulation duty cycle. As the duty cycle of the applied signal is varied, the angular displacement

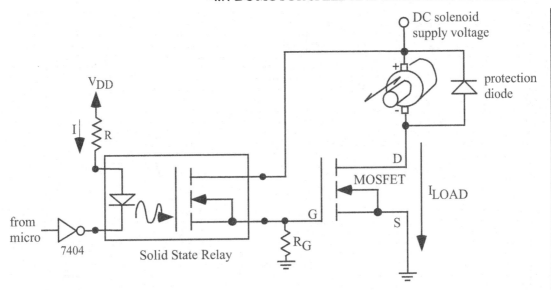

Figure 4.16: Solenoid interface circuit.

of the motor also varies. This type of motor is used to change mechanical positions such as the steering angle of a wheel.

- **Stepper motor:** A stepper motor as its name implies provides an incremental step change in rotation (typically 2.5 degree per step) for a step change in control signal sequence. The motor is typically controlled by a two or four wire interface. For the four wire stepper motor, the microcontroller provides a four bit control sequence to rotate the motor clockwise. To turn the motor counterclockwise, the control sequence is reversed. The low power control signals are interfaced to the motor via MOSFETs or power transistors to provide for the proper voltage and current requirements of the pulse sequence.

4.7.1   DC MOTOR OPERATING PARAMETERS

Space does not allow a full discussion of all motor types. We will concentrate on the DC motor. As previously mentioned, the motor speed may be varied by changing the applied voltage. This is difficult to do with a digital control signal. However, PWM control signal techniques discussed earlier may be combined with a MOSFET interface to precisely control the motor speed. The duty cycle of the PWM signal will also be the percentage of the motor supply voltage applied to the motor, and hence the percentage of rated full speed at which the motor will rotate. The interface circuit to accomplish this type of control is shown in Figure 4.18. Various portions of this interface circuit have been previously discussed. The resistor R_G, typically 10 k ohm, is used to discharge

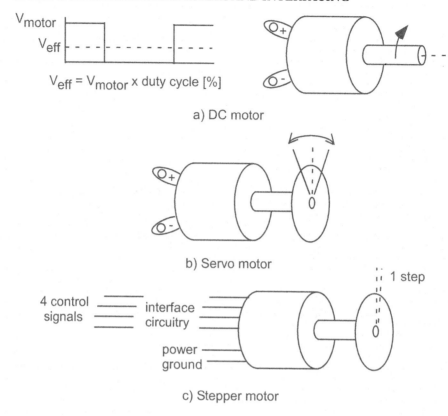

Figure 4.17: Motor types.

the MOSFET gate when no voltage is applied to the gate. For an inductive load, a reversed biased protection diode must be across the load. The interface circuit shown allows the motor to rotate in a given direction. As previously mentioned, to rotate the motor in the opposite direction, the motor polarity must be reversed. This may be accomplished with a high power switching network called an H–bridge specifically designed for this purpose. Reference Pack and Barrett for more information on this topic.

4.7.2   H–BRIDGE DIRECTION CONTROL

For a DC motor to operate in both the clockwise and counter clockwise direction, the polarity of the DC motor supplied must be changed. To operate the motor in the forward direction, the positive battery terminal must be connected to the positive motor terminal while the negative battery terminal must be provided to the negative motor terminal. To reverse the motor direction the motor supply polarity must be reversed. An H–bridge is a circuit employed to perform this polarity switch. The

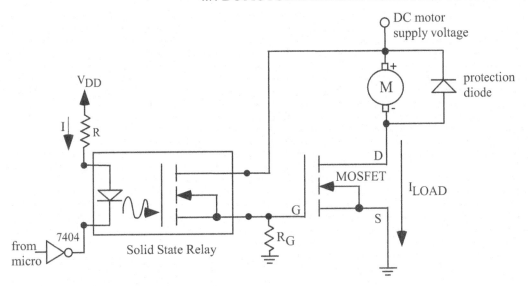

Figure 4.18: DC motor interface.

H–bridge circuit consists of four electronic switches as shown In Figure 4.19. For forward motor direction switches 1 and 4 are closed; whereas, for reverse direction switches 2 and 3 are closed.

Low power H–bridges (500 mA) come in a convenient dual in line package (e.g., 754110). For higher power motors, a H–bridge may be constructed from discrete components as shown in Figure 4.19. The ZTX451 and ZTX551 are NPN and PNP transistors with similar characteristics. The 11DQ06 are Schottky diodes. For driving higher power loads, the switching devices are sized appropriately.

If PWM signals are used to drive the base of the transistors (from microcontroller pins PD4 and PD5), both motor speed and direction may be controlled by the circuit. The transistors used in the circuit must have a current rating sufficient to handle the current requirements of the motor during start and stall conditions.

Example: A linear actuator is a specially designed motor that converts rotary to linear motion. The linear actuator is equipped with a mechanical rod that is extended when asserted in one direction and retracted when the polarity of assertion is reversed. An H–bridge may be used to control a linear actuator as shown in Figure 4.20. In this circuit an enable signal is used to assert the H–bridge while the two pulse width modulation channels are used to extend or retract the H–bridge.

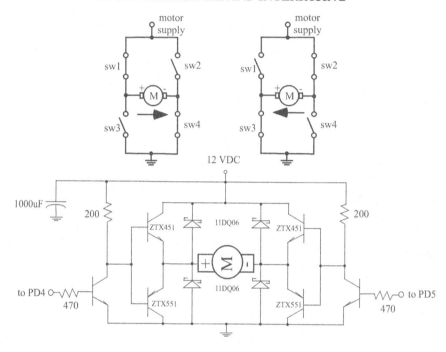

Figure 4.19: H-bridge control circuit.

4.7.3   SERVO MOTOR INTERFACE

The servo motor is used for a precise angular displacement. The displacement is related to the duty cycle of the applied control signal. A servo control circuit and supporting software was provided in Chapter 7.

4.7.4   STEPPER MOTOR CONTROL

Stepper motors are used to provide a discrete angular displacement in response to a control signal step. There are a wide variety of stepper motors including bipolar and unipolar types with different configurations of motor coil wiring. Due to space limitations, we only discuss the unipolar, 5 wire stepper motor. The internal coil configuration for this motor is provided in Figure 4.21(b).

Often, a wiring diagram is not available for the stepper motor. Based on the wiring configuration (Reference Figure 4.21b), one can find out the common line for both coils. It will have a resistance that is one–half of all of the other coils. Once the common connection is found, one can connect the stepper motor into the interface circuit. By changing the other connections, one can determine the correct connections for the step sequence.

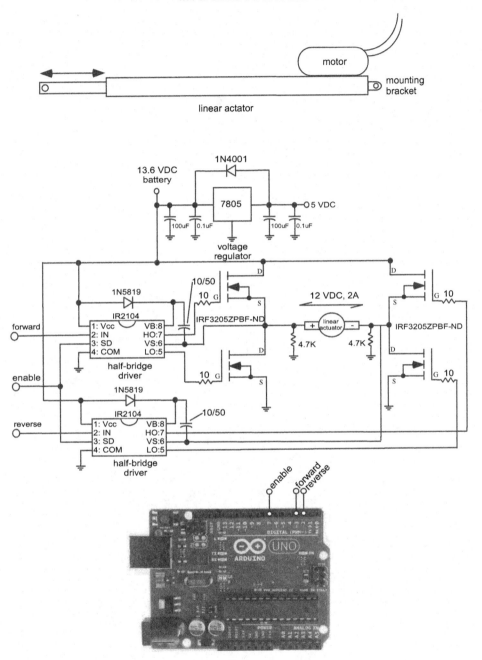

Figure 4.20: Linear actuator control circuit [O'Berto]. (UNO R3 illustration used with permission of the Arduino Team (CC BY–NC–SA) www.arduino.cc).

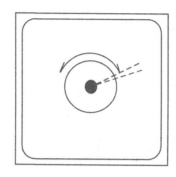

a) a stepper motor rotates a fixed angle per step

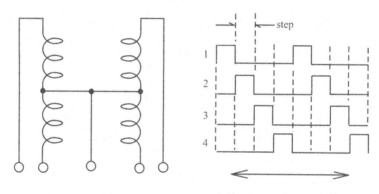

b) coil configuration and step sequence

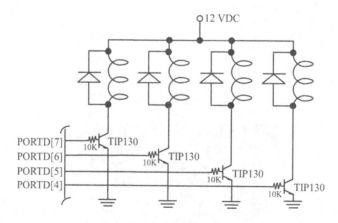

c) stepper motor interface circuit

Figure 4.21: Unipolar stepper motor control circuit.

To rotate the motor either clockwise or counter clockwise, a specific step sequence must be sent to the motor control wires as shown in Figure 4.21(c). As shown in Figure 4.21(c), the control sequence is transmitted by four pins on the microcontroller. In this example, we use PORTD[7:5].

The microcontroller does not have sufficient capability to drive the motor directly. Therefore, an interface circuit is required as shown in Figure 4.21c). For a unipolar stepper motor, we employ a TIP130 power Darlington transistor to drive each coil of the stepper motor. The speed of motor rotation is determined by how fast the control sequence is completed. The TIP 30 must be powered by a supply that has sufficient capability for the stepper motor coils.

Example: An ATmega324 has been connected to a JRP 42BYG016 unipolar, 1.8 degree per step, 12 VDC at 160 mA stepper motor. The interface circuit is shown in Figure 4.22. PORTD pins 7 to 4 are used to provide the step sequence. A one second delay is used between the steps to control motor speed. Pushbutton switches are used on PORTB[1:0] to assert CW and CCW stepper motion. An interface circuit consisting of four TIP130 transistors are used between the microcontroller and the stepper motor to boost the voltage and current of the control signals. Code to provide the step sequence is shown below.

Provided below is a basic function to rotate the stepper motor in the forward or reverse direction.

```
//************************************************************************
//target controller: ATMEL ATmega328
//
//ATMEL AVR ATmega328PV Controller Pin Assignments
//Pin  1 Reset - 1M resistor to Vdd, tact switch to ground, 1.0 uF to ground
//Pin  6 PD4 - to stepper motor coil
//Pin  7 Vdd - 1.0 uF to ground
//Pin  8 Gnd
//Pin 11 PD5 - to stepper motor coil
//Pin 12 PD6 - to stepper motor coil
//Pin 13 PD7 - to stepper motor coil
//Pin 14 PB0 to active high RC debounced switch - CW
//Pin 20 AVcc to Vdd
//Pin 21 ARef to Vcc
//Pin 22 Gnd to Ground
//************************************************************************

//include files*********************************************************

//ATMEL register definitions for ATmega328
#include<iom328pv.h>
#include<macros.h>
                                     //interrupt handler definition
```

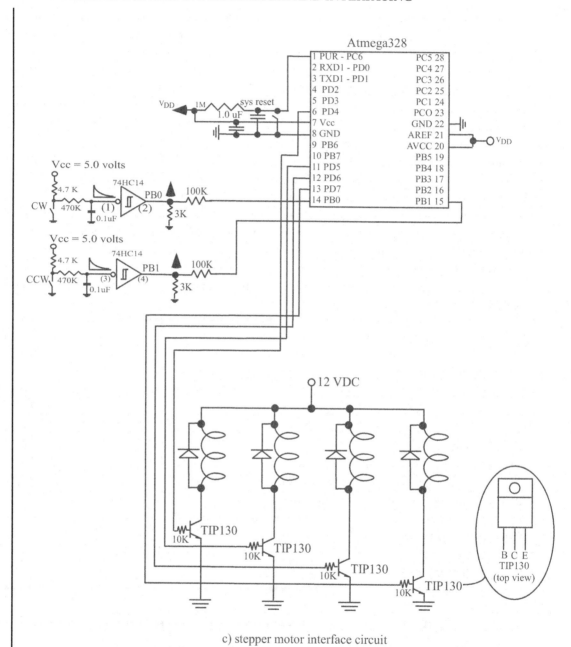

c) stepper motor interface circuit

Figure 4.22: Unipolar stepper motor control circuit.

```
#pragma interrupt_handler timer0_interrupt_isr:17



//function prototypes**********************************************
void initialize_ports(void);          //initializes ports
void read_new_input(void);            //used to read input change on PORTB
void init_timer0_ovf_interrupt(void); //used to initialize timer0 overflow
void timer0_interrupt_isr(void);
void delay(unsigned int);



//main program*****************************************************
//The main program checks PORTB for user input activity. If new activity
//is found, the program responds.

//global variables
unsigned char   old_PORTB = 0x08;  //present value of PORTB
unsigned char   new_PORTB;         //new values of PORTB
unsigned int    input_delay;       //delay counter - increment via Timer0
                                   //overflow interrupt

void main(void)
{
initialize_ports();                //return LED configuration to default
init_timer0_ovf_interrupt();       //used to initialize timer0 overflow

while(1)
  {
  _StackCheck();                   //check for stack overflow
  read_new_input();                //read input status changes on PORTB
  }
}//end main

//Function definitions
//*****************************************************************
//initialize_ports: provides initial configuration for I/O ports
//*****************************************************************

void initialize_ports(void)
```

```
{
//PORTB
DDRB=0xfc;                              //PORTB[7-2] output, PORTB[1:0] input
PORTB=0x00;                             //disable PORTB pull-up resistors

//PORTC
DDRC=0xff;                              //set PORTC[7-0] as output
PORTC=0x00;                             //init low

//PORTD
DDRD=0xff;                              //set PORTD[7-0] as output
PORTD=0x00;                             //initialize low
}


//**************************************************************************
//read_new_input: functions polls PORTB for a change in status. If status
//change has occurred, appropriate function for status change is called
//Pin 1 PB0 to active high RC  debounced switch - CW
//Pin 2 PB1 to active high RC debounced switch  - CCW
//**************************************************************************

void read_new_input(void)
{
new_PORTB = (PINB & 0x03);
if(new_PORTB != old_PORTB){
  switch(new_PORTB){                    //process change in PORTB input

    case 0x01:                          //CW
      while((PINB & 0x03)==0x01)
        {
        PORTD = 0x80;
        delay(15);                      //delay 1s
        PORTD = 0x00;
        delay(1);                       //delay 65 ms

        PORTD = 0x40;
        delay(15);
        PORTD = 0x00;
        delay(1);
```

```
        PORTD = 0x20;
        delay(15);
        PORTD = 0x00;
        delay(1);

        PORTD = 0x10;
        delay(15);
        PORTD = 0x00;
        delay(1);
        }
break;

  case 0x02:                        //CCW
    while((PINB & 0x03)==0x02)
      {
      PORTD = 0x10;
      delay(15);
      PORTD = 0x00;
      delay(1);

      PORTD = 0x20;
      delay(15);
      PORTD = 0x00;
      delay(1);

      PORTD = 0x40;
      delay(15);
      PORTD = 0x00;
      delay(1);

      PORTD = 0x80;
      delay(15);
      PORTD = 0x00;
      delay(1);
}

        break;
```

```
        default:;                  //all other cases
        }                          //end switch(new_PORTB)
    }                              //end if new_PORTB
    old_PORTB=new_PORTB;           //update PORTB
}


//*************************************************************************
//int_timer0_ovf_interrupt(): The Timer0 overflow interrupt is being
//employed as a time base for a master timer for this project. The internal
//oscillator of 8 MHz is divided internally by 8 to provide a 1 MHz time
//base and is divided by 256.  The 8-bit Timer0 register (TCNT0) overflows
//every 256 counts or every 65.5 ms.
//*************************************************************************

void init_timer0_ovf_interrupt(void)
{
TCCR0B = 0x04; //divide timer0 timebase by 256, overfl. occurs every 65.5ms
TIMSK0 = 0x01; //enable timer0 overflow interrupt
asm("SEI");    //enable global interrupt
}


//*************************************************************************
//timer0_interrupt_isr:
//Note: Timer overflow 0 is cleared by hardware when executing the
//corresponding interrupt handling vector.
//*************************************************************************

void timer0_interrupt_isr(void)
{
input_delay++;                     //input delay processing
}


//*************************************************************************
//void delay(unsigned int number_of_65_5ms_interrupts)
//this generic delay function provides the specified delay as the number
//of 65.5 ms "clock ticks" from the Timer0 interrupt.
//Note: this function is only valid when using a 1 MHz crystal or ceramic
//       resonator
//*************************************************************************
```

```
void delay(unsigned int number_of_65_5ms_interrupts)
{
TCNT0 = 0x00;                        //reset timer0
input_delay = 0;
while(input_delay <= number_of_65_5ms_interrupts)
  {
  ;
  }
}


//************************************************************************
```

4.7.5   AC DEVICES

In a similar manner, a high power alternating current (AC) load may be switched on and off using a low power control signal from the microcontroller. In this case, a Solid State Relay is used as the switching device. Solid state relays are available to switch a high power DC or AC load [Crydom]. For example, the Crydom 558–CX240D5R is a printed circuit board mounted, air cooled, single pole single throw (SPST), normally open (NO) solid state relay. It requires a DC control voltage of 3–15 VDC at 15 mA. However, this small microcontroller compatible DC control signal is used to switch 12–280 VAC loads rated from 0.06 to 5 amps [Crydom] as shown in Figure 4.23.

To vary the direction of an AC motor you must use a bi–directional AC motor. A bi–directional motor is equipped with three terminals: common, clockwise, and counterclockwise. To turn the motor clockwise, an AC source is applied to the common and clockwise connections. In like manner, to turn the motor counterclockwise, an AC source is applied to the common and counterclockwise connections. This may be accomplished using two of the Crydom SSRs.

4.8   INTERFACING TO MISCELLANEOUS DEVICES

In this section, we provide a pot pourri of interface circuits to connect a microcontroller to a wide variety of peripheral devices.

4.8.1   SONALERTS, BEEPERS, BUZZERS

In Figure 4.24, we provide several circuits used to interface a microcontroller to a buzzer, beeper or other types of annunciator indexannunciator devices such as a sonalert. It is important that the interface transistor and the supply voltage are matched to the requirements of the sound producing device.

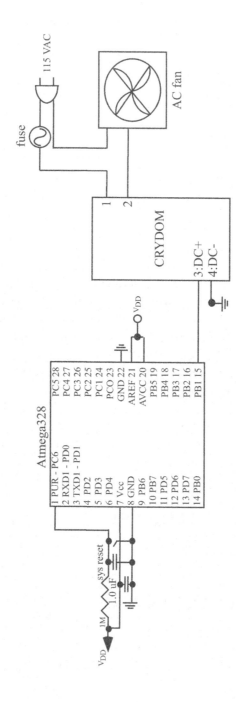

Figure 4.23: AC motor control circuit.

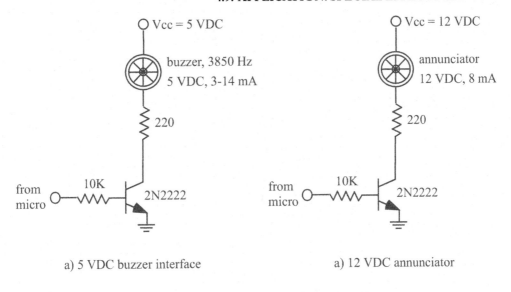

a) 5 VDC buzzer interface   a) 12 VDC annunciator

Figure 4.24: Sonalert, beepers, buzzers.

4.8.2   VIBRATING MOTOR

A vibrating motor is often used to gain one's attention as in a cell phone. These motors are typically rated at 3 VDC and a high current. The interface circuit shown in Figure 4.25 is used to drive the low voltage motor. Note that the control signal provided to the transistor base is 5 VDC. To step the motor supply voltage down to the motor voltage of 3 VDC, two 1N4001 silicon rectifier diodes are used in series. These diodes provide approximately 1.4 to 1.6 VDC voltage drop. Another 1N4001 diode is reversed biased across the motor and the series diode string. The motor may be turned on and off with a 5 VDC control signal from the microcontroller or a PWM signal may be used to control motor speed. It is recommended that a Darlington NPN transistor (TIP 120) be employed in this application.

4.9   APPLICATION: SPECIAL EFFECTS LED CUBE

To illustrate some of the fundamentals of microcontroller interfacing, we construct a three–dimensional LED cube. This design was inspired by an LED cube kit available from Jameco (www.jameco.com). The LED cube consists of four layers of LEDs with 16 layers per LED. Only a single LED is illuminated at a specific time.

Only a single LED is illuminated at a given time. However, different effects may be achieved by how long a specific LED is left illuminated. A specific LED layer is asserted using the layer select pins on the microcontroller using a one–hot–code (a single line asserted while the others are de–asserted). The asserted line is fed through a 74HC244 (three state, octal buffer, line driver)

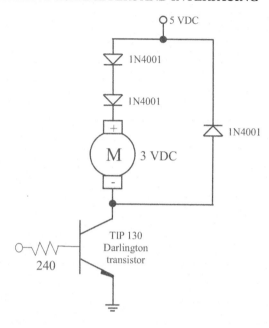

Figure 4.25: Controlling a low voltage motor.

which provides an I_{OH}/I_{OL} current of $\pm$ 35 mA as shown in Figure 4.26. A given output from the 74HC244 is fed to a common anode connections for all 16 LEDs in a layer. All four LEDs in a specific LED position share a common cathode connection. That is, an LED in a specific location within a layer shares a common cathode connection with three other LEDs that share the same position in the other three layers. The common cathode connection from each LED location is fed to a specific output of the 74HC154 4–to–16 decoder. The decoder has a one–cold–code output. To illuminate a specific LED, the appropriate layer select and LED select lines are asserted using the layer_sel[3:0] and led_sel[3:0] lines respectively. This basic design may be easily expanded to a larger LED cube.

4.9.1   CONSTRUCTION HINTS

To limit project costs, low–cost red LEDs (Jameco #333973) were used. The project requires a total of 64 LEDs (4 layers of 16 LEDs each). A LED template pattern was constructed from a 5" by 5" piece of pine wood. A 4–by–4 pattern of holes were drilled into the wood. Holes were spaced 3/4" apart. The hole diameter was slightly smaller than the diameter of the LEDs to allow for a snug LED fit.

The LED array was constructed a layer at a time using the wood template. Each LED was tested before inclusion in the array. A 5 VDC power supply with a series 220 ohm resistor was used

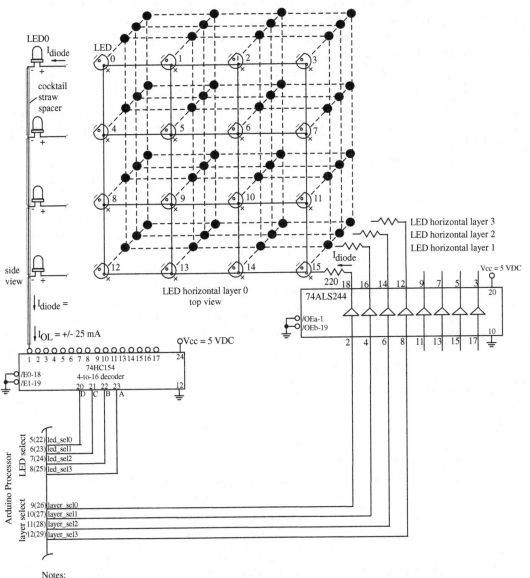

Notes:
1. LED cube consists of 4 layers of 16 LEDs each.
2. Each LED is individually addressed by asserting the appropriate cathode signal (0 to 15) and asserting a specific LED layer.
3. All LEDs in a given layer share a common anode connection.
4. All LEDs in a given position (0 to 15) share a common cathode connection.

Figure 4.26: LED special effects cube.

to insure each LED was fully operational. The LED anodes in a given LED row were then soldered together. A fine tip soldering iron and a small bit of solder were used for each interconnect as shown in Figure 4.27. Cross wires were then used to connect the cathodes of adjacent rows. A 22 gage bare wire was used. Again, a small bit of solder was used for the interconnect points. Four separate LED layers (4 by 4 array of LEDs) were completed.

To assemble the individual layers into a cube, cocktail straw segments were used as spacers between the layers. The straw segments provided spacing between the layers and also offered improved structural stability. The anodes for a given LED position were soldered together. For example, all LEDs in position 0 for all four layers shared a common anode connection.

The completed LED cube was mounted on a perforated printed circuit board (perfboard) to provide a stable base. LED sockets for the 74LS244 and the 74HC154 were also mounted to the perfboard. Connections were routed to a 16 pin ribbon cable connector. The other end of the ribbon cable was interfaced to the appropriate pins of the Arduino processor. The entire LED cube was mounted within a 4" plexiglass cube. The cube is available from the Container Store (www. containerstore.com). A construction diagram is provided in Figure 4.27.

4.9.2 LED CUBE ARDUINO SKETCH CODE

Provided below is the basic code template to illuminate a single LED (LED 0, layer 0). This basic template may be used to generate a number of special effects (e.g. tornado, black hole, etc.). Pin numbers are provided for the Arduino UNO R3. Pin numbers for the Arduino Mega 2560 are provided in the comments.

```
//**********************************************
                           //led select pins
#define led_sel0   5       //Mega2560: pin 22
#define led_sel1   6       //Mega2560: pin 23
#define led_sel2   7       //Mega2560: pin 24
#define led_sel3   8       //Mega2560: pin 25


                           //layer select pins
#define layer_sel0   9     //Mega2560: pin 26
#define layer_sel1   10    //Mega2560: pin 27
#define layer_sel2   11    //Mega2560: pin 28
#define layer_sel3   12    //Mega2560: pin 29

void setup()
{
pinMode(led_sel0, OUTPUT);
pinMode(led_sel1, OUTPUT);
pinMode(led_sel2, OUTPUT);
```

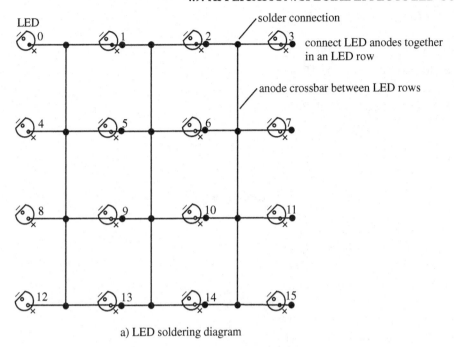

a) LED soldering diagram

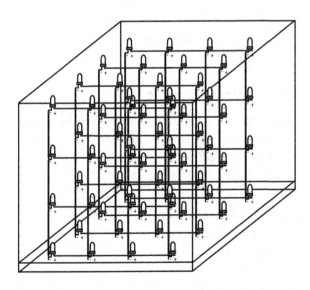

b) 3D LED array mounted within plexiglass cube

Figure 4.27: LED cube construction.

```
pinMode(led_sel3, OUTPUT);

pinMode(layer_sel0, OUTPUT);
pinMode(layer_sel1, OUTPUT);
pinMode(layer_sel2, OUTPUT);
pinMode(layer_sel3, OUTPUT);
}

void loop()
{
                        //illuminate LED 0, layer 0
                        //led select
digitalWrite(led_sel0, LOW);
digitalWrite(led_sel1, LOW);
digitalWrite(led_sel2, LOW);
digitalWrite(led_sel3, LOW);
                        //layer select
digitalWrite(layer_sel0, HIGH);
digitalWrite(layer_sel1, LOW);
digitalWrite(layer_sel2, LOW);
digitalWrite(layer_sel3, LOW);

delay(500);             //delay specified in ms
}


//************************************************
```

In the next example, a function "illuminate_LED" has been added. To illuminate a specific LED, the LED position (0 through 15), the LED layer (0 through 3), and the length of time to illuminate the LED in milliseconds is specified. In this short example, LED 0 is sequentially illuminated in each layer. An LED grid map is provided in Figure 4.28. It is useful for planning special effects.

```
//************************************************
                        //led select pins
#define led_sel0  5     //Mega2560: pin 22
#define led_sel1  6     //Mega2560: pin 23
#define led_sel2  7     //Mega2560: pin 24
#define led_sel3  8     //Mega2560: pin 25

                        //layer select pins
```

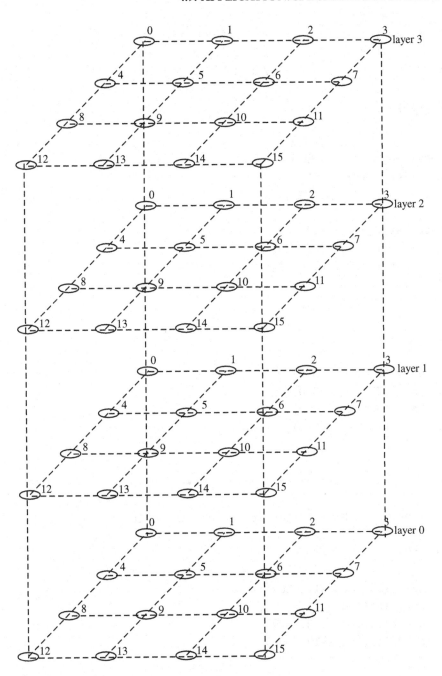

Figure 4.28: LED grid map.

```
#define layer_sel0   9      //Mega2560: pin 26
#define layer_sel1  10      //Mega2560: pin 27
#define layer_sel2  11      //Mega2560: pin 28
#define layer_sel3  12      //Mega2560: pin 29

void setup()
{
pinMode(led_sel0, OUTPUT);
pinMode(led_sel1, OUTPUT);
pinMode(led_sel2, OUTPUT);
pinMode(led_sel3, OUTPUT);

pinMode(layer_sel0, OUTPUT);
pinMode(layer_sel1, OUTPUT);
pinMode(layer_sel2, OUTPUT);
pinMode(layer_sel3, OUTPUT);
}

void loop()
{
illuminate_LED(0, 0, 500);
illuminate_LED(0, 1, 500);
illuminate_LED(0, 2, 500);
illuminate_LED(0, 3, 500);
}

//************************************************

void illuminate_LED(int led, int layer, int delay_time)
{
if(led==0)
  {
  digitalWrite(led_sel0, LOW);
  digitalWrite(led_sel1, LOW);
  digitalWrite(led_sel2, LOW);
  digitalWrite(led_sel3, LOW);
  }
else if(led==1)
  {
```

```
  digitalWrite(led_sel0, HIGH);
  digitalWrite(led_sel1, LOW);
  digitalWrite(led_sel2, LOW);
  digitalWrite(led_sel3, LOW);
  }
else if(led==2)
  {
  digitalWrite(led_sel0, LOW);
  digitalWrite(led_sel1, HIGH);
  digitalWrite(led_sel2, LOW);
  digitalWrite(led_sel3, LOW);
  }
else if(led==3)
  {
  digitalWrite(led_sel0, HIGH);
  digitalWrite(led_sel1, HIGH);
  digitalWrite(led_sel2, LOW);
  digitalWrite(led_sel3, LOW);
  }
else if(led==4)
  {
  digitalWrite(led_sel0, LOW);
  digitalWrite(led_sel1, LOW);
  digitalWrite(led_sel2, HIGH);
  digitalWrite(led_sel3, LOW);
  }
else if(led==5)
  {
  digitalWrite(led_sel0, HIGH);
  digitalWrite(led_sel1, LOW);
  digitalWrite(led_sel2, HIGH);
  digitalWrite(led_sel3, LOW);
  }
else if(led==6)
  {
  digitalWrite(led_sel0, LOW);
  digitalWrite(led_sel1, HIGH);
  digitalWrite(led_sel2, HIGH);
  digitalWrite(led_sel3, LOW);
```

```
   }
else if(led==7)
   {
   digitalWrite(led_sel0, HIGH);
   digitalWrite(led_sel1, HIGH);
   digitalWrite(led_sel2, HIGH);
   digitalWrite(led_sel3, LOW);
   }
if(led==8)
   {
   digitalWrite(led_sel0, LOW);
   digitalWrite(led_sel1, LOW);
   digitalWrite(led_sel2, LOW);
   digitalWrite(led_sel3, HIGH);
   }
else if(led==9)
   {
   digitalWrite(led_sel0, HIGH);
   digitalWrite(led_sel1, LOW);
   digitalWrite(led_sel2, LOW);
   digitalWrite(led_sel3, HIGH);
   }
else if(led==10)
   {
   digitalWrite(led_sel0, LOW);
   digitalWrite(led_sel1, HIGH);
   digitalWrite(led_sel2, LOW);
   digitalWrite(led_sel3, HIGH);
   }
else if(led==11)
   {
   digitalWrite(led_sel0, HIGH);
   digitalWrite(led_sel1, HIGH);
   digitalWrite(led_sel2, LOW);
   digitalWrite(led_sel3, HIGH);
   }
else if(led==12)
   {
   digitalWrite(led_sel0, LOW);
```

```
  digitalWrite(led_sel1, LOW);
  digitalWrite(led_sel2, HIGH);
  digitalWrite(led_sel3, HIGH);
  }
else if(led==13)
  {
  digitalWrite(led_sel0, HIGH);
  digitalWrite(led_sel1, LOW);
  digitalWrite(led_sel2, HIGH);
  digitalWrite(led_sel3, HIGH);
  }
else if(led==14)
  {
  digitalWrite(led_sel0, LOW);
  digitalWrite(led_sel1, HIGH);
  digitalWrite(led_sel2, HIGH);
  digitalWrite(led_sel3, HIGH);
  }
else if(led==15)
  {
  digitalWrite(led_sel0, HIGH);
  digitalWrite(led_sel1, HIGH);
  digitalWrite(led_sel2, HIGH);
  digitalWrite(led_sel3, HIGH);
  }

if(layer==0)
  {
  digitalWrite(layer_sel0, HIGH);
  digitalWrite(layer_sel1, LOW);
  digitalWrite(layer_sel2, LOW);
  digitalWrite(layer_sel3, LOW);
  }
else if(layer==1)
  {
  digitalWrite(layer_sel0, LOW);
  digitalWrite(layer_sel1, HIGH);
  digitalWrite(layer_sel2, LOW);
  digitalWrite(layer_sel3, LOW);
```

```
   }
else if(layer==2)
  {
  digitalWrite(layer_sel0, LOW);
  digitalWrite(layer_sel1, LOW);
  digitalWrite(layer_sel2, HIGH);
  digitalWrite(layer_sel3, LOW);
  }
else if(layer==3)
  {
  digitalWrite(layer_sel0, LOW);
  digitalWrite(layer_sel1, LOW);
  digitalWrite(layer_sel2, LOW);
  digitalWrite(layer_sel3, HIGH);
  }

delay(delay_time);
}


//************************************************
```

In the next example, a "fireworks" special effect is produced. The firework goes up, splits into four pieces, and then falls back down as shown in Figure 4.29. It is useful for planning special effects.

```
//************************************************
                         //led select pins
#define led_sel0   5     //Mega2560: pin 22
#define led_sel1   6     //Mega2560: pin 23
#define led_sel2   7     //Mega2560: pin 24
#define led_sel3   8     //Mega2560: pin 25

                         //layer select pins
#define layer_sel0   9     //Mega2560: pin 26
#define layer_sel1   10    //Mega2560: pin 27
#define layer_sel2   11    //Mega2560: pin 28
#define layer_sel3   12    //Mega2560: pin 29

void setup()
{
pinMode(led_sel0, OUTPUT);
pinMode(led_sel1, OUTPUT);
```

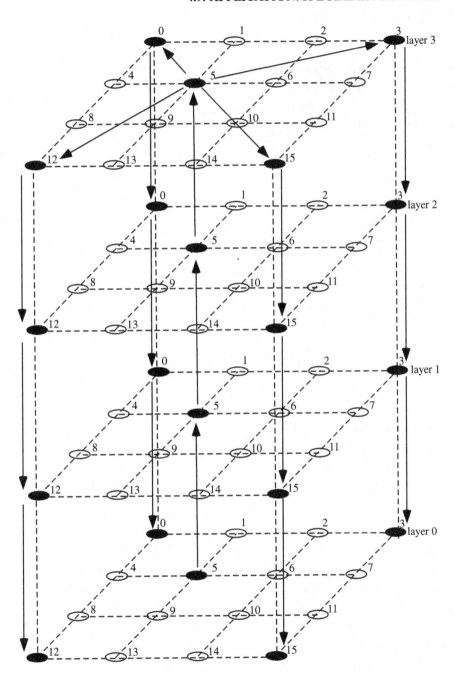

Figure 4.29: LED grid map for a fire work.

```
  pinMode(led_sel2, OUTPUT);
  pinMode(led_sel3, OUTPUT);

  pinMode(layer_sel0, OUTPUT);
  pinMode(layer_sel1, OUTPUT);
  pinMode(layer_sel2, OUTPUT);
  pinMode(layer_sel3, OUTPUT);
  }

  void loop()
  {
  int i;

  //firework going up
  illuminate_LED(5, 0, 100);
  illuminate_LED(5, 1, 100);
  illuminate_LED(5, 2, 100);
  illuminate_LED(5, 3, 100);

  //firework exploding into four pieces
  //at each cube corner
  for(i=0;i<=10;i++)
    {
    illuminate_LED(0,  3, 10);
    illuminate_LED(3,  3, 10);
    illuminate_LED(12, 3, 10);
    illuminate_LED(15, 3, 10);
    delay(10);
    }

  delay(200);

  //firework pieces falling to layer 2
  for(i=0;i<=10;i++)
    {
    illuminate_LED(0,  2, 10);
    illuminate_LED(3,  2, 10);
    illuminate_LED(12, 2, 10);
    illuminate_LED(15, 2, 10);
```

```
  delay(10);
   }

delay(200);

//firework pieces falling to layer 1
for(i=0;i<=10;i++)
   {
   illuminate_LED(0,   1, 10);
   illuminate_LED(3,   1, 10);
   illuminate_LED(12, 1, 10);
   illuminate_LED(15, 1, 10);
   delay(10);
   }

delay(200);

//firework pieces falling to layer 0
for(i=0;i<=10;i++)
   {
   illuminate_LED(0,   0, 10);
   illuminate_LED(3,   0, 10);
   illuminate_LED(12, 0, 10);
   illuminate_LED(15, 0, 10);
   delay(10);
   }

delay(10);
}


//**********************************************

void illuminate_LED(int led, int layer, int delay_time)
{
if(led==0)
   {
   digitalWrite(led_sel0, LOW);
   digitalWrite(led_sel1, LOW);
   digitalWrite(led_sel2, LOW);
```

```
    digitalWrite(led_sel3, LOW);
    }
 else if(led==1)
   {
   digitalWrite(led_sel0, HIGH);
   digitalWrite(led_sel1, LOW);
   digitalWrite(led_sel2, LOW);
   digitalWrite(led_sel3, LOW);
   }
 else if(led==2)
   {
   digitalWrite(led_sel0, LOW);
   digitalWrite(led_sel1, HIGH);
   digitalWrite(led_sel2, LOW);
   digitalWrite(led_sel3, LOW);
   }
 else if(led==3)
   {
   digitalWrite(led_sel0, HIGH);
   digitalWrite(led_sel1, HIGH);
   digitalWrite(led_sel2, LOW);
   digitalWrite(led_sel3, LOW);
   }
 else if(led==4)
   {
   digitalWrite(led_sel0, LOW);
   digitalWrite(led_sel1, LOW);
   digitalWrite(led_sel2, HIGH);
   digitalWrite(led_sel3, LOW);
   }
 else if(led==5)
   {
   digitalWrite(led_sel0, HIGH);
   digitalWrite(led_sel1, LOW);
   digitalWrite(led_sel2, HIGH);
   digitalWrite(led_sel3, LOW);
   }
 else if(led==6)
   {
```

```
digitalWrite(led_sel0, LOW);
digitalWrite(led_sel1, HIGH);
digitalWrite(led_sel2, HIGH);
digitalWrite(led_sel3, LOW);
}
else if(led==7)
{
digitalWrite(led_sel0, HIGH);
digitalWrite(led_sel1, HIGH);
digitalWrite(led_sel2, HIGH);
digitalWrite(led_sel3, LOW);
}
if(led==8)
{
digitalWrite(led_sel0, LOW);
digitalWrite(led_sel1, LOW);
digitalWrite(led_sel2, LOW);
digitalWrite(led_sel3, HIGH);
}
else if(led==9)
{
digitalWrite(led_sel0, HIGH);
digitalWrite(led_sel1, LOW);
digitalWrite(led_sel2, LOW);
digitalWrite(led_sel3, HIGH);
}
else if(led==10)
{
digitalWrite(led_sel0, LOW);
digitalWrite(led_sel1, HIGH);
digitalWrite(led_sel2, LOW);
digitalWrite(led_sel3, HIGH);
}
else if(led==11)
{
digitalWrite(led_sel0, HIGH);
digitalWrite(led_sel1, HIGH);
digitalWrite(led_sel2, LOW);
digitalWrite(led_sel3, HIGH);
```

```
  }
else if(led==12)
  {
  digitalWrite(led_sel0, LOW);
  digitalWrite(led_sel1, LOW);
  digitalWrite(led_sel2, HIGH);
  digitalWrite(led_sel3, HIGH);
  }
else if(led==13)
  {
  digitalWrite(led_sel0, HIGH);
  digitalWrite(led_sel1, LOW);
  digitalWrite(led_sel2, HIGH);
  digitalWrite(led_sel3, HIGH);
  }
else if(led==14)
  {
  digitalWrite(led_sel0, LOW);
  digitalWrite(led_sel1, HIGH);
  digitalWrite(led_sel2, HIGH);
  digitalWrite(led_sel3, HIGH);
  }
else if(led==15)
  {
  digitalWrite(led_sel0, HIGH);
  digitalWrite(led_sel1, HIGH);
  digitalWrite(led_sel2, HIGH);
  digitalWrite(led_sel3, HIGH);
  }

if(layer==0)
  {
  digitalWrite(layer_sel0, HIGH);
  digitalWrite(layer_sel1, LOW);
  digitalWrite(layer_sel2, LOW);
  digitalWrite(layer_sel3, LOW);
  }
else if(layer==1)
  {
```

```
  digitalWrite(layer_sel0, LOW);
  digitalWrite(layer_sel1, HIGH);
  digitalWrite(layer_sel2, LOW);
  digitalWrite(layer_sel3, LOW);
  }
else if(layer==2)
  {
  digitalWrite(layer_sel0, LOW);
  digitalWrite(layer_sel1, LOW);
  digitalWrite(layer_sel2, HIGH);
  digitalWrite(layer_sel3, LOW);
  }
else if(layer==3)
  {
  digitalWrite(layer_sel0, LOW);
  digitalWrite(layer_sel1, LOW);
  digitalWrite(layer_sel2, LOW);
  digitalWrite(layer_sel3, HIGH);
  }

delay(delay_time);
}


//**********************************************
```

4.10   SUMMARY

In this chapter, we discussed the voltage and current operating parameters for the Arduino UNO R3 and Mega 2560 processing board and the Atmel ATmega328 microcontroller. We discussed how this information may be applied to properly design an interface for common input and output circuits. It must be emphasized a properly designed interface allows the microcontroller to operate properly within its parameter envelope. If due to a poor interface design, a microcontroller is used outside its prescribed operating parameter values, spurious and incorrect logic values will result. We provided interface information for a wide range of input and output devices. We also discussed the concept of interfacing a motor to a microcontroller using PWM techniques coupled with high power MOSFET or SSR switching devices.

4.11   REFERENCES

- Pack D, Barrett S (2002) 68HC12 Microcontroller: Theory and Applications. Prentice–Hall Incorporated, Upper Saddle River, NJ.

- Barrett S, Pack D (2004) Embedded Systems Design with the 68HC12 and HCS12. Prentice––Hall Incorporated, Upper Saddle River, NJ.

- Crydom Corporation, 2320 Paseo de las Americas, Suite 201, San Diego, CA (`www.crydom.com`).

- Sick/Stegmann Incorporated, Dayton, OH, (`www.stegmann.com`).

- Images Company, 39 Seneca Loop, Staten Island, NY 10314.

- *Atmel 8–bit AVR Microcontroller with 64/128/256K Bytes In–System Programmable Flash, ATmega640/V, ATmega1280/V, 2560/V* data sheet: 2549P–AVR–10/2012, Atmel Corporation, 2325 Orchard Parkway, San Jose, CA 95131.

- *Atmel 8–bit AVR Microcontroller with 16/32/64K Bytes In–System Programmable Flash, ATmega328P/V, ATmega324P/V, 644P/V* data sheet: 8011I–AVR–05/08, Atmel Corporation, 2325 Orchard Parkway, San Jose, CA 95131.

- Barrett S, Pack D (2006) Microcontrollers Fundamentals for Engineers and Scientists. Morgan and Claypool Publishers. DOI: 10.2200/S00025ED1V01Y200605DCS001

- Barrett S and Pack D (2008) Atmel AVR Microcontroller Primer Programming and Interfacing. Morgan and Claypool Publishers. DOI: 10.2200/S00100ED1V01Y200712DCS015

- Barrett S (2010) Embedded Systems Design with the Atmel AVR Microcontroller. Morgan and Claypool Publishers. DOI: 10.2200/S00225ED1V01Y200910DCS025

- National Semiconductor, *LM34/LM34A/LM34C/LM34CA/LM34D Precision Fahrenheit Temperature Sensor*, 1995.

- Bohm H and Jensen V (1997) Build Your Own Underwater Robot and Other Wet Projects. Westcoast Words.

- seaperch, www.seaperch.org

- SparkFun, www.sparkfun.com

- Pack D and Barrett S (2011) Microcontroller Programming and Interfacing Texas Instruments MSP430. Morgan and Claypool Publishers.

4.12   CHAPTER PROBLEMS

1. What will happen if a microcontroller is used outside of its prescribed operating envelope?

2. Discuss the difference between the terms "sink" and "source" as related to current loading of a microcontroller.

3. Can an LED with a series limiting resistor be directly driven by the Atmel microcontroller? Explain.

4. In your own words, provide a brief description of each of the microcontroller electrical parameters.

5. What is switch bounce? Describe two techniques to minimize switch bounce.

6. Describe a method of debouncing a keypad.

7. What is the difference between an incremental encoder and an absolute encoder? Describe applications for each type.

8. What must be the current rating of the 2N2222 and 2N2907 transistors used in the tri–state LED circuit? Support your answer.

9. Draw the circuit for a six character seven segment display. Fully specify all components. Write a program to display "ATmega328."

10. Repeat the question above for a dot matrix display.

11. Repeat the question above for a LCD display.

12. What is the difference between a unipolar and bipolar stepper motor?

13. What controls the speed of rotation of a stepper motor?

14. A stepper motor provides and angular displacement of 1.8 degrees per step. How can this resolution be improved?

15. Write a function to convert an ASCII numeral representation (0 to 9) to a seven segment display.

16. Why is an interface required between a microcontroller and a stepper motor?

17. How must the LED cube design be modified to incorporate eight layers of LEDs with 16 LEDs per layer?

18. In the LED cube design, what is the maximum amount of forward current that can be safely delivered to a given LED?

CHAPTER 5

Analog to Digital Conversion (ADC)

Objectives: After reading this chapter, the reader should be able to

- Illustrate the analog–to–digital conversion process.

- Assess the quality of analog–to–digital conversion using the metrics of sampling rate, quantization levels, number of bits used for encoding and dynamic range.

- Design signal conditioning circuits to interface sensors to analog–to–digital converters.

- Implement signal conditioning circuits with operational amplifiers.

- Describe the key registers used during an ATmega328 ADC.

- Describe the steps to perform an ADC with the ATmega328.

- Describe the key registers used during an ATmega2560 ADC.

- Describe the steps to perform an ADC with the ATmega2560.

- Program the Arduino UNO R3 and the Arduino Mega 2560 processing boards to perform an ADC using the built–in features of the Arduino Development Environment.

- Program the ATmega328 to perform an ADC in C.

- Program the ATmega2560 to perform an ADC in C.

- Describe the operation of a digital–to–analog converter (DAC).

5.1   OVERVIEW

A microcontroller is used to process information from the natural world, decide on a course of action based on the information collected, and then issue control signals to implement the decision. Since the information from the natural world, is analog or continuous in nature, and the microcontroller is a digital or discrete based processor, a method to convert an analog signal to a digital form is required. An ADC system performs this task while a digital to analog converter (DAC) performs the conversion in the opposite direction. We will discuss both types of converters in this chapter.

Most microcontrollers are equipped with an ADC subsystem; whereas, DACs must be added as an external peripheral device to the controller.

In this chapter, we discuss the ADC process in some detail. In the first section, we discuss the conversion process itself, followed by a presentation of the successive–approximation hardware implementation of the process. We then review the basic features of the ATmega328 and the ATmega2560 ADC systems followed by a system description and a discussion of key ADC registers. We conclude our discussion of the analog–to–digital converter with several illustrative code examples. We show how to program the ADC using the built–in features of the Arduino Development Environment and C. We conclude the chapter with a discussion of the DAC process and interface a multi–channel DAC to the ATmega328. We also discuss the Arduino Development Environment built–in features that allow generation of an output analog signal via pulse width modulation (PWM) techniques. Throughout the chapter, we provide detailed examples.

5.2   SAMPLING, QUANTIZATION AND ENCODING

In this subsection, we provide an abbreviated discussion of the ADC process. This discussion was condensed from "Atmel AVR Microcontroller Primer Programming and Interfacing." The interested reader is referred to this text for additional details and examples [Barrett and Pack]. We present three important processes associated with the ADC: sampling, quantization, and encoding.

Sampling. We first start with the subject of sampling. Sampling is the process of taking 'snap shots' of a signal over time. When we sample a signal, we want to sample it in an optimal fashion such that we can capture the essence of the signal while minimizing the use of resources. In essence, we want to minimize the number of samples while retaining the capability to faithfully reconstruct the original signal from the samples. Intuitively, the rate of change in a signal determines the number of samples required to faithfully reconstruct the signal, provided that all adjacent samples are captured with the same sample timing intervals.

Sampling is important since when we want to represent an analog signal in a digital system, such as a computer, we must use the appropriate sampling rate to capture the analog signal for a faithful representation in digital systems. Harry Nyquist from Bell Laboratory studied the sampling process and derived a criterion that determines the minimum sampling rate for any continuous analog signals. His, now famous, minimum sampling rate is known as the Nyquist sampling rate, which states that one must sample a signal at least twice as fast as the highest frequency content of the signal of interest. For example, if we are dealing with the human voice signal that contains frequency components that span from about 20 Hz to 4 kHz, the Nyquist sample theorem requires that we must sample the signal at least at 8 kHz, 8000 'snap shots' every second. Engineers who work for telephone companies must deal with such issues. For further study on the Nyquist sampling rate, refer to Pack and Barrett listed in the References section.

When a signal is sampled a low pass anti–aliasing filter must be employed to insure the Nyquist sampling rate is not violated. In the example above, a low pass filter with a cutoff frequency of 4 KHz would be used before the sampling circuitry for this purpose.

Quantization. Now that we understand the sampling process, let's move on to the second process of the analog–to–digital conversion, quantization. Each digital system has a number of bits it uses as the basic unit to represent data. A bit is the most basic unit where single binary information, one or zero, is represented. A nibble is made up of four bits put together. A byte is eight bits.

We have tacitly avoided the discussion of the form of captured signal samples. When a signal is sampled, digital systems need some means to represent the captured samples. The quantization of a sampled signal is how the signal is represented as one of the quantization levels. Suppose you have a single bit to represent an incoming signal. You only have two different numbers, 0 and 1. You may say that you can distinguish only low from high. Suppose you have two bits. You can represent four different levels, 00, 01, 10, and 11. What if you have three bits? You now can represent eight different levels: 000, 001, 010, 011, 100, 101, 110, and 111. Think of it as follows. When you had two bits, you were able to represent four different levels. If we add one more bit, that bit can be one or zero, making the total possibilities eight. Similar discussion can lead us to conclude that given n bits, we have 2^n unique numbers or levels one can represent.

Figure 5.1 shows how n bits are used to quantize a range of values. In many digital systems, the incoming signals are voltage signals. The voltage signals are first obtained from physical signals (pressure, temperature, etc.) with the help of transducers, such as microphones, angle sensors, and infrared sensors. The voltage signals are then conditioned to map their range with the input range of a digital system, typically 0 to 5 volts. In Figure 5.1, n bits allow you to divide the input signal range of a digital system into 2^n different quantization levels. As can be seen from the figure, the more quantization levels means the better mapping of an incoming signal to its true value. If we only had a single bit, we can only represent level 0 and level 1. Any analog signal value in between the range had to be mapped either as level 0 or level 1, not many choices. Now imagine what happens as we increase the number of bits available for the quantization levels. What happens when the available number of bits is 8? How many different quantization levels are available now? Yes, 256. How about 10, 12, or 14? Notice also that as the number of bits used for the quantization levels increases for a given input range the 'distance' between two adjacent levels decreases accordingly.

Finally, the encoding process involves converting a quantized signal into a digital binary number. Suppose again we are using eight bits to quantize a sampled analog signal. The quantization levels are determined by the eight bits and each sampled signal is quantized as one of 256 quantization levels. Consider the two sampled signals shown in Figure 5.1. The first sample is mapped to quantization level 2 and the second one is mapped to quantization level 198. Note the amount of quantization error introduced for both samples. The quantization error is inversely proportional to the number of bits used to quantize the signal.

Encoding. Once a sampled signal is quantized, the encoding process involves representing the quantization level with the available bits. Thus, for the first sample, the encoded sampled value is 0000_0001, while the encoded sampled value for the second sample is 1100_0110. As a result of the encoding process, sampled analog signals are now represented as a set of binary numbers. Thus,

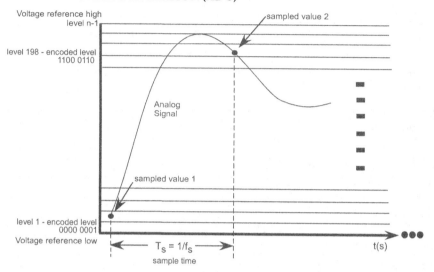

Figure 5.1: Sampling, quantization, and encoding.

the encoding is the last necessary step to represent a sampled analog signal into its corresponding digital form, shown in Figure 5.1.

5.2.1   RESOLUTION AND DATA RATE

Resolution. Resolution is a measure used to quantize an analog signal. In fact, resolution is nothing more than the voltage 'distance' between two adjacent quantization levels we discussed earlier. Suppose again we have a range of 5 volts and one bit to represent an analog signal. The resolution in this case is 2.5 volts, a very poor resolution. You can imagine how your TV screen will look if you only had only two levels to represent each pixel, black and white. The maximum error, called the resolution error, is 2.5 volts for the current case, 50 % of the total range of the input signal. Suppose you now have four bits to represent quantization levels. The resolution now becomes 1.25 volts or 25 % of the input range. Suppose you have 20 bits for quantization levels. The resolution now becomes 4.77×10^{-6} volts, 9.54×10^{-5}% of the total range. The discussion we presented simply illustrates that as we increase the available number of quantization levels within a fixed voltage range, the distance between adjacent levels decreases, reducing the quantization error of a sampled signal. As the number grows, the error decreases, making the representation of a sampled analog signal more accurate in the corresponding digital form. The number of bits used for the quantization is directly proportional to the resolution of a system. You now should understand the technical background when you watch high definition television broadcasting. In general, resolution may be defined as:

$$resolution = (voltage\ span)/2^b = (V_{ref\ high} - V_{ref\ low})/2^b$$

for the ATmega328, the resolution is:

$$resolution = (5 - 0)/2^{10} = 4.88\,mV$$

Data rate. The definition of the data rate is the amount of data generated by a system per some time unit. Typically, the number of bits or the number of bytes per second is used as the data rate of a system. We just saw that the more bits we use for the quantization levels, the more accurate we can represent a sampled analog signal. Why not use the maximum number of bits current technologies can offer for all digital systems, when we convert analog signals to digital counterparts? It has to do with the cost involved. In particular, suppose you are working for a telephone company and your switching system must accommodate 100,000 customers. For each individual phone conversation, suppose the company uses an 8KHz sampling rate (f_s) and you are using 10 bits for the quantization levels for each sampled signal.[1] This means the voice conversation will be sampled every 125 microseconds (T_s) due to the reciprocal relationship between (f_s) and (T_s). If all customers are making out of town calls, what is the number of bits your switching system must process to accommodate all calls? The answer will be 100,000 x 8000 x 10 or eight billion bits per every second! You will need some major computing power to meet the requirement for processing and storage of the data. For such reasons, when designers make decisions on the number of bits used for the quantization levels and the sampling rate, they must consider the computational burden the selection will produce on the computational capabilities of a digital system versus the required system resolution.

Dynamic range. You will also encounter the term "dynamic range" when you consider finding appropriate analog–to–digital converters. The dynamic range is a measure used to describe the signal to noise ratio. The unit used for the measurement is Decibel (dB), which is the strength of a signal with respect to a reference signal. The greater the dB number, the stronger the signal is compared to a noise signal. The definition of the dynamic range is 20 $log\,2^b$ where b is the number of bits used to convert analog signals to digital signals. Typically, you will find 8 to 12 bits used in commercial analog–to–digital converters, translating the dynamic range from 20 $log\,2^8$ dB to 20 $log\,2^{12}$ dB.

5.3   ANALOG–TO–DIGITAL CONVERSION (ADC) PROCESS

The goal of the ADC process is to accurately represent analog signals as digital signals. Toward this end, three signal processing procedures, sampling, quantization, and encoding, described in the previous section must be combined together. Before the ADC process takes place, we first need to convert a physical signal into an electrical signal with the help of a transducer. A transducer is an electrical and/or mechanical system that converts physical signals into electrical signals or electrical signals to physical signals. Depending on the purpose, we categorize a transducer as an input transducer or an output transducer. If the conversion is from physical to electrical, we call it an input transducer. The mouse, the keyboard, and the microphone for your personal computer all fall under this category. A camera, an infrared sensor, and a temperature sensor are also input transducers.

---

[1] For the sake of our discussion, we ignore other overheads involved in processing a phone call such as multiplexing, de–multiplexing, and serial–to–parallel conversion.

The output transducer converts electrical signals to physical signals. The computer screen and the printer for your computer are output transducers. Speakers and electrical motors are also output transducers. Therefore, transducers play the central part for digital systems to operate in our physical world by transforming physical signals to and from electrical signals. It is important to carefully design the interface between transducers and the microcontroller to insure proper operation. A poorly designed interface could result in improper embedded system operation or failure. Interface techniques are discussed in detail in Chapter 8.

5.3.1   TRANSDUCER INTERFACE DESIGN (TID) CIRCUIT

In addition to transducers, we also need a signal conditioning circuitry before we apply the ADC. The signal conditioning circuitry is called the transducer interface. The objective of the transducer interface circuit is to scale and shift the electrical signal range to map the output of the input transducer to the input range of the analog–to–digital converter which is typically 0 to 5 VDC. Figure 5.2 shows the transducer interface circuit using an input transducer.

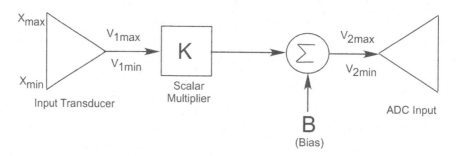

Figure 5.2: A block diagram of the signal conditioning for an analog–to–digital converter. The range of the sensor voltage output is mapped to the analog–to–digital converter input voltage range. The scalar multiplier maps the magnitudes of the two ranges and the bias voltage is used to align two limits.

The output of the input transducer is first scaled by constant K. In the figure, we use a microphone as the input transducer whose output ranges from -5 VDC to + 5 VDC. The input to the analog–to–digital converter ranges from 0 VDC to 5 VDC. The box with constant K maps the output range of the input transducer to the input range of the converter. Naturally, we need to multiply all input signals by 1/2 to accommodate the mapping. Once the range has been mapped, the signal now needs to be shifted. Note that the scale factor maps the output range of the input transducer as -2.5 VDC to +2.5 VDC instead of 0 VDC to 5 VDC. The second portion of the circuit shifts the range by 2.5 VDC, thereby completing the correct mapping. Actual implementation of the TID circuit components is accomplished using operational amplifiers.

In general, the scaling and bias process may be described by two equations:

$$V_{2max} = (V_{1max} \times K) + B$$

$$V_{2min} = (V_{1min} \times K) + B$$

The variable V_{1max} represents the maximum output voltage from the input transducer. This voltage occurs when the maximum physical variable (X_{max}) is presented to the input transducer. This voltage must be scaled by the scalar multiplier (K) and then have a DC offset bias voltage (B) added to provide the voltage V_{2max} to the input of the ADC converter [USAFA].

Similarly, The variable V_{1min} represents the minimum output voltage from the input transducer. This voltage occurs when the minimum physical variable (X_{min}) is presented to the input transducer. This voltage must be scaled by the scalar multiplier (K) and then have a DC offset bias voltage (B) added to produce voltage V_{2min} to the input of the ADC converter.

Usually, the values of V_{1max} and V_{1min} are provided with the documentation for the transducer. Also, the values of V_{2max} and V_{2min} are known. They are the high and low reference voltages for the ADC system (usually 5 VDC and 0 VDC for a microcontroller). We thus have two equations and two unknowns to solve for K and B. The circuits to scale by K and add the offset B are usually implemented with operational amplifiers.

Example: A photodiode is a semiconductor device that provides an output current corresponding to the light impinging on its active surface. The photodiode is used with a transimpedance amplifier to convert the output current to an output voltage. A photodiode/transimpedance amplifier provides an output voltage of 0 volts for maximum rated light intensity and -2.50 VDC output voltage for the minimum rated light intensity. Calculate the required values of K and B for this light transducer so it may be interfaced to a microcontroller's ADC system.

$$V_{2max} = (V_{1max} \times K) + B$$

$$V_{2min} = (V_{1min} \times K) + B$$

$$5.0\ V = (0\ V \times K) + B$$

$$0\ V = (-2.50\ V \times K) + B$$

The values of K and B may then be determined to be 2 and 5 VDC, respectively.

5.3.2   OPERATIONAL AMPLIFIERS

In the previous section, we discussed the transducer interface design (TID) process. Going through this design process yields a required value of gain (K) and DC bias (B). Operational amplifiers (op amps) are typically used to implement a TID interface. In this section, we briefly introduce operational amplifiers including ideal op amp characteristics, classic op amp circuit configurations, and an example to illustrate how to implement a TID with op amps. Op amps are also used in a wide variety of other applications including analog computing, analog filter design, and a myriad of other applications. We do not have the space to investigate all of these related applications. The interested reader is referred to the References section at the end of the chapter for pointers to some excellent texts on this topic.

5.3.2.1   The ideal operational amplifier

A generic ideal operational amplifier is illustrated in Figure 5.3. An ideal operational does not exist in the real world. However, it is a good first approximation for use in developing op amp application circuits.

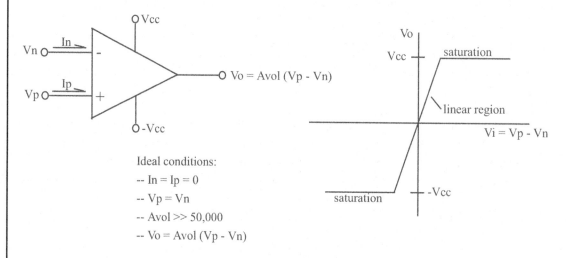

Figure 5.3: Ideal operational amplifier characteristics.

The op amp is an active device (requires power supplies) equipped with two inputs, a single output, and several voltage source inputs. The two inputs are labeled Vp, or the non–inverting input, and Vn, the inverting input. The output of the op amp is determined by taking the difference between Vp and Vn and multiplying the difference by the open loop gain (A_{vol}) of the op amp which is typically a large value much greater than 50,000. Due to the large value of A_{vol}, it does not take much of a difference between Vp and Vn before the op amp will saturate. When an op amp saturates, it does not damage the op amp, but the output is limited to the supply voltages $\pm V_{cc}$. This will clip the output, and hence distort the signal, at levels slightly less than $\pm V_{cc}$. Op amps are typically used in a closed loop, negative feedback configuration. A sample of classic operational amplifier configurations with negative feedback are provided in Figure 5.4 [Faulkenberry].

It should be emphasized that the equations provided with each operational amplifier circuit are only valid if the circuit configurations are identical to those shown. Even a slight variation in the circuit configuration may have a dramatic effect on circuit operation. It is important to analyze each operational amplifier circuit using the following steps:

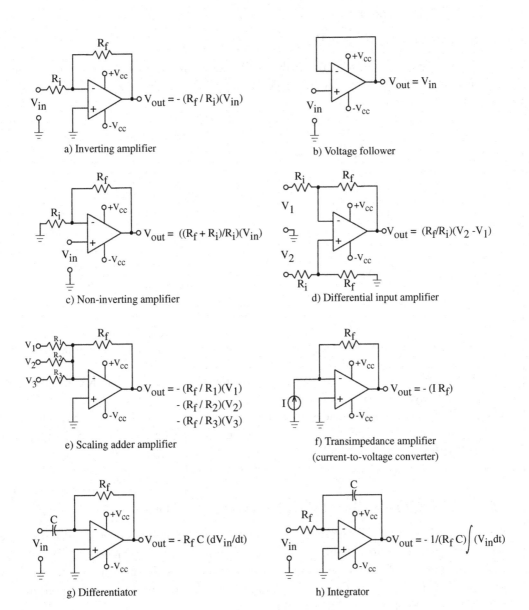

Figure 5.4: Classic operational amplifier configurations. Adapted from [Faulkenberry].

- Write the node equation at Vn for the circuit.

- Apply ideal op amp characteristics to the node equation.

- Solve the node equation for Vo.

As an example, we provide the analysis of the non–inverting amplifier circuit in Figure 5.5. This same analysis technique may be applied to all of the circuits in Figure 5.4 to arrive at the equations for Vout provided.

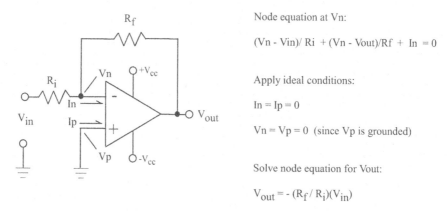

Node equation at Vn:

$(Vn - Vin)/\ Ri\ + (Vn - Vout)/Rf\ +\ In\ = 0$

Apply ideal conditions:

$In = Ip = 0$

$Vn = Vp = 0$ (since Vp is grounded)

Solve node equation for Vout:

$V_{out} = -\ (R_f/\ R_i)(V_{in})$

Figure 5.5: Operational amplifier analysis for the non–inverting amplifier. Adapted from [Faulkenberry].

Example: In the previous section, it was determined that the values of K and B were 2 and 5 VDC, respectively. The two–stage op amp circuitry provided in Figure 5.6 implements these values of K and B. The first stage provides an amplification of -2 due to the use of the non–inverting amplifier configuration. In the second stage, a summing amplifier is used to add the output of the first stage with a bias of − 5 VDC. Since this stage also introduces a minus sign to the result, the overall result of a gain of 2 and a bias of +5 VDC is achieved.

5.4   ADC CONVERSION TECHNOLOGIES

The ATmega328 and the ATmega2560 use a successive–approximation converter technique to convert an analog sample into a 10–bit digital representation. In this section, we will discuss this type of conversion process. For a review of other converter techniques, the interested reader is referred to "Atmel AVR Microcontroller Primer: Programming and Interfacing." In certain applications, you are required to use converter technologies external to the microcontroller.

5.4.1   SUCCESSIVE–APPROXIMATION

The ATmega328 and the ATmega2560 microcontrollers are equipped with a successive–approximation ADC type converter. The successive–approximation technique uses a digital–to–

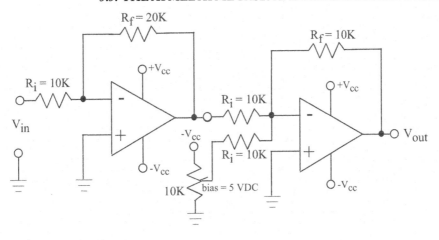

Figure 5.6: Operational amplifier implementation of the transducer interface design (TID) example circuit.

analog converter, a controller, and a comparator to perform the ADC process. Starting from the most significant bit down to the least significant bit, the controller turns on each bit at a time and generates an analog signal, with the help of the digital–to–analog converter, to be compared with the original input analog signal. Based on the result of the comparison, the controller changes or leaves the current bit and turns on the next most significant bit. The process continues until decisions are made for all available bits. Figure 5.7 shows the architecture of this type of converter. The advantage of this technique is that the conversion time is uniform for any input, but the disadvantage of the technology is the use of complex hardware for implementation.

5.5   THE ATMEL ATMEGA328 AND ATMEGA2560 ADC SYSTEM

The Atmel ATmega328 and the ATmega2560 microcontrollers are equipped with a flexible and powerful ADC system. It has the following features [Atmel]:

- 10–bit resolution

- ± 2 least significant bit (LSB) absolute accuracy

- 13 ADC clock cycle conversion time

- **ATmega328:** 6 multiplexed single ended input channels

- **ATmega2560:** 16 multiplexed single ended input channels

- Selectable right or left result justification

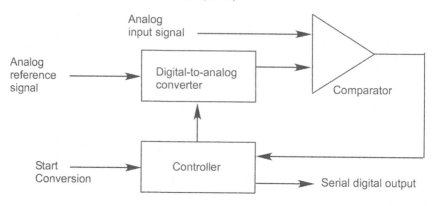

Figure 5.7: Successive-approximation ADC.

- 0 to Vcc ADC input voltage range

Let's discuss each feature in turn. The first feature of discussion is "10–bit resolution." Resolution is defined as:

$$Resolution = (V_{RH} - V_{RL})/2^b$$

V_{RH} and V_{RL} are the ADC high and low reference voltages. Whereas, "b" is the number of bits available for conversion. For the ATmega processor with reference voltages of 5 VDC, 0 VDC, and 10–bits available for conversion, resolution is 4.88 mV. Absolute accuracy specified as ± 2 LSB is then ± 9.76 mV at this resolution [Atmel].

It requires 13 analog–to–digital clock cycles to perform an ADC conversion. The ADC system may be run at a slower clock frequency than the main microcontroller clock source. The main microcontroller clock is divided down using the ADC Prescaler Select (ADPS[2:0]) bits in the ADC Control and Status Register A (ADCSRA). A slower ADC clock results in improved ADC accuracy at higher controller clock speeds.

The ADC is equipped with a single successive–approximation converter. Only a single ADC channel may be converted at a given time. The input of the ADC is equipped with a multiple input (ATmega328:6/Atemega2560:16) analog multiplexer. The analog input for conversion is selected using the MUX[3:0] bits in the ADC Multiplexer Selection Register (ADMUX).

The 10–bit result from the conversion process is placed in the ADC Data Registers, ADCH and ADCL. These two registers provide 16 bits for the 10–bit result. The result may be left justified by setting the ADLAR (ADC Left Adjust Result) bit of the ADMUX register. Right justification is provided by clearing this bit.

The analog input voltage for conversion must be between 0 and Vcc volts. If this is not the case, external circuitry must be used to insure the analog input voltage is within these prescribed bounds as discussed earlier in the chapter.

5.5.1   BLOCK DIAGRAM

The block diagram for the ATmega328 and ATmega2560 ADC conversion system is provided in Figure 5.8. The left edge of the diagram provides the external microcontroller pins to gain access to the ADC. For the ATmega328, the six analog input channels are provided at pins ADC[5:0]; whereas, for the ATmega2560 the sixteen analog input channels are provided at pins ADC[15:0]. The ADC reference voltage pins are provided at AREF and AVCC. The key features and registers of the ADC system previously discussed are included in the diagram.

5.5.2   ATMEGA328 ADC REGISTERS

The key registers for the ATmega328 ADC system are shown in Figure 5.9. It must be emphasized that the ADC system has many advanced capabilities that we do not discuss here. Our goal is to review the basic ADC conversion features of this powerful system. We have already discussed many of the register setting already. We will discuss each register in turn [Atmel].

5.5.2.1   ATmega328 ADC Multiplexer Selection Register (ADMUX)

As previously discussed, the ADMUX register contains the ADLAR bit to select left or right justification and the MUX[3:0] bits to determine which analog input will be provided to the analog--to--digital converter for conversion. To select a specific input for conversion is accomplished when a binary equivalent value is loaded into the MUX[3:0] bits. For example, to convert channel ADC7, "0111" is loaded into the ADMUX register. This may be accomplished using the following C instruction:

ADMUX = 0x07;

The REFS[1:0] bits of the ADMUX register are also used to determine the reference voltage source for the ADC system. These bits may be set to the following values:

- REFS[0:0] = 00: AREF used for ADC voltage reference

- REFS[0:1] = 01: AVCC with external capacitor at the AREF pin

- REFS[1:0] = 10: Reserved

- REFS[1:1] = 11: Internal 1.1 VDC voltage reference with an external capacitor at the AREF pin

5.5.2.2   ATmega328 ADC Control and Status Register A (ADCSRA)

The ADCSRA register contains the ADC Enable (ADEN) bit. This bit is the "on/off" switch for the ADC system. The ADC is turned on by setting this bit to a logic one. The ADC Start

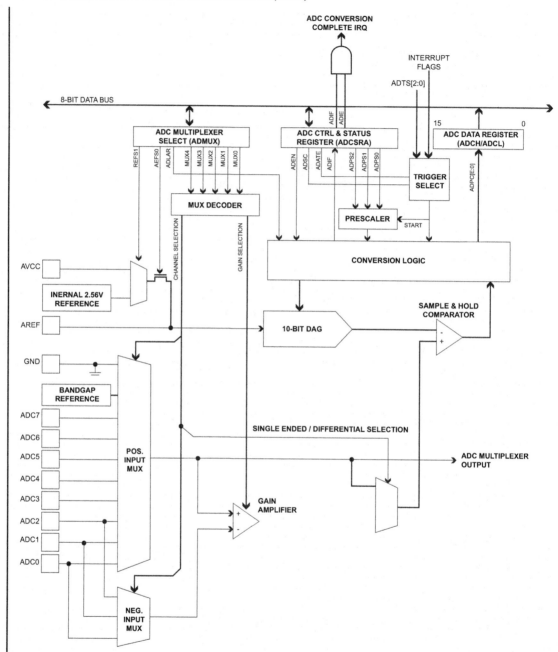

Figure 5.8: Atmel AVR ATmega328 ADC block diagram. (Figure used with permission of Atmel, Incorporated.)

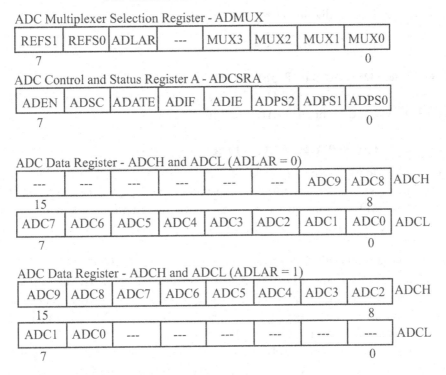

ADC Multiplexer Selection Register - ADMUX

REFS1	REFS0	ADLAR	---	MUX3	MUX2	MUX1	MUX0
7							0

ADC Control and Status Register A - ADCSRA

ADEN	ADSC	ADATE	ADIF	ADIE	ADPS2	ADPS1	ADPS0
7							0

ADC Data Register - ADCH and ADCL (ADLAR = 0)

---	---	---	---	---	---	ADC9	ADC8	ADCH
15							8	

ADC7	ADC6	ADC5	ADC4	ADC3	ADC2	ADC1	ADC0	ADCL
7							0	

ADC Data Register - ADCH and ADCL (ADLAR = 1)

ADC9	ADC8	ADC7	ADC6	ADC5	ADC4	ADC3	ADC2	ADCH
15							8	

ADC1	ADC0	---	---	---	---	---	---	ADCL
7							0	

Figure 5.9: ATmega328 ADC Registers. Adapted from Atmel.

Conversion (ADSC) bit is also contained in the ADCSRA register. Setting this bit to logic one initiates an ADC. The ADCSRA register also contains the ADC Interrupt flag (ADIF) bit. This bit sets to logic one when the ADC is complete. The ADIF bit is reset by writing a logic one to this bit.

The ADC Prescaler Select (ADPS[2:0]) bits are used to set the ADC clock frequency. The ADC clock is derived from dividing down the main microcontroller clock. The ADPS[2:0] may be set to the following values:

- ADPS[2:0] = 000: division factor 2

- ADPS[2:0] = 001: division factor 2

- ADPS[2:0] = 010: division factor 4

- ADPS[2:0] = 011: division factor 8

- ADPS[2:0] = 100: division factor 16

- ADPS[2:0] = 101: division factor 32

- ADPS[2:0] = 110: division factor 64

- ADPS[2:0] = 111: division factor 128

5.5.2.3 ATmega328 ADC Data Registers (ADCH, ADCL)

As previously discussed, the ADC Data Register contains the result of the ADC. The results may be left (ADLAR=1) or right (ADLAR=0) justified.

5.5.3 ATMEGA2560 ADC REGISTERS

The key registers for the ATmega2560 ADC system are shown in Figure 5.10. It must be emphasized that the ADC system has many advanced capabilities that we do not discuss here. Our goal is to review the basic ADC conversion features of this powerful system. We have already discussed many of the register setting already. We will discuss each register in turn [Atmel].

5.5.3.1 ATmega2560 ADC Multiplexer Selection Register (ADMUX)

As previously discussed, the ADMUX register contains the ADLAR bit to select left or right justification and the MUX[4:0] bits to determine which analog input will be provided to the analog-–to–digital converter for conversion. There is another MUX bit (MUX5) contained in the ADC Control and Status Register B (ADCSRB).

There is a wide variety of single–ended, differential input, and gain combinations that may be selected for ADC conversion. Complete details are provided in the ATmega2560 data sheet (www.atmel.com). For example, to select a specific single–ended input (0 through 7) for conversion, the binary equivalent value of the channel number is loaded into the MUX[5:0] bits. For example, to convert channel ADC7, "00_0111" is loaded into the ADMUX register and the MUX5 bit of the ADCSRB register. This may be accomplished using the following C instructions:

ADMUX = 0x07;
ADCSRB = 0x00;

For a single–ended conversion on ADC channels 8 through 15, the following bits must be loaded to ADMUX bits 5 to 0:

- ADC 8: 10_0000

- ADC 9: 10_0001

- ADC 10: 10_0010

- ADC 11: 10_0011

- ADC 12: 10_0100

- ADC 13: 10_0101

- ADC 14: 10_0110

ADC Multiplexer Selection Register - ADMUX

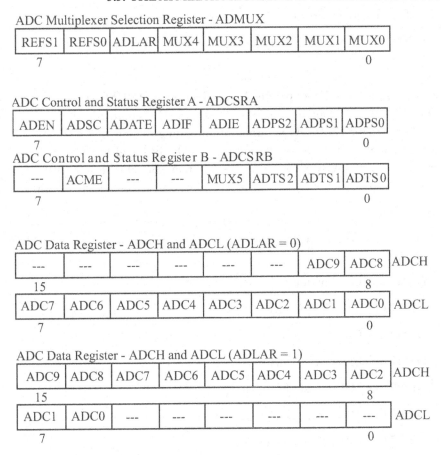

REFS1	REFS0	ADLAR	MUX4	MUX3	MUX2	MUX1	MUX0
7							0

ADC Control and Status Register A - ADCSRA

ADEN	ADSC	ADATE	ADIF	ADIE	ADPS2	ADPS1	ADPS0
7							0

ADC Control and Status Register B - ADCSRB

---	ACME	---	---	MUX5	ADTS2	ADTS1	ADTS0
7							0

ADC Data Register - ADCH and ADCL (ADLAR = 0)

---	---	---	---	---	---	ADC9	ADC8	ADCH
15							8	

ADC7	ADC6	ADC5	ADC4	ADC3	ADC2	ADC1	ADC0	ADCL
7							0	

ADC Data Register - ADCH and ADCL (ADLAR = 1)

ADC9	ADC8	ADC7	ADC6	ADC5	ADC4	ADC3	ADC2	ADCH
15							8	

ADC1	ADC0	---	---	---	---	---	---	ADCL
7							0	

Figure 5.10: ATmega2560 ADC Registers. Adapted from Atmel.

- ADC 15: 10_0111

For example, to convert channel ADC8, "10_0000" is loaded into the ADMUX register and the MUX5 bit of the ADCSRB register. This may be accomplished using the following C instructions:

```
ADMUX = 0x00;
ADCSRB = 0x08;
```

The REFS[1:0] bits of the ADMUX register are also used to determine the reference voltage source for the ADC system. These bits may be set to the following values:

- REFS[0:0] = 00: AREF used for ADC voltage reference

- REFS[0:1] = 01: AVCC with external capacitor at the AREF pin

- REFS[1:0] = 10: Reserved

- REFS[1:1] = 11: Internal 1.1 VDC voltage reference with an external capacitor at the AREF pin

5.5.3.2   ATmega2560 ADC Control and Status Register A (ADCSRA)

The ADCSRA register contains the ADC Enable (ADEN) bit. This bit is the "on/off" switch for the ADC system. The ADC is turned on by setting this bit to a logic one. The ADC Start Conversion (ADSC) bit is also contained in the ADCSRA register. Setting this bit to logic one initiates an ADC. The ADCSRA register also contains the ADC Interrupt flag (ADIF) bit. This bit sets to logic one when the ADC is complete. The ADIF bit is reset by writing a logic one to this bit.

The ADC Prescaler Select (ADPS[2:0]) bits are used to set the ADC clock frequency. The ADC clock is derived from dividing down the main microcontroller clock. The ADPS[2:0] may be set to the following values:

- ADPS[2:0] = 000: division factor 2

- ADPS[2:0] = 001: division factor 2

- ADPS[2:0] = 010: division factor 4

- ADPS[2:0] = 011: division factor 8

- ADPS[2:0] = 100: division factor 16

- ADPS[2:0] = 101: division factor 32

- ADPS[2:0] = 110: division factor 64

- ADPS[2:0] = 111: division factor 128

5.5.3.3   ATmega2560 ADC Data Registers (ADCH, ADCL)

As previously discussed, the ADC Data Register contains the result of the ADC. The results may be left (ADLAR=1) or right (ADLAR=0) justified.

5.6   PROGRAMMING THE ADC USING THE ARDUINO DEVELOPMENT ENVIRONMENT

The Arduino Development Environment has the built–in function analogRead to perform an ADC conversion. The format for the analogRead function is:

```
unsigned int   return_value;


return_value = analogRead(analog_pin_read);
```

The function returns an unsigned integer value from 0 to 1023, corresponding to the voltage span from 0 to 5 VDC.

Recall that we introduced the use of this function in the Application section at the end of Chapter 3. The analogRead function was used to read the analog output values from the three IR sensors on the Blinky 602A robot.

5.7   ATMEGA328: PROGRAMMING THE ADC IN C

Provided below are two functions to operate the ATmega328 ADC system. The first function "InitADC( )" initializes the ADC by first performing a dummy conversion on channel 0. In this particular example, the ADC prescalar is set to 8 (the main microcontroller clock was operating at 10 MHz).

The function then enters a while loop waiting for the ADIF bit to set indicating the conversion is complete. After conversion the ADIF bit is reset by writing a logic one to it.

The second function, "ReadADC(unsigned char)," is used to read the analog voltage from the specified ADC channel. The desired channel for conversion is passed in as an unsigned character variable. The result is returned as a left justified, 10 bit binary result. The ADC prescalar must be set to 8 to slow down the ADC clock at higher external clock frequencies (10 MHz) to obtain accurate results. After the ADC is complete, the results in the eight bit ADCL and ADCH result registers are concatenated into a 16–bit unsigned integer variable and returned to the function call.

```
//**************************************************************
//InitADC: initialize analog-to-digital converter
//**************************************************************

void InitADC( void)
{
ADMUX = 0;                      //Select channel 0
ADCSRA = 0xC3;                  //Enable ADC & start 1st dummy
                                //conversion

                                //Set ADC module prescalar to 8

                                //critical for accurate ADC results
while (!(ADCSRA & 0x10));       //Check if conversation is ready
ADCSRA |= 0x10;                 //Clear conv rdy flag - set the bit
}


//**************************************************************
//ReadADC: read analog voltage from analog-to-digital converter -
//the desired channel for conversion is passed in as an unsigned
```

```
//character variable.
//The result is returned as a right justified, 10 bit binary result.
//The ADC prescalar must be set to 8 to slow down the ADC clock at higher
//external clock frequencies (10 MHz) to obtain accurate results.
//****************************************************************************



unsigned int ReadADC(unsigned char channel)
{
unsigned int binary_weighted_voltage, binary_weighted_voltage_low;
                                    //weighted binary voltage
unsigned int binary_weighted_voltage_high;

ADMUX = channel;                        //Select channel
ADCSRA |= 0x43;                         //Start conversion
                                        //Set ADC module prescalar to 8
                                        //critical for accurate ADC results
while (!(ADCSRA & 0x10));                //Check if conversion is ready
ADCSRA |= 0x10;                         //Clear Conv rdy flag - set the bit
binary_weighted_voltage_low = ADCL;     //Read 8 low bits first (important)
                                        //Read 2 high bits, multiply by 256
binary_weighted_voltage_high = ((unsigned int)(ADCH << 8));
binary_weighted_voltage = binary_weighted_voltage_low |
                          binary_weighted_voltage_high;
return binary_weighted_voltage;         //ADCH:ADCL
}


//****************************************************************************
```

5.8   ATMEGA2560: PROGRAMMING THE ADC IN C

Provided below are two functions to operate the ATmega2560 ADC system. The first function "InitADC( )" initializes the ADC by first performing a dummy conversion on channel 0. In this particular example, the ADC prescalar is set to 8 (the main microcontroller clock was operating at 10 MHz).

The function then enters a while loop waiting for the ADIF bit to set indicating the conversion is complete. After conversion the ADIF bit is reset by writing a logic one to it.

The second function, "ReadADC(unsigned char)," is used to read the analog voltage from the specified ADC channel. The desired channel for conversion is passed in as an unsigned character variable. The result is returned as a left justified, 10 bit binary result. The ADC prescalar must be set to 8 to slow down the ADC clock at higher external clock frequencies (10 MHz) to obtain accurate results. After the ADC is complete, the results in the eight bit ADCL and ADCH result registers are concatenated into a 16–bit unsigned integer variable and returned to the function call.

```c
//*************************************************************************
//InitADC: initialize analog-to-digital converter
//*************************************************************************

void InitADC( void)
{
ADMUX = 0;                          //Select channel 0
ADCSRA = 0xC3;                      //Enable ADC & start 1st dummy
                                    //conversion

                                    //Set ADC module prescalar to 8

                                    //critical for accurate ADC results
while (!(ADCSRA & 0x10));           //Check if conversation is ready
ADCSRA |= 0x10;                     //Clear conv rdy flag - set the bit
}


//*************************************************************************
//ReadADC: read analog voltage from analog-to-digital converter -
//the desired channel for conversion is passed in as an unsigned
//character variable.
//The result is returned as a right justified, 10 bit binary result.
//The ADC prescalar must be set to 8 to slow down the ADC clock at higher
//external clock frequencies (10 MHz) to obtain accurate results.
//*************************************************************************



unsigned int ReadADC(unsigned char channel)
{
unsigned int binary_weighted_voltage, binary_weighted_voltage_low;
                                    //weighted binary voltage
unsigned int binary_weighted_voltage_high;

if(channel <= 7)
  {
```

```
  ADMUX = channel;                        //Select channel
  ADCSRB = 0x00;                          //Configure ADMUX[5]
  }
else
  {
  channel = (unsigned char)(channel - 8);
  ADMUX = channel;
  ADCSRB = 0x08;                          //Configure ADMUX[5]
  }

ADCSRA |= 0x43;                           //Start conversion
                                          //Set ADC module prescalar to 8
                                          //critical for accurate ADC results
while (!(ADCSRA & 0x10));                 //Check if conversion is ready
ADCSRA |= 0x10;                           //Clear Conv rdy flag - set the bit
binary_weighted_voltage_low = ADCL;       //Read 8 low bits first (important)
                                          //Read 2 high bits, multiply by 256
binary_weighted_voltage_high = ((unsigned int)(ADCH << 8));
binary_weighted_voltage = binary_weighted_voltage_low |
                          binary_weighted_voltage_high;
return binary_weighted_voltage;           //ADCH:ADCL
}


//************************************************************************
```

5.9   EXAMPLE: ADC RAIN GAGE INDICATOR WITH THE ARDUINO UNO R3

In this example, we construct a rain gage type level display using small light emitting diodes. The rain gage indicator consists of a panel of eight light emitting diodes. The gage may be constructed from individual diodes or from an LED bar containing eight elements. Whichever style is chosen, the interface requirements between the processor and the LEDs are the same.

The requirement for this project is to use the analog–to–digital converter to illuminate up to eight LEDs based on the input voltage. A 10k trimmer potentiometer is connected to the ADC channel to vary the input voltage. We first provide a solution using the Arduino Development Environment with the Arduino UNO R3 processing board. Then a solution employing the ATmega328 programmed in C is provided.

5.9.1   ADC RAIN GAGE INDICATOR USING THE ARDUINO DEVELOPMENT ENVIRONMENT

The circuit configuration employing the Arduino UNO R3 processing board is provided in Figure 5.11. The DIGITAL pins of the microcontroller are used to communicate with the LED interface circuit. We describe the operation of the LED interface circuit in Chapter 8.

The sketch to implement the project requirements is provided below. As in previous examples, we define the Arduino UNO R3 pins, set them for output via the setup() function, and write the loop() function. In this example, the loop() function senses the voltage from the 10K trimmer potentiometer and illuminates a series of LEDs corresponding to the sensed voltage levels.

```
//**************************************************************************
#define trim_pot   A0                   //analog input pin

                                        //digital output pins
                                        //LED indicators 0 - 7
#define LED0        0                    //digital pin
#define LED1        1                    //digital pin
#define LED2        2                    //digital pin
#define LED3        3                    //digital pin
#define LED4        4                    //digital pin
#define LED5        5                    //digital pin
#define LED6        6                    //digital pin
#define LED7        7                    //digital pin


int trim_pot_reading;                   //declare variable for trim pot

void setup()
  {
                                        //LED indicators - wall detectors
  pinMode(LED0, OUTPUT);                //configure pin 0 for digital output
  pinMode(LED1, OUTPUT);                //configure pin 1 for digital output
  pinMode(LED2, OUTPUT);                //configure pin 2 for digital output
  pinMode(LED3, OUTPUT);                //configure pin 3 for digital output
  pinMode(LED4, OUTPUT);                //configure pin 4 for digital output
  pinMode(LED5, OUTPUT);                //configure pin 5 for digital output
  pinMode(LED6, OUTPUT);                //configure pin 6 for digital output
  pinMode(LED7, OUTPUT);                //configure pin 7 for digital output
  }
```

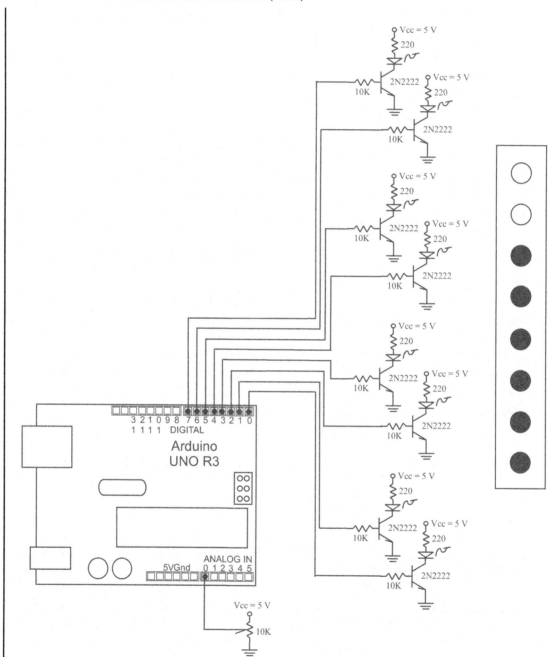

Figure 5.11: ADC with rain gage level indicator.

```
void loop()
  {
                                  //read analog output from trim pot
    trim_pot_reading = analogRead(trim_pot);

    if(trim_pot_reading < 128)
      {
      digitalWrite(LED0, HIGH);
      digitalWrite(LED1, LOW);
      digitalWrite(LED2, LOW);
      digitalWrite(LED3, LOW);
      digitalWrite(LED4, LOW);
      digitalWrite(LED5, LOW);
      digitalWrite(LED6, LOW);
      digitalWrite(LED7, LOW);
      }
    else if(trim_pot_reading < 256)
      {
      digitalWrite(LED0, HIGH);
      digitalWrite(LED1, HIGH);
      digitalWrite(LED2, LOW);
      digitalWrite(LED3, LOW);
      digitalWrite(LED4, LOW);
      digitalWrite(LED5, LOW);
      digitalWrite(LED6, LOW);
      digitalWrite(LED7, LOW);
      }
    else if(trim_pot_reading < 384)
      {
      digitalWrite(LED0, HIGH);
      digitalWrite(LED1, HIGH);
      digitalWrite(LED2, HIGH);
      digitalWrite(LED3, LOW);
      digitalWrite(LED4, LOW);
      digitalWrite(LED5, LOW);
      digitalWrite(LED6, LOW);
      digitalWrite(LED7, LOW);
      }
    else if(trim_pot_reading < 512)
```

```
       {
     digitalWrite(LED0, HIGH);
     digitalWrite(LED1, HIGH);
     digitalWrite(LED2, HIGH);
     digitalWrite(LED3, HIGH);
     digitalWrite(LED4, LOW);
     digitalWrite(LED5, LOW);
     digitalWrite(LED6, LOW);
     digitalWrite(LED7, LOW);
       }
   else if(trim_pot_reading < 640)
       {
     digitalWrite(LED0, HIGH);
     digitalWrite(LED1, HIGH);
     digitalWrite(LED2, HIGH);
     digitalWrite(LED3, HIGH);
     digitalWrite(LED4, HIGH);
     digitalWrite(LED5, LOW);
     digitalWrite(LED6, LOW);
     digitalWrite(LED7, LOW);
       }
   else if(trim_pot_reading < 768)
     {
     digitalWrite(LED0, HIGH);
     digitalWrite(LED1, HIGH);
     digitalWrite(LED2, HIGH);
     digitalWrite(LED3, HIGH);
     digitalWrite(LED4, HIGH);
     digitalWrite(LED5, HIGH);
     digitalWrite(LED6, LOW);
     digitalWrite(LED7, LOW);
     }
   else if(trim_pot_reading < 896)
     {
     digitalWrite(LED0, HIGH);
     digitalWrite(LED1, HIGH);
     digitalWrite(LED2, HIGH);
     digitalWrite(LED3, HIGH);
     digitalWrite(LED4, HIGH);
```

```
        digitalWrite(LED5, HIGH);
        digitalWrite(LED6, HIGH);
        digitalWrite(LED7, LOW);

        }
     else
        {
        digitalWrite(LED0, HIGH);
        digitalWrite(LED1, HIGH);
        digitalWrite(LED2, HIGH);
        digitalWrite(LED3, HIGH);
        digitalWrite(LED4, HIGH);
        digitalWrite(LED5, HIGH);
        digitalWrite(LED6, HIGH);
        digitalWrite(LED7, HIGH);

        }
delay(500);                          //delay 500 ms
}


//**************************************************************************
```

5.9.2 ADC RAIN GAGE INDICATOR IN C

We now implement the rain gage indicator with the control algorithm (sketch) programmed in C for the ATmega328. The circuit configuration employing the ATmega328 is provided in Figure 5.12. PORT D of the microcontroller is used to communicate with the LED interface circuit. In this example, we extend the requirements of the project:

- Write a function to display an incrementing binary count from 0 to 255 on the LEDs.

- Write a function to display a decrementing binary count from 255 to 0 on the LEDs.

- Use the analog–to–digital converter to illuminate up to eight LEDs based on the input voltage. A 10k trimmer potentiometer is connected to the ADC channel to vary the input voltage.

The algorithm for the project was written by Anthony (Tony) Kunkel, MSEE and Geoff Luke, MSEE, at the University of Wyoming for an Industrial Controls class assignment. A 30 ms delay is provided between PORTD LED display updates. This prevents the display from looking as a series of LEDs that are always illuminated.

Note: The delay function provided in this example is not very accurate. It is based on counting the number of assembly language no operation (NOPs). If a single NOP requires a single clock cycle

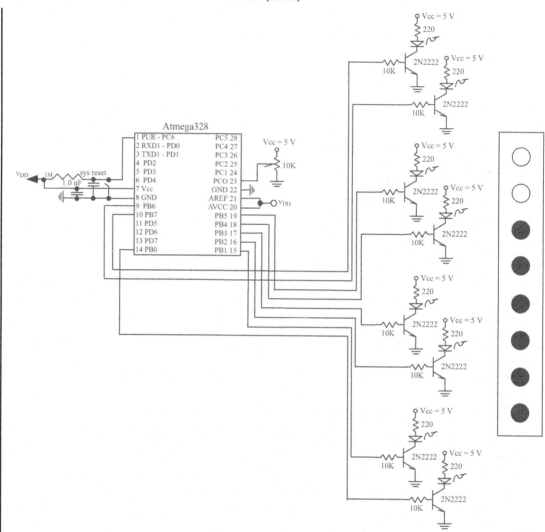

Figure 5.12: ADC with rain gage level indicator.

to execute. A rough (and inaccurate) estimation of expended time may be calculated. The actual delay depends on the specific resolution of clock source used with the microcontroller, the clock frequency of the clock source and the specific compiler used to implement the code. In this specific example an accurate delay is not required since we are simply trying to slow down the code execution time so the LED changes may be observed. In the next chapter we provide a more accurate method of providing a time delay based on counting interrupts.

```
//*********************************************************************
//Tony Kunkel and Geoff Luke
//University of Wyoming
//*********************************************************************
//This program calls four functions:
//First: Display an incrementing binary count on PORTD from 0-255
//Second: Display a decrementing count on PORTD from 255-0
//Third: Display rain gauge info on PORTD
//Fourth: Delay when ever PORTD is updated
//*********************************************************************

//ATMEL register definitions for ATmega328
#include<iom328pv.h>

//function prototypes
void display_increment(void);        //Function
to display increment to PORTD
void display_decrement(void);        //Function
to display decrement to PORTD
void rain_gage(void)
void InitADC(void);                  //Initialize ADC converter
unsigned int ReadADC();              //Read specified ADC channel
void delay(void);                    //Function to delay


//*********************************************************************

int main(void)
{
display_increment();                 //Display incrementing binary on

                                     //PORTD from 0-255
delay();                             //Delay

display_decrement();                 //Display decrementing binary on

                                     //PORTD from 255-0
delay();                             //Delay

InitADC();
```

```
while(1)
  {
  rain_gage();                  //Display gage info on PORTD
  delay();                      //Delay
  }

return 0;
}




//*******************************************************************
//function definitions
//*******************************************************************


//*******************************************************************
//delay
//Note: The delay function provided in this example is not very accurate.
//It is based on counting the number of assembly language no operation
//(NOPs).  If a single NOP requires a single clock cycle to execute.  A
//rough (and inaccurate) estimation of expended time may be calculated.
//The actual delay depends on the specific resolution of clock source
//used with the microcontroller, the clock frequency of the clock source
//and the specific compiler used to implement the code.  In this specific
//example an accurate delay is not required since we are simply trying to
//slow down the code execution time so the LED changes may be observed.
//*******************************************************************

void delay(void)
{
int i, k;

for(i=0; i<400; i++)
  {
  for(k=0; k<300; k++)
      {
      asm("nop");               //assembly language nop, requires 2 cycles
      }
```

```
   }
}


//********************************************************************
//Displays incrementing binary count from 0 to 255
//********************************************************************

void display_increment(void)
{
int i;
unsigned char j = 0x00;

DDRD = 0xFF;                     //set PORTD to output

for(i=0; i<255; i++)
  {
  j++;                           //increment j
  PORTD = j;                     //assign j to data port
  delay();                       //wait
  }
}


//********************************************************************
//Displays decrementing binary count from 255 to 0
//********************************************************************

void display_decrement(void)
{
int i;
unsigned char j = 0xFF;

DDRD = 0xFF;                     //set PORTD to output

for(i=0; i<256; i++)
  {
  j=(j-0x01);                    //decrement j by one
  PORTD = j;                     //assign char j to data port
  delay();                       //wait
  }
```

```
}


//*********************************************************************
//Initializes ADC
//*********************************************************************

void InitADC(void)
{
ADMUX = 0;                    //Select channel 0
ADCSRA = 0xC3;                //Enable ADC & start dummy conversion
                              //Set ADC module prescalar
                              //to 8 critical for
                              //accurate ADC results
while (!(ADCSRA & 0x10));      //Check if conversation is ready
ADCSRA |= 0x10;               //Clear conv rdy flag - set the bit
}


//*********************************************************************
//ReadADC: read analog voltage from analog-to-digital converter -
//the desired channel for conversion is passed in as an unsigned
//character variable. The result is returned as a right justified,
//10 bit binary result.
//The ADC prescalar must be set to 8 to slow down the ADC clock at
//higher external clock frequencies (10 MHz) to obtain accurate results.
//*********************************************************************

unsigned int ReadADC(unsigned char channel)
{
unsigned int binary_weighted_voltage, binary_weighted_voltage_low;
unsigned int binary_weighted_voltage_high; //weighted binary voltage

ADMUX = channel;                     //Select channel
ADCSRA |= 0x43;                      //Start conversion
                                     //Set ADC module prescalar
                                     //to 8 critical for
                                     //accurate ADC results
while (!(ADCSRA & 0x10));             //Check if conversion is ready
ADCSRA |= 0x10;                      //Clear Conv rdy flag - set the bit
```

```
binary_weighted_voltage_low = ADCL; //Read 8 low bits first-(important)
                                    //Read 2 high bits, multiply by 256
binary_weighted_voltage_high = ((unsigned int)(ADCH << 8));
binary_weighted_voltage = binary_weighted_voltage_low |
                          binary_weighted_voltage_high;
return binary_weighted_voltage;      //ADCH:ADCL
}


//***********************************************************************
//Displays voltage magnitude as LED level on PORTB
//***********************************************************************

void rain_gage(void)
{
unsigned int ADCValue;

ADCValue = readADC(0x00);

DDRD = 0xFF;                    //set PORTD to output

if(ADCValue < 128)
  {
  PORTD = 0x01;
  }
else if(ADCValue < 256)
  {
  PORTD = 0x03;
  }
else if(ADCValue < 384)
  {
  PORTD = 0x07;
  }
else if(ADCValue < 512)
  {
  PORTD = 0x0F;
  }
else if(ADCValue < 640)
  {
    PORTD = 0x1F;
```

```
  }
else if(ADCValue < 768)
  {
  PORTD = 0x3F;
  }
else if(ADCValue < 896)
  {
  PORTD = 0x7F;
  }
else
  {
  PORTD = 0xAA;
  }
}
//*********************************************************************
//*********************************************************************
```

5.9.3   ADC RAIN GAGE USING THE ARDUINO DEVELOPMENT ENVIRONMENT–REVISITED

If you carefully compare the two implementations of the rain gage indicator provided in the two pre-vious examples, you will note that some activities are easier to perform in the Arduino Development Environment while others are easier to accomplish in C. Is it possible to mix the two techniques for a more efficient sketch? The answer is "yes!"

In this example, we revise the earlier Arduino Development Environment sketch using por-tions of the original sketch with the ATmega328 control algorithm code to significantly shorten the program. In particular, we will address PORTD directly in the sketch to shorten up the code when the LEDs are illuminated. Recall in Chapter 1, we provided the open source schematic of the Arduino UNO R3 processing board (Figure 1.7). Careful study of the schematic reveals that DIGITAL pins 7 to 0 are connected to PORTD of the Arduino UNO R3 processing board. This allows direct configuration of PORTD using C language constructs.

```
//*********************************************************************
#define trim_pot    0                    //analog input pin

int trim_pot_reading;                     //declare variable for trim pot

void setup()
  {
  DDRD = 0xFF;                            //Set PORTD (DIGITAL pins 7 to 0)
  }                                       //as output
```

```
void loop()
  {
                                        //read analog output from trim pot
   trim_pot_reading = analogRead(trim_pot);

   if(trim_pot_reading < 128)
     {
     PORTD = 0x01;                      //illuminate LED 0
     }
   else if(trim_pot_reading < 256)
     {
     PORTD = 0x03;                      //illuminate LED 0-1
     }
   else if(trim_pot_reading < 384)
     {
     PORTD = 0x07;                      //illuminate LED 0-2
     }
   else if(trim_pot_reading < 512)
     {
     PORTD = 0x0F;                      //illuminate LED 0-3
     }
   else if(trim_pot_reading < 640)
     {
     PORTD = 0x1F;                      //illuminate LED 0-4
     }
   else if(trim_pot_reading < 768)
     {
     PORTD = 0x3F;                      //illuminate LED 0-5
     }
   else if(trim_pot_reading < 896)
     {
     PORTD = 0x7F;                      //illuminate LED 0-6
     }
   else
     {
     PORTD = 0xFF;                      //illuminate LED 0-7
     }
 delay(500);                           //delay 500 ms
```

```
}


//*******************************************************************************
```

5.10   ONE–BIT ADC – THRESHOLD DETECTOR

A threshold detector circuit or comparator configuration contains an operational amplifier employed in the open loop configuration. That is, no feedback is provided from the output back to the input to limit gain. A threshold level is applied to one input of the op amp. This serves as a comparison reference for the signal applied to the other input. The two inputs are constantly compared to one another. When the input signal is greater than the set threshold value, the op amp will saturate to a value slightly less than +Vcc as shown in Figure 5.13a). When the input signal falls below the threshold the op amp will saturate at a voltage slightly greater than −Vcc. If a single–sided op amp is used in the circuit (e.g., LM324), the −Vcc supply pin may be connected to ground. In this configuration, the op map provides for a one–bit ADC circuit.

A bank of threshold detectors may be used to construct a multi–channel threshold detector as shown in Figure 5.13c). This provides a flash converter type ADC. It is a hardware version of a rain gage indicator. In Chapter 1, we provided a 14–channel version for use in a laboratory instrumentation project.

5.11   DIGITAL–TO–ANALOG CONVERSION (DAC)

Once a signal is acquired to a digital system with the help of the analog–to digital conversion process and has been processed, frequently the processed signal is converted back to another analog signal. A simple example of such a conversion occurs in digital audio processing. Human voice is converted to a digital signal, modified, processed, and converted back to an analog signal for people to hear. The process to convert digital signals to analog signals is completed by a digital–to–analog converter. The most commonly used technique to convert digital signals to analog signals is the summation method shown in Figure 5.14.

With the summation method of digital–to–analog conversion, a digital signal, represented by a set of ones and zeros, enters the digital–to–analog converter from the most significant bit to the least significant bit. For each bit, a comparator checks its logic state, high or low, to produce a clean digital bit, represented by a voltage level. Typically, in a microcontroller context, the voltage level is +5 or 0 volts to represent logic one or logic zero, respectively. The voltage is then multiplied by a scalar value based on its significant position of the digital signal as shown in Figure 5.14. Once all bits for the signal have been processed, the resulting voltage levels are summed together to produce the final analog voltage value. Notice that the production of a desired analog signal may involve further signal conditioning such as a low pass filter to 'smooth' the quantized analog signal and a transducer interface circuit to match the output of the digital–to–analog converter to the input of an output transducer.

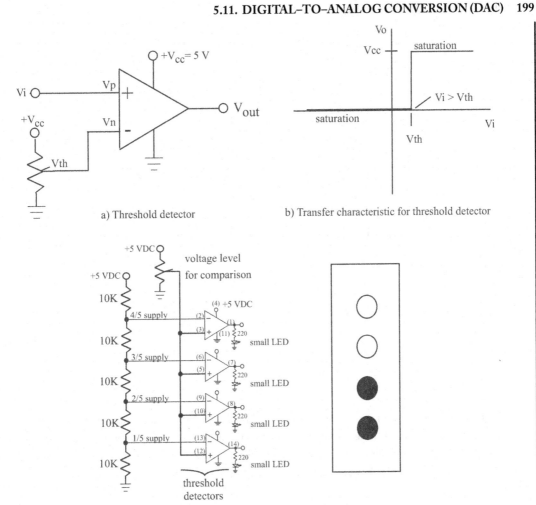

a) Threshold detector

b) Transfer characteristic for threshold detector

c) 4-channel threshold detector

Figure 5.13: One–bit ADC threshold detector.

5.11.1 DAC WITH THE ARDUINO DEVELOPMENT ENVIRONMENT

The analogWrite command within the Arduino Development Environment issues a signal from 0 to 5 VDC by sending a constant from 0 to 255 using pulse width modulation (PWM) techniques. This signal, when properly filtered, serves as a DC signal. The Blinky 602A control sketch provided in Chapter 3 used the analogWrite command to issue drive signals to the robot motors to navigate the robot through the maze.

The form of the analogWrite command is the following:

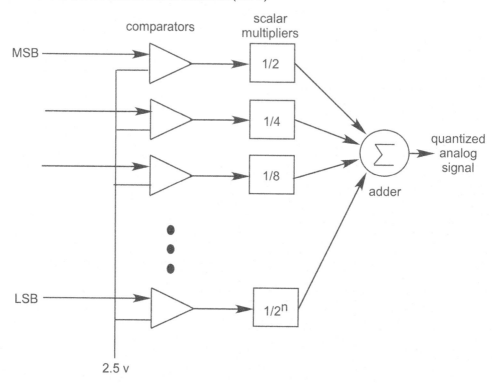

Figure 5.14: A summation method to convert a digital signal into a quantized analog signal. Comparators are used to clean up incoming signals and the resulting values are multiplied by a scalar multiplier and the results are added to generate the output signal. For the final analog signal, the quantized analog signal should be connected to a low pass filter followed by a transducer interface circuit.

```
analogWrite(output pin, value);
```

5.11.2 DAC WITH EXTERNAL CONVERTERS

A microcontroller can be equipped with a wide variety of DAC configurations including:

- Single channel, 8–bit DAC connected via a parallel port (e.g., Motorola MC1408P8)

- Quad channel, 8–bit DAC connected via a parallel port (e.g., Analog Devices AD7305)

- Quad channel, 8–bit DAC connected via the SPI (e.g., Analog Devices AD7304)

- Octal channel, 8–bit DAC connected via the SPI (e.g., Texas Instrument TLC5628)

Space does not allow an in depth look at each configuration, but we will examine the TLC5628 in more detail.

5.11.3 OCTAL CHANNEL, 8–BIT DAC VIA THE SPI

The Texas Instruments (TI) TLC5628 is an eight channel, 8–bit DAC connected to the microcontroller via the SPI. Reference Figure 5.15a). It has a wide range of features packed into a 16–pin chip including [Texas Instrument]:

- Eight individual, 8–bit DAC channels,

- Operation from a single 5 VDC supply,

- Data load via the SPI interface,

- Programmable gain (1X or 2X), and

- A 12–bit command word to program each DAC.

Figure 5.15b) provides the interconnection between the ATmega328 and the TLC5628. The ATmega328 SPI's SCK line is connected to the TLC5628: the MOSI to the serial data line and PORTB[2] to the Load line. As can be seen in the timing diagram, two sequential bytes are sent from the ATmega328 to select the appropriate DAC channel and to provide updated data to the DAC. The Load line is pulsed low to update the DAC. In this configuration, the LDAC line is tied low. The function to transmit data to the DAC is left as an assignment for the reader at the end of the chapter.

5.12 APPLICATION: ART PIECE ILLUMINATION SYSTEM – REVISITED

In Chapter 2, we investigated an illumination system for a painting. The painting was to be illuminated via high intensity white LEDs. The LEDs could be mounted in front of or behind the painting as the artist desired. We equipped the lighting system with an IR sensor to detect the presence of someone viewing the piece. We also wanted to adjust the intensity of the lighting based on how the close viewer was to the art piece. A circuit diagram of the system is provided in Figure 5.16.

The Arduino Development Environment sketch to sense how away the viewer is and issue a proportional intensity control signal to illuminate the LED is provided below. The analogRead function is used to obtain the signal from the IR sensor. The analogWrite function is used to issue a proportional signal.

```
//************************************************************************
                                        //analog input pins
#define viewer_sensor        A5        //analog pin - left IR sensor

                                        //digital output pins
#define illumination_output  0         //illumination output pin
```

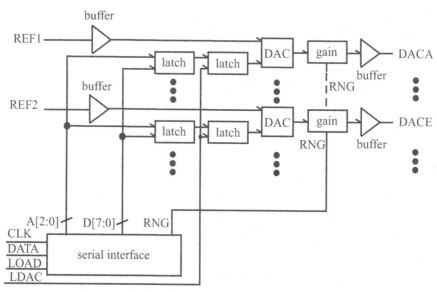

a) TLC5628 octal 8-bit DACs

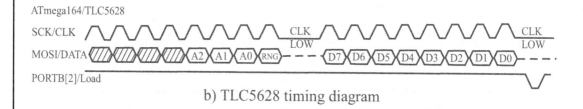

b) TLC5628 timing diagram

A[2:0]	DAC Updated	D[7:0]	Output Voltage
000	DACA	0000_0000	GND
001	DACB	0000_0001	(1/256) x REF(1+RNG)
010	DACC	:	:
011	DACD	:	:
100	DACE	:	:
101	DACF	:	:
110	DACG	:	:
111	DACH	1111_1111	(255/256) x REF (1+RNG)

c) TLC5628 bit assignments.

Figure 5.15: Eight channel DAC. Adapted from [Texas Instrument].

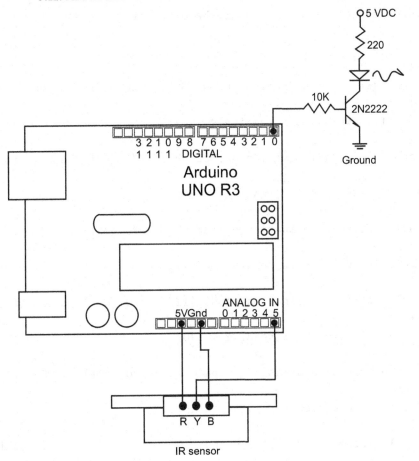

Figure 5.16: IR sensor interface.

```
unsigned int viewer_sensor_reading;        //current value of sensor output

void setup()
  {
  pinMode(illumination_output, OUTPUT);     //config. pin 0 for dig. output
  }

void loop()
  {                                         //read analog output from
                                            //IR sensors
```

```
    viewer_sensor_reading  = analogRead(viewer_sensor);

    if(viewer_sensor_reading < 128)
      {
      analogWrite(illumination_output,31);   //0 (off) to 255 (full speed)
      }
    else if(viewer_sensor_reading < 256)
      {
      analogWrite(illumination_output,63);   //0 (off) to 255 (full speed)
      }
    else if(viewer_sensor_reading < 384)
      {
      analogWrite(illumination_output,95);   //0 (off) to 255 (full speed)
      }
    else if(viewer_sensor_reading < 512)
      {
      analogWrite(illumination_output,127); //0 (off) to 255 (full speed)
      }
    else if(viewer_sensor_reading < 640)
      {
      analogWrite(illumination_output,159); //0 (off) to 255 (full speed)
      }
    else if(viewer_sensor_reading < 768)
      {
      analogWrite(illumination_output,191); //0 (off) to 255 (full speed)
      }
    else if(viewer_sensor_reading < 896)
      {
      analogWrite(illumination_output,223); //0 (off) to 255 (full speed)
      }
    else
      {
      analogWrite(illumination_output,255); //0 (off) to 255 (full speed)
      }
delay(500);                                 //delay 500 ms
}

//******************************************************************************
```

5.13 ARDUINO MEGA 2560 EXAMPLE: KINESIOLOGY AND HEALTH LABORATORY INSTRUMENTATION

In this example an Arduino Mega 2560 is used to control a piece of laboratory equipment. The Kinesiology and Health (KNH) Department required assistance in developing a piece of laboratory research equipment. An overview of the equipment is provided in Figure 5.17. This example was originally presented in "Embedded Systems Design with the Atmel AVR Microcontroller [Barrett]."

The KNH researchers needed a display panel containing two columns of large (10 mm diameter) red LEDs. The LEDs needed to be viewable at a distance of approximately 5 meters. The right column of LEDs would indicate the desired level of exertion for the subject under test. This LED array would be driven by an external signal generator using a low frequency ramp signal. The left column of LEDs would indicate actual subject exertion level. This array would be driven by a powered string potentiometer.

As its name implies, a string potentiometer is equipped with a string pull. The resistance provided by the potentiometer is linearly related to the string displacement. Once powered, the string pot provides an output voltage proportional to the string pull length. The end of the string would be connected to a displacing arm on a piece of exercise equipment (e.g. a leg lift apparatus).

The following features were required:

- The ability to independently drive a total of 28 large (10 mm diameter) LEDs.

- An interface circuit to drive each LED.

- Two analog–to–digital channels to convert the respective string potentiometer and signal generator inputs into digital signals.

- Input buffering circuitry to isolate and protect the panel from the string pot and signal generator and to guard against accidental setting overload.

- An algorithm to link the analog input signals to the appropriate LED activation signals.

Due to the high number of output pins required to drive the LEDs, the Arduino Mega 2560 microcontroller is a good choice for this project. The overall project may be partitioned using the structure chart illustrated in Figure 5.18.

The full up, two array, 14–bit output driving a total of 28 large LEDs circuit diagram is provided in Figure 5.19. The LED circuit was described in detail in Chapter 4.

As shown in Figure 5.20, any rest period and exercise period may be selected by careful choice of the signal parameters applied to the desired exertion array. In the example shown, a rest period of 2s and an exercise period of 1s are shown. To obtain this parameter setting, the signal generator is set for a ramp signal with the parameters shown. The input protection circuitry limits the excursion of the input signal to approximately 0 to 4 volts.

Provided below is the template for the KNH control software. The "analogRead" function returns a value from 0 to 1023 as determined by the analog value read from 0 to 5 VDC. The 5 VDC

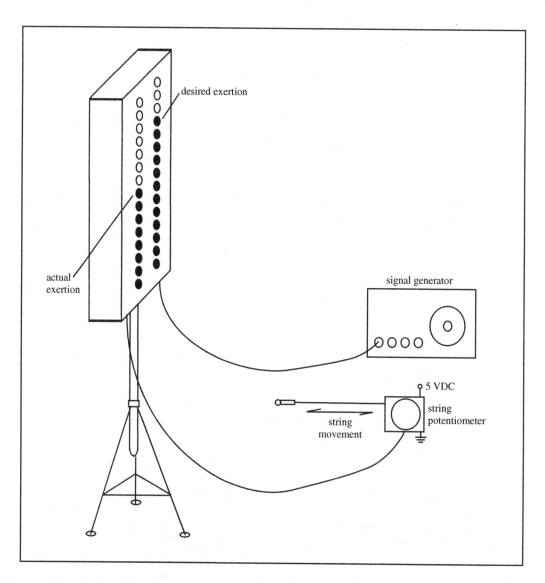

Figure 5.17: KNH project overview.

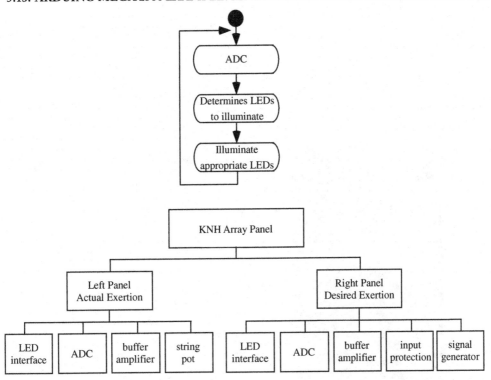

Figure 5.18: UML activity diagram and structure chart for KNH arrays.

range has been divided into thirteen equal spans of 78 ADC levels each ($1024/13 = 78$). In the code template, an LED is illuminated in each array corresponding to the analog value sensed from the string potentiometer and the signal generator.

```
//**************************************************************************
                                 //analog input pins
#define string_pot      A0       //analog in - string potentiometer
#define signal_gen      A1       //analog in - signal generator


                                 //digital output pins
                                 //LED array indicators
                                 //string potentiometer array
#define string_pot_1    21       //digital pin - string pot 1
   :
   :
#define string_pot_14   34       //digital pin - string pot 14
```

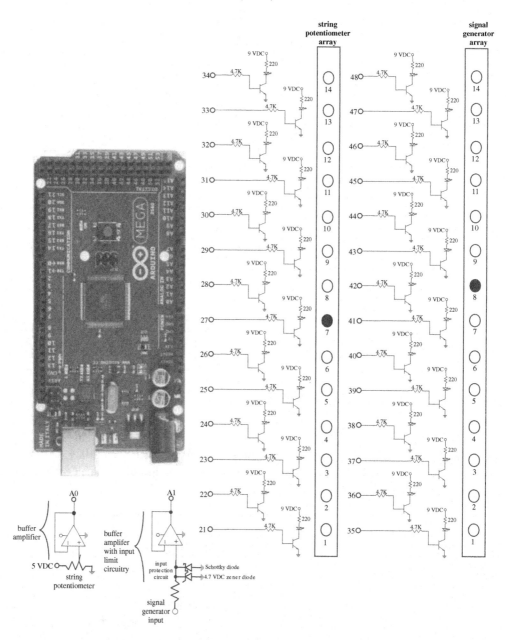

Figure 5.19: Circuit diagram for the KNH arrays. (left) Actual exertion array and (right) desired exertion array. (UNO R3 illustration used with permission of the Arduino Team (CC BY–NC–SA) www. arduino.cc).

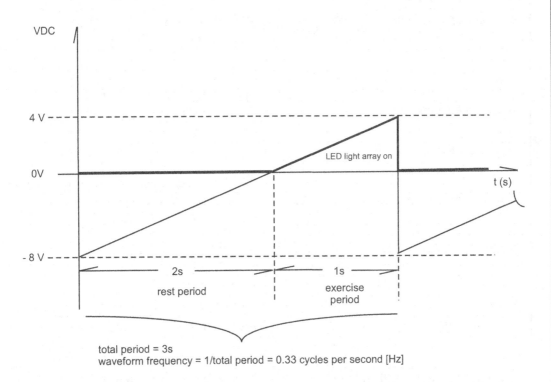

Figure 5.20: KNH array settings. LEDs will illuminate for input voltages between 0 and 4.7 VDC. More LEDs incrementally illuminate as input voltage increases in response to signal generator input signal. The input to the array is protected by a 4.7 VDC zener diode and a Schottky diode to limit voltage input to levels between approximately 0 and 4.7 VDC.

```
                                     //digital output pins
                                     //LED array indicators
                                     //signal generator array
#define sig_gen_1        35          //digital pin - sig gen 1
   :
   :
#define sig_gen_14       48          //digital pin - sig gen 14



int string_pot_value;                //declare variable string pot
int signal_gen_value;                //declare signal generator

void setup()
  {
                                     //LED indicators - string pot array
  pinMode(string_pot_1,   OUTPUT);   //string pot 1 output
     :
     :
  pinMode(string_pot_14, OUTPUT);    //string pot 14 output


                                     //LED indicators - signal gen array
  pinMode(sig_gen_1,   OUTPUT);      //signal gen 1 output
     :
     :
  pinMode(sig_gen_14, OUTPUT);       //signal gen 14 output



  }

void loop()
  {
                                        //read analog inputs
                                        // - string potentiometer
                                        // - signal generator
    string_pot_value = analogRead(string_pot);
```

```
signal_gen_value = analogRead(signal_gen);

//illuminate appropriate LED in string pot array
if(string_pot_value > 0)&&(string_pot_value <= 78)
  {
  digitalWrite(string_pot_1,HIGH);   //turn LED on
  digitalWrite(string_pot_2, LOW);   //turn LED off
    :

    :

  digitalWrite(string_pot_14, LOW); //turn LED off
  }

 elseif(string_pot_value > 78)&&(string_pot_value <= 156)
  {
  digitalWrite(string_pot_1, LOW);   //turn LED off
  digitalWrite(string_pot_2,HIGH);   //turn LED on
    :

    :

  digitalWrite(string_pot_14, LOW); //turn LED off
  }


    :

    :


 elseif(string_pot_value > 936)&&(string_pot_value <= 1023)
  {
  digitalWrite(string_pot_1, LOW);    //turn LED off
  digitalWrite(string_pot_2, LOW);    //turn LED off
    :

    :

  digitalWrite(string_pot_14,HIGH);   //turn LED on
  }


//illuminate appropriate LED in string pot array
if(signal_gen_value > 0)&&(signal_gen_value <= 78)
  {
  digitalWrite(sig_gen_1,HIGH);        //turn LED on
```

```
      digitalWrite(sig_gen_2, LOW);        //turn LED off
        :
        :

      digitalWrite(ig_gen_14, LOW);        //turn LED off
      }


   elseif(signal_gen_value > 78)&&(signal_gen_value <= 156)
      {
      digitalWrite(sig_gen_1, LOW);        //turn LED off
      digitalWrite(sig_gen_2,HIGH);        //turn LED on
        :
        :

      digitalWrite(sig_gen_14,LOW);        //turn LED off
      }


        :
        :



   elseif(signal_gen_value > 936)&&(signal_gen_value <= 1023)
      {
      digitalWrite(sig_gen_1, LOW);          //turn LED off
      digitalWrite(sig_gen_2, LOW);          //turn LED off
        :
        :

      digitalWrite(sig_gen_14,HIGH);         //turn LED on
      }


}
//*****************************************************************
```

5.14 SUMMARY

In this chapter, we presented the differences between analog and digital signals and used this knowledge to discuss three sub–processing steps involved in analog to digital converters: sampling, quantization, and encoding. We also presented the quantization errors and the data rate associated with the ADC process. The dynamic range of an analog–to–digital converter, one of the measures to describe a conversion process, was also presented. We then presented the successive–approximation converter. Transducer interface design concepts were then discussed along with supporting information on operational amplifier configurations. We then reviewed the operation, registers, and actions required to program the ADC system aboard the ATmega328 and the ATmega2560. We concluded the chapter with a discussion of the ADC process and an implementation using a multi–channel DAC connected to the ATmega328 SPI system and an instrumentation example using the Arduino Mega 2560.

5.15 REFERENCES

- *Atmel 8–bit AVR Microcontroller with 16K Bytes In–System Programmable Flash, ATmega328, ATmega328L,* data sheet: 2466L–AVR–06/05, Atmel Corporation, 2325 Orchard Parkway, San Jose, CA 95131.

- *Atmel 8-bit AVR Microcontroller with 4/8/16/32K Bytes In-System Programmable Flash, ATmega48PA, 88PA, 168PA, 328P* data sheet: 8171D-AVR-05/11, Atmel Corporation, 2325 Orchard Parkway, San Jose, CA 95131.

- *Atmel 8-bit AVR Microcontroller with 64/128/256K Bytes In-System Programmable Flash, ATmega640/V, ATmega1280/V, 2560/V* data sheet: 2549P-AVR-10/2012, Atmel Corporation, 2325 Orchard Parkway, San Jose, CA 95131.

- Barrett S, Pack D (2006) Microcontrollers Fundamentals for Engineers and Scientists. Morgan and Claypool Publishers. DOI: 10.2200/S00025ED1V01Y200605DCS001

- Barrett S and Pack D (2008) Atmel AVR Microcontroller Primer Programming and Interfacing. Morgan and Claypool Publishers. DOI: 10.2200/S00100ED1V01Y200712DCS015

- Barrett S (2010) Embedded Systems Design with the Atmel AVR Microcontroller. Morgan and Claypool Publishers. DOI: 10.2200/S00225ED1V01Y200910DCS025

- Roland Thomas and Albert Rosa, *The Analysis and Design of Linear Circuits, Fourth Edition,* Wiley & Sons, Inc., New York, 2003.

- M.A. Hollander, editor, *Electrical Signals and Systems, Fourth Edition,* McGraw-Hill Companies, Inc, 1999.

- Daniel Pack and Steven Barrett, *Microcontroller Theory and Applications: HC12 and S12,* Prentice Hall, 2ed, Upper Saddle River, New Jersey 07458, 2008.

- Alan Oppenheim and Ronald Schafer, *Discrete-time Signal Processing, Second Edition,* Prentice Hall, Upper Saddle River, New Jersey, 1999.

- John Enderle, Susan Blanchard, and Joseph Bronzino, *Introduction to Biomedical Engineering,* Academic Press, 2000.

- L. Faulkenberry, *An Introduction to Operational Amplifiers,* John Wiley & Sons, New York, 1977.

• P. Horowitz and W. Hill, *The Art of Electronics*, Cambridge University Press, 1989.

• L. Faulkenberry, *Introduction to Operational Amplifiers with Linear Integrated Circuit Applications*, 1982.

• D. Stout and M. Kaufman, *Handbook of Operational Amplifier Circuit Design* McGraw-Hill Book Company, 1976.

• S. Franco, *Design with Operational Amplifiers and Analog Integrated Circuits, third edition*, McGraw-Hill Book Company, 2002.

• *TLC5628C, TLC5628I Octal 8-bit Digital-to-Analog Converters*, Texas Instruments, Dallas, TX, 1997.

5.16 CHAPTER PROBLEMS

1. Given a sinusoid with 500 Hz frequency, what should be the minimum sampling frequency for an analog-to-digital converter, if we want to faithfully reconstruct the analog signal after the conversion?

2. If 12 bits are used to quantize a sampled signal, what is the number of available quantized levels? What will be the resolution of such a system if the input range of the analog-to-digital converter is 10V?

3. Given the 12 V input range of an analog-to-digital converter and the desired resolution of 0.125 V, what should be the minimum number of bits used for the conversion?

4. Investigate the analog-to-digital converters in your audio system. Find the sampling rate, the quantization bits, and the technique used for the conversion.

5. A flex sensor provides 10K ohm of resistance for 0 degrees flexure and 40K ohm of resistance for 90 degrees of flexure. Design a circuit to convert the resistance change to a voltage change (Hint: consider a voltage divider). Then design a transducer interface circuit to convert the output from the flex sensor circuit to voltages suitable for the ATmega328 and the ATmega2560 ADC systems.

6. If an analog signal is converted by an analog-to-digital converter to a binary representation and then back to an analog voltage using a DAC, will the original analog input voltage be the same as the resulting analog output voltage? Explain.

7. Derive each of the characteristic equations for the classic operation amplifier configurations provided in Figure 5.4.

8. If a resistor was connected between the non-inverting terminal and ground in the inverting amplifier configuration of Figure 5.4a), how would the characteristic equation change?

9. A photodiode provides a current proportional to the light impinging on its active area. What classic operational amplifier configuration should be used to current the diode output to a voltage?

10. Does the time to convert an analog input signal to a digital representation vary in a successive-approximation converter relative to the magnitude of the input signal?

11. Calculate the signal parameters required to drive the KNH array with a 1s rest time and a 2s exercise time?

12. What is the purpose of the input protection circuitry for the desired exertion array in the KNH panel?

CHAPTER 6

Interrupt Subsystem

Objectives: After reading this chapter, the reader should be able to

- Understand the need of a microcontroller for interrupt capability.

- Describe the general microcontroller interrupt response procedure.

- Describe the ATmega328 and the ATmega2560 interrupt features.

- Properly configure and program an interrupt event for the ATmega328 and the ATmega2560 in C.

- Properly configure and program an interrupt event for the Arduino UNO R3 and the Arduino Mega 2560 using built–in features of the Arduino Development Environment.

- Use the interrupt system to implement a real time clock.

- Employ the interrupt system as a component in an embedded system.

6.1 OVERVIEW

A microcontroller normally executes instructions in an orderly fetch–decode–execute sequence as dictated by a user–written program as shown in Figure 6.1. However, the microcontroller must be equipped to handle unscheduled (although planned), higher priority events that might occur inside or outside the microcontroller. To process such events, a microcontroller requires an interrupt system.

The interrupt system onboard a microcontroller allows it to respond to higher priority events. Appropriate responses to these events may be planned, but we do not know when these events will occur. When an interrupt event occurs, the microcontroller will normally complete the instruction it is currently executing and then transition program control to interrupt event specific tasks. These tasks, which resolve the interrupt event, are organized into a function called an interrupt service routine (ISR). Each interrupt will normally have its own interrupt specific ISR. Once the ISR is complete, the microcontroller will resume processing where it left off before the interrupt event occurred.

In this chapter, we discuss the ATmega328 and the ATmega2560 interrupt system in detail. We provide several examples on how to program an interrupt in C and also using the built–in features of the Arduino Development Environment.

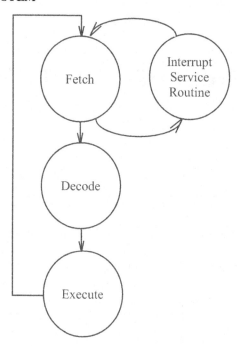

Figure 6.1: Microcontroller Interrupt Response.

6.1.1 ATMEGA328 INTERRUPT SYSTEM

The ATmega328 is equipped with a powerful and flexible complement of 26 interrupt sources. Two of the interrupts originate from external interrupt sources while the remaining 24 interrupts support the efficient operation of peripheral subsystems aboard the microcontroller. The ATmega328 interrupt sources are shown in Figure 6.2. The interrupts are listed in descending order of priority. As you can see, the RESET has the highest priority, followed by the external interrupt request pins INT0 (pin 4) and INT1 (pin 5). The remaining interrupt sources are internal to the ATmega328.

6.1.2 ATMEGA2560 INTERRUPT SYSTEM

The ATmega2560 is equipped with a powerful and flexible complement of 57 interrupt sources. Eight of the interrupts originate from external interrupt sources including three pin change interrupts. The remaining 46 interrupts support the efficient operation of peripheral subsystems aboard the microcontroller. The ATmega2560 interrupt sources are shown in Figure 6.3. The interrupts are listed in descending order of priority. As you can see, the RESET has the highest priority, followed by the external interrupt request pins INT0 through INT7.

Vector	Address	Source	Definition	AVR-GCC ISR name
1	0x0000	RESET	External pin, power-on reset, brown-out reset, watchdog system reset	
2	0x0002	INT0	External interrupt request 0	INT0_vect
3	0x0004	INT1	External interrupt request 1	INT1_vect
4	0x0006	PCINT0	Pin change interrupt request 0	PCINT0 vect
5	0x0008	PCINT1	Pin change interrupt request 1	PCINT1_vect
6	0x000A	PCINT2	Pin change interrupt request 2	PCINT2_vect
7	0x000C	WDT	Watchdog time-out interrupt	WDT_vect
8	0x000E	TIMER2 COMPA	Timer/Counter2 Compare Match A	TIMER_2_COMPA_vect
9	0x0010	TIMER2 COMPB	Timer/Counter2 Compare Match B	TIMER_2_COMPB_vect
10	0x0012	TIMER2 OVF	Timer/Counter2 Overflow	TIMER_2_OVF_vect
11	0x0014	TIMER1 CAPT	Timer/Counter1 Capture Event	TIMER_1_CAPT_vect
12	0x0016	TIMER1 COMPA	Timer/Counter1 Compare Match A	TIMER_1_COMPA_vect
13	0x0018	TIMER1 COMPB	Timer/Counter1 Compare Match B	TIMER_1_COMPB_vect
14	0x001A	TIMER1 OVF	Timer/Counter1 Overflow	TIMER_1_OVF_vect
15	0x001C	TIMER0 COMPA	Timer/Counter0 Compare Match A.	TIMER_0_COMPA_vect
16	0x001E	TIMER0 COMPB	Timer/Counter0 Compare Match B	TIMER_0_COMPB_vect
17	0x0020	TIMER0 OVF	Timer/Counter0 Overflow	TIMER_0_OVF vect
18	0x0022	SPI, STC	SPI Serial Transfer Complete	SPI_STC_vect
19	0x0024	USART, RX	USART Rx Complete	USART_RX_vect
20	0x0026	USART, UDRE	USART, Data Register Empty	USART_UDRE_vect
21	0x0028	USART, TX	USART, Tx Complete	USART_TX_vect
22	0x002A	ADC	ADC Conversiono Complete	ADC_vect
23	0x002C	EE READY	EEPROM Ready	EE_READY_vect
24	0x002E	ANALOG COMP	Analog Comparator	ANALOG_COMP_vect
25	0x0030	TWI	2-wire Serial Interface	TWI_vect
26	0x0032	SPM READY	Store Program Memory Ready	SPM_ready_vect

Figure 6.2: Atmel AVR ATmega328 Interrupts. (Adapted from figure used with permission of Atmel, Incorporated.)

Vector	Address	Source	Definition	AVR-GCC ISR name
1	0x0000	RESET	External pin, power-on reset, brown-out reset, watchdog system reset	
2	0x0002	INT0	External interrupt request 0	INT0_vect
3	0x0004	INT1	External interrupt request 1	INT1_vect
4	0x0006	INT2	External interrupt request 2	INT2_vect
5	0x0008	INT3	External interrupt request 3	INT3_vect
6	0x000A	INT4	External interrupt request 4	INT4_vect
7	0x000C	INT5	External interrupt request 5	INT5_vect
8	0x000E	INT6	External interrupt request 6	INT6_vect
9	0x0010	INT7	External interrupt request 7	INT7_vect
10	0x0012	PCINT0	Pin change interrupt request 0	PCINT0_vect
11	0x0014	PCINT1	Pin change interrupt request 1	PCINT1_vect
12	0x0016	PCINT2	Pin change interrupt request 2	PCINT2_vect
13	0x0018	WDT	Watchdog time-out interrupt	WDT_vect
14	0x001A	TIMER2 COMPA	Timer/Counter2 Compare Match A	TIMER2_COMPA_vect
15	0x001C	TIMER2 COMPB	Timer/Counter2 Compare Match B	TIMER2_COMPB_vect
16	0x001E	TIMER2 OVF	Timer/Counter2 Overflow	TIMER2_OVF_vect
17	0x0020	TIMER1 CAPT	Timer/Counter1 Capture Event	TIMER1_CAPT_vect
18	0x0022	TIMER1 COMPA	Timer/Counter1 Compare Match A	TIMER1_COMPA_vect
19	0x0024	TIMER1 COMPB	Timer/Counter1 Compare Match B	TIMER1_COMPB_vect
20	0x0026	TIMER1 COMPC	Timer/Counter1 Compare Match C	TIMER1_COMPC_vect
21	0x0028	TIMER1 OVF	Timer/Counter1 Overflow	TIMER1_OVF_vect
22	0x002A	TIMER0 COMPA	Timer/Counter0 Compare Match A	TIMER0_COMPA_vect
23	0x002C	TIMER0 COMPB	Timer/Counter0 Compare Match B	TIMER0_COMPB_vect
24	0x002E	TIMER0 OVF	Timer/Counter0 Overflow	TIMER0_OVF_vect
25	0x0030	SPI, STC	SPI Serial Transfer Complete	SPI_STC_vect
26	0x0032	USART0, RX	USART0, Rx Complete	USART0_RX_vect
27	0x0034	USART0, UDRE	USART0, Data Register Empty	USART0_UDRE_vect
28	0x0036	USART0, TX	USART0, Tx Complete	USART0_TX_vect
29	0x0038	ANALOG COMP	Analog Comparator	ANALOG_COMP_vect
30	0X003A	ADC	ADC Conversion Complete	ADC_vect
31	0x003C	EE READY	EEROM Ready	EE_READY_vect
32	0x003E	TIMER3 CAPT	Timer/Counter3 Capture Event	TIMER3_CAPT_vect
33	0x0040	TIMER3 COMPA	Timer/Counter3 Compare Match A	TIMER3_COMPA_vect
34	0x0042	TIMER3 COMPB	Timer/Counter3 Compare Match B	TIMER3_COMPB_vect
35	0x0044	TIMER3 COMPC	Timer/Counter3 Compare Match C	TIMER3_COMPC_vect
36	0x0046	TIMER3 OVF	Timer/Counter3 Overflow	TIMER3_OVF_vect
37	0x0048	USART1, RX	USART1, Rx Complete	USART1_RX_vect
38	0x004A	USART1, UDRE	USART1, Data Register Empty	USART1_UDRE_vect
39	0x004C	USART1, TX	USART1, Tx Complete	USART1_TX_vect
40	0x004E	TWI	2-wire Serial Interface	TWI_vect

Figure 6.3: Atmel AVR ATmega2560 Interrupts. (Adapted from figure used with permission of Atmel, Incorporated.) *(Continues.)*

41	0x0050	SPM READY	Store Program Memory Ready	SPM_ready_vect
42	0x0052	TIMER 4 CAPT	Timer/Counter4 Capture Event	TIMER 4_CAPT_vect
43	0x0054	TIMER 4 COMPA	Timer/Counter4 Compare Match A	TIMER 4_COMPA_vect
44	0x0056	TIMER 4 COMPB	Timer/Counter4 Compare Match B	TIMER 4_COMPB_vect
45	0x0058	TIMER 4 COMPC	Timer/Counter4 Compare Match C	TIMER 4_COMPC_vect
46	0x005A	TIMER 4 OVF	Timer/Counter4 Overflow	TIMER 4_OVF_vect
47	0x005C	TIMER 5 CAPT	Timer/Counter5 Capture Event	TIMER 5_CAPT_vect
48	0x005E	TIMER 5 COMPA	Timer/Counter5 Compare Match A	TIMER 5_COMPA_vect
49	0x0060	TIMER 5 COMPB	Timer/Counter5 Compare Match B	TIMER 5_COMPB_vect
50	0x0062	TIMER 5 COMPC	Timer/Counter5 Compare Match C	TIMER 5_COMPC_vect
51	0x0064	TIMER 5 OVF	Timer/Counter5 Overflow	TIMER 5_OVF_vect
52	0x0066	USART 2, RX	USART 2, Rx Complete	USART 2_RX_vect
53	0x0068	USART 2, UDRE	USART 2, Data Register Empty	USART 2_UDRE_vect
54	0x006A	USART 2, TX	USART 2, Tx Complete	USART 2_TX_vect
55	0x006C	USART 3, RX	USART 3, Rx Complete	USART 3_RX_vect
56	0x006E	USART 3, UDRE	USART 3, Data Register Empty	USART 3_UDRE_vect
57	0x0070	USART 3, TX	USART 3, Tx Complete	USART 3_TX_vect

Figure 6.3: *(Continued.)* Atmel AVR ATmega2560 Interrupts. (Adapted from figure used with permission of Atmel, Incorporated.)

6.1.3   GENERAL INTERRUPT RESPONSE

When an interrupt occurs, the microcontroller completes the current instruction, stores the address of the next instruction on the stack, and starts executing instructions in the designated interrupt service routine (ISR) corresponding to the particular interrupt source. It also turns off the interrupt system to prevent further interrupts while one is in progress. The execution of the ISR is performed by loading the beginning address of the interrupt service routine specific for that interrupt into the program counter. The interrupt service routine will then commence. Execution of the ISR continues until the return from interrupt instruction (reti) is encountered. Program control then reverts back to the main program.

6.2   INTERRUPT PROGRAMMING OVERVIEW

To program an interrupt the user is responsible for the following actions:

- Ensure the interrupt service routine for a specific interrupt is tied to the correct interrupt vector address, which points to the starting address of the interrupt service routine.

- Ensure the interrupt system has been globally enabled. This is accomplished with the assembly language instruction SEI.

- Ensure the specific interrupt subsystem has been locally enabled.

- Ensure the registers associated with the specific interrupt have been configured correctly.

In the next two examples that follow, we illustrate how to accomplish these steps. We use the ImageCraft ICC AVR compiler which contains excellent support for interrupts. Other compilers have similar features.

6.3 PROGRAMMING ATMEGA328 INTERRUPTS IN C AND THE ARDUINO DEVELOPMENT ENVIRONMENT

In this section, we provide two representative examples of writing interrupts. We provide both an externally generated interrupt event and also one generated from within the microcontroller. For each type of interrupt, we illustrate how to program it in C and also with the Arduino Development Environment built–in features. For the C examples, we use the ImageCraft ICC AVR compiler.

The ImageCraft ICC AVR compiler uses the following syntax to link an interrupt service routine to the correct interrupt vector address:

```
#pragma interrupt_handler timer_handler:4

void timer_handler(void)
{
    :
    :
}
```

As you can see, the #pragma with the reserved word **interrupt_handler** is used to communicate to the compiler that the routine name that follows is an interrupt service routine. The number that follows the ISR name corresponds to the interrupt vector number in Figure 6.2. The ISR is then written like any other function. It is important that the ISR name used in the #pragma instruction identically matches the name of the ISR in the function body. Since the compiler knows the function is an ISR, it will automatically place the assembly language RETI instruction at the end of the ISR.

6.3.1 EXTERNAL INTERRUPT PROGRAMMING–ATMEGA328

The external interrupts INT0 (pin 4) and INT1 (pin 5) trigger an interrupt within the ATmega328 when an user–specified external event occurs at the pin associated with the specific interrupt. Interrupts INT0 and INT1 may be triggered with a falling or rising edge or a low level signal. The specific settings for each interrupt is provided in Figure 6.4.

6.3.1.1 Programming external interrupts in C

Provided below is the code snapshot to configure an interrupt for INT0. In this specific example, an interrupt will occur when a positive edge transition occurs on the ATmega328 INT0 external interrupt pin.

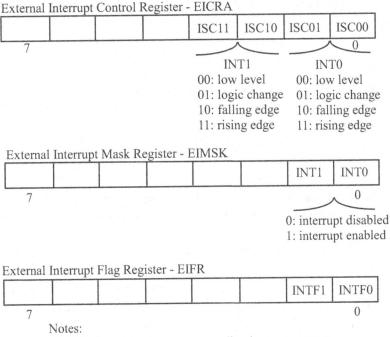

External Interrupt Control Register - EICRA

				ISC11	ISC10	ISC01	ISC00

7   0

INT1
00: low level
01: logic change
10: falling edge
11: rising edge

INT0
00: low level
01: logic change
10: falling edge
11: rising edge

External Interrupt Mask Register - EIMSK

						INT1	INT0

7   0

0: interrupt disabled
1: interrupt enabled

External Interrupt Flag Register - EIFR

						INTF1	INTF0

7   0

Notes:
- INTFx flag sets when corresponding interrupt occurs.
- INTFx flag reset by executing ISR or writing logic one to flag.

Figure 6.4: ATmega328 Interrupt INT0 and INT1 Registers.

```
//interrupt handler definition
#pragma interrupt_handler int0_ISR:2

//function prototypes
void int0_ISR(void);
void initialize_interrupt0(void);


//***********************************************************************

//The following function call should be inserted in the main program to
//initialize the INT0 interrupt to respond to a positive edge trigger.
//This function should only be called once.

:
```

```
initialize_interrupt_int0();
:




//****************************************************************************

//function definitions

//****************************************************************************
//initialize_interrupt_int0:  initializes interrupt INT0.
//Note: stack is automatically initialized by the compiler
//****************************************************************************

void initialize_interrupt_int0(void)    //initialize interrupt INT0
{
DDRD = 0xFB;                             //set PD2 (int0) as input
PORTD &= ~0x04;                          //disable pullup resistor PD2
EIMSK = 0x01;                            //enable INT0
EICRA = 0x03;                            //set for positive edge trigger
asm("SEI");                              //global interrupt enable
}




//****************************************************************************
//int0_ISR: interrupt service routine for INT0
//****************************************************************************

void int0_ISR(void)
{

//Insert interrupt specific actions here.

}
```

The INT0 interrupt is reset by executing the associated interrupt service routine or writing a logic one to the INTF0 bit in the External Interrupt Flag Register (EIFR).

6.3.1.2 Programming external interrupts using the Arduino Development Environment built–in features–Atmega328

The Arduino Development Environment has four built–in functions to support external the INT0 and INT1 external interrupts [www.arduino.cc].

These are the four functions:

- **interrupts()**. This function enables interrupts.

- **noInterrupts()**. This function disables interrupts.

- **attachInterrupt(interrupt, function, mode)**. This function links the interrupt to the appropriate interrupt service routine.

- **detachInterrupt(interrupt)**. This function turns off the specified interrupt.

The Arduino UNO R3 processing board is equipped with two external interrupts: INT0 on DIGITAL pin 2 and INT1 on DIGITAL pin 3. The **attachInterrupt(interrupt, function, mode)** function is used to link the hardware pin to the appropriate interrupt service pin. The three arguments of the function are configured as follows:

- **interrupt.** Interrupt specifies the INT interrupt number: either 0 or 1.

- **function.** Function specifies the name of the interrupt service routine.

- **mode.** Mode specifies what activity on the interrupt pin will initiate the interrupt: **LOW** level on pin, **CHANGE** in pin level, **RISING** edge, or **FALLING** edge.

To illustrate the use of these built–in Arduino Development Environment features, we revisit the previous example.

```
//***************************************************************

void setup()
{
attachInterrupt(0, int0_ISR, RISING);
}

void loop()
{

//wait for interrupts

}

//***************************************************************
```

```
//int0_ISR: interrupt service routine for INT0
//*********************************************************************

void int0_ISR(void)
{

//Insert interrupt specific actions here.

}
//*********************************************************************
```

6.3.2   ATMEGA328 INTERNAL INTERRUPT PROGRAMMING

In this example, we use Timer/Counter0 as a representative example on how to program internal interrupts. In the example that follows, we use Timer/Counter0 to provide prescribed delays within our program.

We discuss the ATmega328 timer system in detail in the next chapter. Briefly, the Timer/Counter0 is an eight bit timer. It rolls over every time it receives 256 timer clock "ticks." There is an interrupt associated with the Timer/Counter0 overflow. If activated, the interrupt will occur every time the contents of the Timer/Counter0 transitions from 255 back to 0 count. We can use this overflow interrupt as a method of keeping track of real clock time (hours, minutes, and seconds) within a program. In this specific example, we use the overflow to provide precision program delays.

6.3.2.1   Programming an internal interrupt in C–Atmega328

In this example, the ATmega328 is being externally clocked by a 10 MHz ceramic resonator. The resonator frequency is further divided by 256 using the clock select bits CS[2:1:0] in Timer/Counter Control Register B (TCCR0B). When CS[2:1:0] are set for [1:0:0], the incoming clock source is divided by 256. This provides a clock "tick" to Timer/Counter0 every 25.6 microseconds. Therefore, the eight bit Timer/Counter0 will rollover every 256 clock "ticks" or every 6.55 ms.

To create a precision delay, we write a function called delay. The function requires an unsigned integer parameter value indicating how many 6.55 ms interrupts the function should delay. The function stays within a while loop until the desired number of interrupts has occurred. For example, to delay one second the function would be called with the parameter value "153." That is, it requires 153 interrupts occurring at 6.55 ms intervals to generate a one second delay.

The code snapshots to configure the Time/Counter0 Overflow interrupt is provided below along with the associated interrupt service routine and the delay function.

```
//function prototypes*************************************************
                                        //delay specified number 6.55ms
void delay(unsigned int number_of_6_55ms_interrupts);
```

```
void init_timer0_ovf_interrupt(void);   //initialize timer0 overf.interrupt


//interrupt handler definition*********************************************
                                        //interrupt handler definition
#pragma interrupt_handler timer0_interrupt_isr:17



//global variables*********************************************************
unsigned int    input_delay;            //counts number of Timer/Counter0
                                        //Overflow interrupts


//main program*************************************************************

void main(void)
{
init_timer0_ovf_interrupt();            //initialize Timer/Counter0 Overflow

                                        //interrupt - call once at beginning
                                        //of program
:
:
delay(153);                             //1 second delay

}

//*************************************************************************
//int_timer0_ovf_interrupt(): The Timer/Counter0 Overfl. interrupt is being
//employed as a time base for a master timer for this project. The ceramic
//resonator  operating at 10 MHz is divided by 256.
//The 8-bit Timer0 register
//(TCNT0) overflows every 256 counts or every 6.55 ms.
//*************************************************************************

void init_timer0_ovf_interrupt(void)
{
TCCR0B = 0x04;                          //divide timer0 timebase by 256,
                                        //overflow occurs every 6.55ms
```

```
TIMSK0 = 0x01;                          //enable timer0 overflow interrupt
asm("SEI");                             //enable global interrupt
}


//*****************************************************************************
//timer0_interrupt_isr:
//Note: Timer overflow 0 is cleared automatically
//when executing the corresponding interrupt handling vector.
//*****************************************************************************

void timer0_interrupt_isr(void)
{
input_delay++;                          //increment overflow counter
}


//*****************************************************************************
//delay(unsigned int num_of_6_55ms_interrupts): this generic delay function
//provides the specified delay as the number of 6.55 ms "clock ticks"
//from the Timer/Counter0 Overflow interrupt.
//
//Note: this function is only valid when using a 10 MHz crystal or ceramic
//resonator. If a different source frequency is used, the clock
//tick delay value must be recalculated.
//*****************************************************************************

void delay(unsigned int number_of_6_55ms_interrupts)
{
TCNT0 = 0x00;                           //reset timer0
input_delay = 0;                        //reset timer0 overflow counter

while(input_delay <= number_of_6_55ms_interrupts)
  {
  ;                                     //wait for spec. number of interpts.
  }
}


//*****************************************************************************
```

6.3.2.2 Programming an internal interrupt using the Arduino Development Environment–Arduino UNO R3

The Arduino Development Environment uses the GNU tool chain and the AVR Libc to compile programs. Internal interrupt configuration uses AVR–GCC conventions. To tie the interrupt event to the correct interrupt service routine, the AVR–GCC interrupt name must be used. These vector names are provided in the right column of Figure 6.2.

In the following sketch, the previous example is configured for use with the Arduino Development Environment using AVR–GCC conventions. Also, the timing functions in the previous example assumed a time base of 10 MHz. The Arduino UNO R3 is clocked with a 16 MHz crystal. Therefore, some of the parameters in the sketch were adjusted to account for this difference in time base.

```
//*********************************************************************
#include <avr/interrupt.h>

unsigned int    input_delay;            //counts number of Timer/Counter0
                                        //Overflow interrupts


void setup()
{
init_timer0_ovf_interrupt();            //initialize Timer/Counter0 Overfl.
}

void loop()
{

  :

delay(244);                             //1 second delay

  :

}


//*********************************************************************
// ISR(TIMER0_OVF_vect) - increments counter on every interrupt.
//*********************************************************************

ISR(TIMER0_OVF_vect)
{
input_delay++;                          //increment overflow counter
```

```
}

//*****************************************************************************
//int_timer0_ovf_interrupt(): The Timer/Counter0 Overflow interrupt is
//being employed as a time base for a master timer for this project.
//The crystal //resonator  operating at 16 MHz is divided by 256.
//The 8-bit Timer0 register (TCNT0) overflows every 256 counts or every
//4.1 ms.
//*****************************************************************************

void init_timer0_ovf_interrupt(void)
{
TCCR0B = 0x04; //divide timer0 timebase by 256, overflow occurs every 4.1 ms
TIMSK0 = 0x01; //enable timer0 overflow interrupt
asm("SEI");    //enable global interrupt
}


//*****************************************************************************
//delay(unsigned int num_of_4_1ms_interrupts): this generic delay function
//provides the specified delay as the number of 4.1 ms "clock ticks" from
//the Timer/Counter0 Overflow interrupt.
//
//Note: this function is only valid when using a 16 MHz crystal or ceramic
//resonator. If a different source frequency is used, the clock
//tick delay value must be recalculated.
//*****************************************************************************

void delay(unsigned int number_of_4_1ms_interrupts)
{
TCNT0 = 0x00;                            //reset timer0
input_delay = 0;                         //reset timer0 overflow counter

while(input_delay <= number_of_4_1ms_interrupts)
  {
  ;                                      //wait for spec. number of intrpts.
  }
}


//*****************************************************************************
```

6.4 PROGRAMMING ATMEGA2560 INTERRUPTS IN C AND THE ARDUINO DEVELOPMENT ENVIRONMENT

In this section, we provide two representative examples of writing interrupts. We provide both an externally generated interrupt event and also one generated from within the microcontroller. For each type of interrupt, we illustrate how to program it in C and also with the Arduino Development Environment built–in features. For the C examples, we use the ImageCraft ICC AVR compiler.

The ImageCraft ICC AVR compiler uses the following syntax to link an interrupt service routine to the correct interrupt vector address:

```
#pragma interrupt_handler timer_handler:4

void timer_handler(void)
{
:
:
}
```

As you can see, the #pragma with the reserved word **interrupt_handler** is used to communicate to the compiler that the routine name that follows is an interrupt service routine.

The number that follows the ISR name corresponds to the interrupt vector number in Figure 6.3. The ISR is then written like any other function. It is important that the ISR name used in the #pragma instruction identically matches the name of the ISR in the function body. Since the compiler knows the function is an ISR, it will automatically place the assembly language RETI instruction at the end of the ISR.

6.4.1 EXTERNAL INTERRUPT PROGRAMMING–ATMEGA2560

The external interrupts INT0 through INT7 trigger an interrupt within the ATmega2560 when an user–specified external event occurs at the pin associated with the specific interrupt. Interrupts INT0 through INT7 may be triggered with a falling or rising edge or a low level signal. The specific settings for each interrupt is provided in Figure 6.5.

6.4.1.1 Programming external interrupts in C

Provided below is the code snapshot to configure an interrupt for INT0. In this specific example, an interrupt will occur when a positive edge transition occurs on the ATmega2560 INT0 external interrupt pin.

```
//interrupt handler definition
#pragma interrupt_handler int0_ISR:2

//function prototypes
void int0_ISR(void);
```

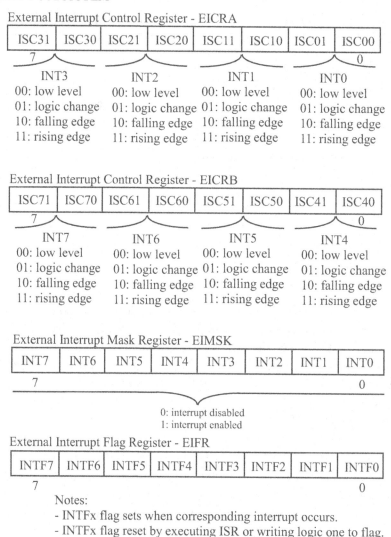

Figure 6.5: ATmega2560 Interrupt Registers.

```
void initialize_interrupt0(void);
```

```
//**********************************************************************
```

```
//The following function call should be inserted in the main program to
```

```
//initialize the INT0 interrupt to respond to a positive edge trigger.
//This function should only be called once.


 :
initialize_interrupt_int0();
 :




//*************************************************************************

//function definitions

//*************************************************************************
//initialize_interrupt_int0:  initializes interrupt INT0.
//Note: stack is automatically initialized by the compiler
//*************************************************************************

void initialize_interrupt_int0(void)    //initialize interrupt INT0
{
DDRD = 0xFB;                             //set PD2 (int0) as input
PORTD &= ~0x04;                          //disable pullup resistor PD2
EIMSK = 0x01;                            //enable INT0
EICRA = 0x03;                            //set for positive edge trigger
asm("SEI");                              //global interrupt enable
}



//*************************************************************************
//int0_ISR: interrupt service routine for INT0
//*************************************************************************

void int0_ISR(void)
{

//Insert interrupt specific actions here.

}
```

The INT0 interrupt is reset by executing the associated interrupt service routine or writing a logic one to the INTF0 bit in the External Interrupt Flag Register (EIFR).

6.4.1.2 Programming external interrupts using the Arduino Development Environment built–in features–Atmega2560

The Arduino Development Environment has four built–in functions to support some of the external interrupts (INT0 through INT7) [www.arduino.cc].

These are the four functions:

- **interrupts().** This function enables interrupts.

- **noInterrupts().** This function disables interrupts.

- **attachInterrupt(interrupt, function, mode).** This function links the interrupt to the appropriate interrupt service routine.

- **detachInterrupt(interrupt).** This function turns off the specified interrupt.

The Arduino Mega 2560 processing board is equipped with six external interrupts:

- INT0, pin 2

- INT1, pin 3

- INT2, pin 21

- INT3, pin 1

- INT4, pin 19

- INT5, pin 18

The **attachInterrupt(interrupt, function, mode)** function is used to link the hardware pin to the appropriate interrupt service pin. The three arguments of the function are configured as follows:

- **interrupt.** Interrupt specifies the INT interrupt number: either 0 or 1.

- **function.** Function specifies the name of the interrupt service routine.

- **mode.** Mode specifies what activity on the interrupt pin will initiate the interrupt: **LOW** level on pin, **CHANGE** in pin level, **RISING** edge, or **FALLING** edge.

To illustrate the use of these built–in Arduino Development Environment features, we revisit the previous example.

```
//***************************************************************************

void setup()
{
attachInterrupt(0, int0_ISR, RISING);
}

void loop()
{

//wait for interrupts

}


//***************************************************************************
//int0_ISR: interrupt service routine for INT0
//***************************************************************************

void int0_ISR(void)
{

//Insert interrupt specific actions here.

}
//***************************************************************************
```

6.4.2   ATMEGA2560 INTERNAL INTERRUPT PROGRAMMING

In this example, we use Timer/Counter0 as a representative example on how to program internal interrupts. In the example that follows, we use Timer/Counter0 to provide prescribed delays within our program.

We discuss the ATmega2560 timer system in detail in the next chapter. Briefly, the Timer/Counter0 is an eight bit timer. It rolls over every time it receives 256 timer clock "ticks." There is an interrupt associated with the Timer/Counter0 overflow. If activated, the interrupt will occur every time the contents of the Timer/Counter0 transitions from 255 back to 0 count. We can use this overflow interrupt as a method of keeping track of real clock time (hours, minutes, and seconds) within a program. In this specific example, we use the overflow to provide precision program delays.

6.4.2.1 Programming an internal interrupt in C–Atmega2560

In this example, the ATmega2560 is being externally clocked by a 10 MHz ceramic resonator. The resonator frequency is further divided by 256 using the clock select bits CS[2:1:0] in Timer/Counter Control Register B (TCCR0B). When CS[2:1:0] are set for [1:0:0], the incoming clock source is divided by 256. This provides a clock "tick" to Timer/Counter0 every 25.6 microseconds. Therefore, the eight bit Timer/Counter0 will rollover every 256 clock "ticks" or every 6.55 ms.

To create a precision delay, we write a function called delay. The function requires an unsigned integer parameter value indicating how many 6.55 ms interrupts the function should delay. The function stays within a while loop until the desired number of interrupts has occurred. For example, to delay one second the function would be called with the parameter value "153." That is, it requires 153 interrupts occurring at 6.55 ms intervals to generate a one second delay.

The code snapshots to configure the Time/Counter0 Overflow interrupt is provided below along with the associated interrupt service routine and the delay function.

```
//function prototypes***************************************************
                                        //delay specified number 6.55ms
void delay(unsigned int number_of_6_55ms_interrupts);
void init_timer0_ovf_interrupt(void);   //initialize timer0 overf.interrupt



//interrupt handler definition******************************************
                                        //interrupt handler definition
#pragma interrupt_handler timer0_interrupt_isr:24



//global variables******************************************************
unsigned int   input_delay;             //counts number of Timer/Counter0
                                        //Overflow interrupts

//main program**********************************************************

void main(void)
{
init_timer0_ovf_interrupt();            //initialize Timer/Counter0 Overflow

                                        //interrupt - call once at beginning
                                        //of program
:
:
```

```
delay(153);                               //1 second delay

}


//************************************************************************
//int_timer0_ovf_interrupt(): The Timer/Counter0 Overflow interrupt is being
//employed as a time base for a master timer for this project. The ceramic
//resonator  operating at 10 MHz is divided by 256.
//The 8-bit Timer0 register
//(TCNT0) overflows every 256 counts or every 6.55 ms.
//************************************************************************

void init_timer0_ovf_interrupt(void)
{
TCCR0B = 0x04;                            //divide timer0 timebase by 256,
                                          //overflow occurs every 6.55ms

TIMSK0 = 0x01;                            //enable timer0 overflow interrupt
asm("SEI");                               //enable global interrupt
}


//************************************************************************
//timer0_interrupt_isr:
//Note: Timer overflow 0 is cleared automatically
//when executing the corresponding interrupt handling vector.
//************************************************************************

void timer0_interrupt_isr(void)
{
input_delay++;                            //increment overflow counter
}


//************************************************************************
//delay(unsigned int num_of_6_55ms_interrupts): this generic delay function
//provides the specified delay as the number of 6.55 ms "clock ticks"
//from the Timer/Counter0 Overflow interrupt.
//
//Note: this function is only valid when using a 10 MHz crystal or ceramic
//resonator. If a different source frequency is used, the clock
//tick delay value must be recalculated.
```

```
//********************************************************************

void delay(unsigned int number_of_6_55ms_interrupts)
{
TCNT0 = 0x00;                          //reset timer0
input_delay = 0;                       //reset timer0 overflow counter

while(input_delay <= number_of_6_55ms_interrupts)
  {
  ;                                    //wait for spec. number of interpts.
  }
}


//********************************************************************
```

6.4.2.2 Programming an internal interrupt using the Arduino Development Environment–Arduino Mega 2560

The Arduino Development Environment uses the GNU tool chain and the AVR Libc to compile programs. Internal interrupt configuration uses AVR–GCC conventions. To tie the interrupt event to the correct interrupt service routine, the AVR–GCC interrupt name must be used. These vector names are provided in the right column of Figure 6.3.

In the following sketch, the previous example is configured for use with the Arduino Development Environment using AVR–GCC conventions. Also, the timing functions in the previous example assumed a time base of 10 MHz. The Arduino Mega 2560 is clocked with a 16 MHz crystal. Therefore, some of the parameters in the sketch were adjusted to account for this difference in time base.

```
//********************************************************************
#include <avr/interrupt.h>

unsigned int   input_delay;            //counts number of Timer/Counter0
                                       //Overflow interrupts

void setup()
{
init_timer0_ovf_interrupt();           //initialize Timer/Counter0 Overfl.
}

void loop()
{
```

```
:

delay(244);                                    //1 second delay

:

}


//****************************************************************************
// ISR(TIMER0_OVF_vect) - increments counter on every interrupt.
//****************************************************************************

ISR(TIMER0_OVF_vect)
{
input_delay++;                                 //increment overflow counter
}


//****************************************************************************
//int_timer0_ovf_interrupt(): The Timer/Counter0 Overflow interrupt is
//being employed as a time base for a master timer for this project.
//The crystal //resonator  operating at 16 MHz is divided by 256.
//The 8-bit Timer0 register (TCNT0) overflows every 256 counts or every
//4.1 ms.
//****************************************************************************

void init_timer0_ovf_interrupt(void)
{
TCCR0B = 0x04; //divide timer0 timebase by 256, overflow occurs every 4.1 ms
TIMSK0 = 0x01; //enable timer0 overflow interrupt
asm("SEI");    //enable global interrupt
}


//****************************************************************************
//delay(unsigned int num_of_4_1ms_interrupts): this generic delay function
//provides the specified delay as the number of 4.1 ms "clock ticks" from
//the Timer/Counter0 Overflow interrupt.
//
//Note: this function is only valid when using a 16 MHz crystal or ceramic
//resonator. If a different source frequency is used, the clock
```

```
//tick delay value must be recalculated.
//**********************************************************************

void delay(unsigned int number_of_4_1ms_interrupts)
{
TCNT0 = 0x00;                         //reset timer0
input_delay = 0;                      //reset timer0 overflow counter

while(input_delay <= number_of_4_1ms_interrupts)
  {
  ;                                   //wait for spec. number of intrpts.
  }
}


//**********************************************************************
```

6.5   FOREGROUND AND BACKGROUND PROCESSING

A microcontroller can only process a single instruction at a time. It processes instructions in a fetch––decode–execute sequence as determined by the program and its response to external events. In many cases, a microcontroller has to process multiple events seemingly simultaneously. How is this possible with a single processor?

Normal processing accomplished by the microcontroller is called foreground processing. An interrupt may be used to periodically break into foreground processing, 'steal' some clock cycles to accomplish another event called background processing, and then return processor control back to the foreground process.

As an example, a microcontroller controlling access for an electronic door must monitor input commands from a user and generate the appropriate pulse width modulation (PWM) signals to open and close the door. Once the door is in motion, the controller must monitor door motor operation for obstructions, malfunctions, and other safety related parameters. This may be accomplished using interrupts. In this example, the microcontroller is responding to user input status in the foreground while monitoring safety related status in the background using interrupts as illustrated in Figure 6.6.

Example: This example illustrates foreground and background processing. We use a green LED to indicate when the microcontroller is processing in the foreground and a flashing red LED indicates background processing. A switch is connected to an external interrupt pin (INT0, pin 2). When the switch is depressed, the microcontroller executes the associated interrupt service routine to flash the red LED. The circuit configuration is provided in Figure 6.7.

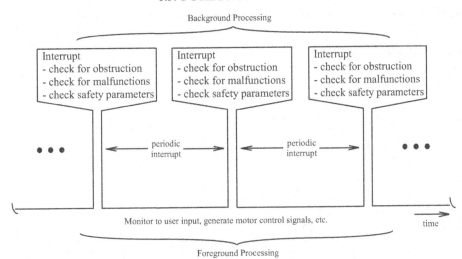

Figure 6.6: Interrupt used for background processing. The microcontroller responds to user input status in the foreground while monitoring safety related status in the background using interrupts.

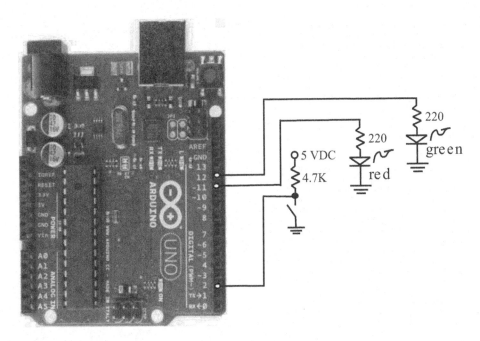

Figure 6.7: Foreground background processing. (UNO R3 illustration used with permission of the Arduino Team (CC BY–NC–SA) www.arduino.cc).

```
//*********************************************************

#define green_LED 12
#define red_LED   11
#define ext_sw     2

int switch_value;

void setup()
{
pinMode(green_LED, OUTPUT);
pinMode(red_LED,   OUTPUT);
pinMode(ext_sw,    INPUT);
                //trigger interrupt on INT0 for falling edge
attachInterrupt(0, background, FALLING);
}

void loop()
{
digitalWrite(green_LED, HIGH);    //foreground processing
digitalWrite(red_LED,   LOW);
}

void background()                 //background processing
{
unsigned int i;

digitalWrite(green_LED, LOW);
digitalWrite(red_LED,   HIGH);

for (i=0; i<=64000; i++)          //delay
  {
  asm("nop");
  }

digitalWrite(red_LED,   LOW);

for (i=0; i<=64000; i++)          //delay
  {
```

```
    asm("nop");
    }

digitalWrite(red_LED,    HIGH);

for (i=0; i<=64000; i++)          //delay
    {
    asm("nop");
    }

digitalWrite(red_LED,    LOW);

for (i=0; i<=64000; i++)          //delay
    {
    asm("nop");
    }

digitalWrite(red_LED,    HIGH);
}
//**********************************************************
```

6.6   INTERRUPT EXAMPLES

In this section, we provide several varied examples on using interrupts internal and external to the microcontroller.

6.6.1   APPLICATION 1: REAL TIME CLOCK IN C

A microcontroller only 'understands' elapsed time in reference to its timebase clock ticks. To keep track of clock time in seconds, minutes, hours etc., a periodic interrupt may be generated for use as a 'clock tick' for a real time clock. In this example, we use the Timer 0 overflow to generate a periodic clock tick very 6.55 ms. The ticks are counted in reference to clock time variables and may be displayed on a liquid crystal display. This is also a useful technique for generating very long delays in a microcontroller.

```
//function prototypes**********************************************
                                    //delay specified number 6.55ms

void delay(unsigned int number_of_6_55ms_interrupts);
void init_timer0_ovf_interrupt(void);//initialize timer0 overflow interrupt
```

```
//interrupt handler definition*****************************************
                                //interrupt handler definition
#pragma interrupt_handler timer0_interrupt_isr:17



//global variables****************************************************
unsigned int days_ctr, hrs_ctr, mins_ctr, sec_ctr, ms_ctr;


//main program*******************************************************

void main(void)
{
day_ctr = 0; hr_ctr = 0; min_ctr = 0; sec_ctr = 0; ms_ctr = 0;



init_timer0_ovf_interrupt();          //initialize Timer/Counter0 Overflow

  //interrupt - call once at beginning
                                      //of program
while(1)
  {
  ;                                   //wait for interrupts
  }
}


//*********************************************************************
//int_timer0_ovf_interrupt(): The Timer/Counter0 Overflow interrupt is
//being employed as a time base for a master timer for this project.
//The ceramic resonator  operating at 10 MHz is divided by 256.
//The 8-bit Timer0 register (TCNT0) overflows every 256 counts or
//every 6.55 ms.
//*********************************************************************

void init_timer0_ovf_interrupt(void)
{
TCCR0B = 0x04; //divide timer0 timebase by 256, overfl. occurs every 6.55ms

TIMSK0 = 0x01; //enable timer0 overflow interrupt
asm("SEI");    //enable global interrupt
```

```
}

//***************************************************************************
//timer0_interrupt_isr:
//Note: Timer overflow 0 is cleared by hardware when executing the
//corresponding interrupt handling vector.
//***************************************************************************

void timer0_interrupt_isr(void)
{

//Update millisecond counter
ms_ctr = ms_ctr + 1;                        //increment ms counter


                                            //Update second counter
if(ms_ctr == 154)                           //counter equates to 1000 ms at 154
  {
  ms_ctr = 0;                               //reset ms counter
  sec_ctr = sec_ctr + 1;                    //increment second counter
  }

//Update minute counter
if(sec_ctr == 60)
  {
  sec_ctr = 0;                              //reset sec counter
  min_ctr = min_ctr + 1;                    //increment min counter
  }

//Update hour counter
if(min_ctr == 60)
  {
  min_ctr = 0;                              //reset min counter
  hr_ctr  = hr_ctr + 1;                     //increment hr counter
  }

//Update day counter
if(hr_ctr == 24)
  {
  hr_ctr = 0;                               //reset hr counter
```

```
    day_ctr  = day_ctr + 1;                    //increment day counter
    }
}
```

//**

6.6.2   APPLICATION 2: REAL TIME CLOCK USING THE ARDUINO DEVELOPMENT ENVIRONMENT

In this example, we reconfigure the previous example using the Arduino Development Environment. The timing functions in the previous example assumed a time base of 10 MHz. The Arduino UNO R3 is clocked with a 16 MHz crystal. Therefore, some of the parameters in the sketch are adjusted to account for this difference in time base.

```
//********************************************************************************
#include <avr/interrupt.h>

//global variables***************************************************************
unsigned int days_ctr, hrs_ctr, mins_ctr, sec_ctr, ms_ctr;

void setup()
{
day_ctr = 0; hr_ctr = 0; min_ctr = 0; sec_ctr = 0; ms_ctr = 0;
init_timer0_ovf_interrupt();              //init. Timer/Counter0 Overflow
}

void loop()
{
:
:                                         //wait for interrupts
:
}

//********************************************************************************
// ISR(TIMER0_OVF_vect) Timer0 interrupt service routine.
//
//Note: Timer overflow 0 is cleared by hardware when executing the
//corresponding interrupt handling vector.
//********************************************************************************

ISR(TIMER0_OVF_vect)
```

```c
{
//Update millisecond counter
ms_ctr = ms_ctr + 1;                        //increment ms counter

                                            //Update second counter
                                            //counter equates to 1000 ms at 244

if(ms_ctr == 244)                           //each clock tick is 4.1 ms
  {
  ms_ctr = 0;                               //reset ms counter
  sec_ctr = sec_ctr + 1;                    //increment second counter
  }

//Update minute counter
if(sec_ctr == 60)
  {
  sec_ctr = 0;                              //reset sec counter
  min_ctr = min_ctr + 1;                    //increment min counter
  }

//Update hour counter
if(min_ctr == 60)
  {
  min_ctr = 0;                              //reset min counter
  hr_ctr  = hr_ctr + 1;                     //increment hr counter
  }

//Update day counter
if(hr_ctr == 24)
  {
  hr_ctr = 0;                               //reset hr counter
  day_ctr  = day_ctr + 1;                   //increment day counter
  }
}


//*************************************************************************
//int_timer0_ovf_interrupt(): The Timer/Counter0
//Overflow interrupt is being employed as a time base for a master timer
//for this project. The ceramic resonator  operating at 16 MHz is
```

```
//divided by 256.
//The 8-bit Timer0 register (TCNT0) overflows every 256 counts or
//every 4.1 ms.
//***********************************************************************

void init_timer0_ovf_interrupt(void)
{
TCCR0B = 0x04; //divide timer0 timebase by 256, overfl. occurs every 4.1 ms
TIMSK0 = 0x01; //enable timer0 overflow interrupt
asm("SEI");    //enable global interrupt
}
//***********************************************************************
```

6.6.3   APPLICATION 3: INTERRUPT DRIVEN USART IN C

In Chapter 4, we discussed the serial communication capability of the USART in some detail. In the following example, we revisit the USART and use it in an interrupt driven mode.

Example. You have been asked to evaluate a new positional encoder technology. The encoder provides 12–bit resolution. The position data is sent serially at 9600 Baud as two sequential bytes as shown in Figure 6.8. The actual encoder is new technology and production models are not available for evaluation.

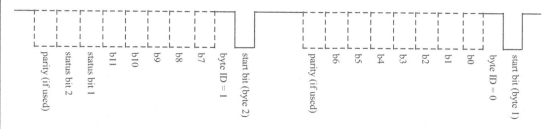

Figure 6.8: Encoder data format. The position data is sent serially at 9600 Baud as two sequential bytes.

Since the actual encoder is not available for evaluation, another Atmel ATmega328 will be used to send signals in identical format and Baud rate as the encoder. The test configuration is illustrated in Figure 6.9. The ATmega328 on the bottom serves as the positional encoder. The microcontroller is equipped with two pushbuttons at PD2 and PD3. The pushbutton at PD2 provides a debounced input to open a simulated door and increment the positional encoder. The pushbutton at PD3 provides a debounced input to close a simulated door and decrement the positional encoder. The current count of the encoder (eight most significant bits) is fed to a digital–to–analog converter (DAC0808) to provide an analog representation.

The positional data from the encoder is sent out via the USART in the format described in Figure 6.8. The top ATmega328 receives the positional data using interrupt driven USART techniques. The current position is converted to an analog signal via the DAC. The transmitted and received signals may be compared at the respective DAC outputs.

Provided below is the code for the ATmega328 that serves as the encoder simulator followed by the code for receiving the data.

```
//**********************************************************************
//author: Steven Barrett, Ph.D., P.E.
//last revised: April 15, 2010
//file: encode.c
//target controller: ATMEL ATmega328
//
//ATmega328 clock source: internal 8 MHz clock
//
//ATMEL AVR ATmega328PV Controller Pin Assignments
//Chip Port Function I/O Source/Dest Asserted Notes
//
//Pin  1 to system reset circuitry
//Pin  2 PD0: USART receive pin (RXD)
//Pin  3 PD1: USART transmit pin (TXD)
//Pin  4 PD2 to active high RC debounced switch - Open
//Pin  5 PD3 to active high RC debounced switch - Close
//Pin  7 Vcc - 1.0 uF to ground
//Pin  8 Gnd
//Pin  9 PB6 to pin A7(11) on DAC0808
//Pin 10 PB7 to pin A8(12) on DAC0808
//Pin 14 PB0 to pin A1(5)  on DAC0808
//Pin 15 PB1 to pin A2(6)  on DAC0808
//Pin 16 PB2 to pin A3(7)  on DAC0808
//Pin 17 PB3 to pin A4(8)  on DAC0808
//Pin 18 PB4 to pin A5(9)  on DAC0808
//Pin 19 PB5 to pin A6(10) on DAC0808
//Pin 20 AVCC to 5 VDC
//Pin 21 AREF to 5 VDC
//Pin 22 Ground
//**********************************************************************

//include files***********************************************************
#include<iom328v.h>
#include<macros.h>
```

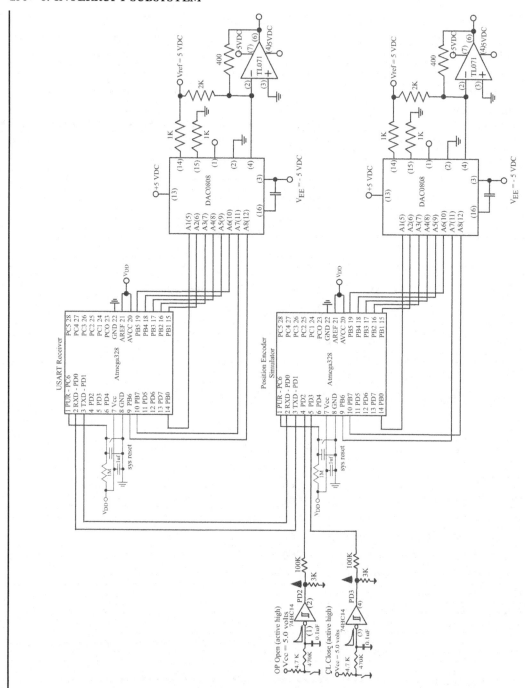

Figure 6.9: Encoder test configuration.

```c
//function prototypes*************************************************

 //delay specified number 6.55ms int
void initialize_ports(void);            //initializes ports
void USART_init(void);
void USART_TX(unsigned char data);


//main program*******************************************************
//global variables
unsigned char   old_PORTD = 0x08;      //present value of PORTD
unsigned char   new_PORTD;             //new values of PORTD
unsigned int    door_position = 0;

void main(void)
{
initialize_ports();                    //return LED configuration to default
USART_init();

//main activity loop - checks PORTD to see if either PD2 (open)
//or PD3 (close) was depressed.
//If either was depressed the program responds.


while(1)
  {
  _StackCheck();                       //check for stack overflow
  read_new_input();
  //read input status changes on PORTB
  }
}//end main

//Function definitions
//*****************************************************************
//initialize_ports: provides initial configuration for I/O ports
//*****************************************************************

void initialize_ports(void)
{
```

```
//PORTB
DDRB=0xff;                              //PORTB[7-0] output
PORTB=0x00;                             //initialize low


//PORTC
DDRC=0xff;                              //set PORTC[7-0] as output
PORTC=0x00;                             //initialize low


//PORTD
DDRD=0xf2;                              //set PORTD[7-4, 0] as output
PORTD=0x00;                             //initialize low
}


//**********************************************************************
//**********************************************************************
//read_new_input: functions polls PORTD for a change in status. If status
//change has occurred, appropriate function for status change is called
//Pin  4 PD2 to active high RC debounced switch - Open
//Pin  5 PD3 to active high RC debounced switch - Close
//**********************************************************************

void read_new_input(void)
{
unsigned int   gate_position;      //measure instantaneous position of gate
unsigned int i;
unsigned char ms_door_position, ls_door_position, DAC_data;

new_PORTD = (PIND & 0x0c);
 //mask all pins but PORTD[3:2]
if(new_PORTD != old_PORTD){
  switch(new_PORTD){                //process change in PORTD input

    case 0x01:                      //Open
      while(PIND == 0x04)
        {
        //split into two bytes
        ms_door_position=(unsigned char)(((door_position >> 6)
                              &(0x00FF))|0x01);
        ls_door_position=(unsigned char)(((door_position << 1)
```

```
                                     &(0x00FF))&0xFE);

   //TX data to USART
   USART_TX(ms_door_position);
   USART_TX(ls_door_position);

   //format data for DAC and send to DAC on PORTB
   DAC_data=(unsigned char)((door_position >> 4)&(0x00FF));
   PORTB = DAC_data;

   //increment position counter
   if(door_position >= 4095)
      door_position = 4095;
   else
      door_position++;
   }
 break;

case 0x02:                          //Close
  while(PIND == 0x02)
    {
    //split into two bytes
    ms_door_position=(unsigned char)(((door_position >>6)
                                  &(0x00FF))|0x01);
    ls_door_position=(unsigned char)(((door_position <<1)
                                  &(0x00FF))&0xFE);
    //TX data to USART
    USART_TX(ms_door_position);
    USART_TX(ls_door_position);

    //format data for DAC and send to DAC on PORTB
    DAC_data=(unsigned char)((door_position >> 4)&(0x00FF));

    PORTB = DAC_data;

    //decrement position counter
    if(door_position <= 0)
      door_position = 0;
    else
      door_position-;
```

```
        }
        break;

        default:;                        //all other cases
      }                                  //end switch(new_PORTD)
  }                                      //end if new_PORTD
  old_PORTD = new_PORTD;                 //update PORTD
}


//**************************************************************************
//USART_init: initializes the USART system
//
//Note: ATmega328 clocked by internal 8 MHz clock
//**************************************************************************

void USART_init(void)
{
UCSRA = 0x00;                   //control
register initialization
UCSRB = 0x08;                   //enable transmitter
UCSRC = 0x86;                   //async, no parity, 1 stop bit, 8 data bits
                                //Baud Rate initialization
                                //8 MHz clock requires UBRR value of 51
                                // or 0x0033 to achieve 9600 Baud rate
UBRRH = 0x00;
UBRRL = 0x33;
}


//**************************************************************************
//USART_transmit: transmits single byte of data
//**************************************************************************

void USART_transmit(unsigned char data)
{
while((UCSRA & 0x20)==0x00)  //wait for UDRE flag
  {
  ;
  }
UDR = data;                     //load data to UDR for transmission
```

```
}

//**********************************************************************
      Receive ATmega328 code follows.
//**********************************************************************
//author: Steven Barrett, Ph.D., P.E.
//last revised: April 15, 2010
//file: receive.c
//target controller: ATMEL ATmega328
//
//ATmega328 clock source: internal 8 MHz clock
//
//ATMEL AVR ATmega328PV Controller Pin Assignments
//Chip Port Function I/O Source/Dest Asserted Notes
//
//Pin  1 to system reset circuitry
//Pin  2 PD0: USART receive pin (RXD)
//Pin  3 PD1: USART transmit pin (TXD)
//Pin  4 PD2 to active high RC debounced switch - Open
//Pin  5 PD3 to active high RC debounced switch - Close
//Pin  7 Vcc - 1.0 uF to ground
//Pin  8 Gnd
//Pin  9 PB6 to pin A7(11) on DAC0808
//Pin 10 PB7 to pin A8(12) on DAC0808
//Pin 14 PB0 to pin A1(5)  on DAC0808
//Pin 15 PB1 to pin A2(6)  on DAC0808
//Pin 16 PB2 to pin A3(7)  on DAC0808
//Pin 17 PB3 to pin A4(8)  on DAC0808
//Pin 18 PB4 to pin A5(9)  on DAC0808
//Pin 19 PB5 to pin A6(10) on DAC0808
//Pin 20 AVCC to 5 VDC
//Pin 21 AREF to 5 VDC
//Pin 22 Ground
//**********************************************************************

//include files*********************************************************
#include<iom328v.h>
#include<macros.h>
#include<eeprom.h>                     //EEPROM support functions
```

```
#pragma data: eeprom
unsigned int door_position_EEPROM
#pragma data:data


//function prototypes***********************************************
void initialize_ports(void);              //initializes ports
void InitUSART(void);
unsigned char USART_RX(void);


                                          //interrupt handler definition
#pragma interrupt_handler USART_RX_interrupt_isr: 19

//main program*****************************************************
unsigned int    door_position = 0;
unsigned char   data_rx;
unsigned int    dummy1 = 0x1234;
unsigned int    keep_going =1;
unsigned int    loop_counter = 0;
unsigned int    ms_position, ls_position;

void main(void)
{
initialize_ports();                       //return LED configuration to deflt.
USART_init();

                                          //limited startup features
                                          //main activity loop - processor will
                                          //continually cycle through loop
                                          //waiting for USART data

while(1)
  {
                                          //continuous loop waiting for
                                          //interrupts

  _StackCheck();                          //check for stack overflow
  }
}//end main
```

```
//Function definitions
//**************************************************************************
//initialize_ports: provides initial configuration for I/O ports
//**************************************************************************

void initialize_ports(void)
{
//PORTB
DDRB=0xff;                          //PORTB[7-0] output
PORTB=0x00;                         //initialize low

//PORTC
DDRC=0xff;                          //set PORTC[7-0] as output
PORTC=0x00;                         //initialize low

//PORTD
DDRD=0xff;                          //set PORTD[7-0] as output
PORTD=0x00;                         //initialize low
}


//**************************************************************************
//USART_init: initializes the USART system
//
//Note: ATmega328 clocked by internal 8 MHz clock
//**************************************************************************

void USART_init(void)
{
UCSRA = 0x00;                //control
register initialization
UCSRB = 0x08;                //enable transmitter
UCSRC = 0x86;                //async, no parity, 1 stop bit, 8 data bits
                             //Baud Rate initialization
                             //8 MHz clock requires UBRR value of 51
                             // or 0x0033 to achieve 9600 Baud rate

UBRRH = 0x00;
UBRRL = 0x33;
}
```

```
//****************************************************************************
//USART_RX_interrupt_isr
//****************************************************************************

void USART_RX_interrupt_isr(void)
{
unsigned char data_rx, DAC_data;
unsigned int  ls_position, ms_position;

//Receive USART data
data_rx = UDR;

//Retrieve door position data from EEPROM
EEPROM_READ((int) &door_position_EEPROM, door_position);

//Determine which byte to update
if((data_rx & 0x01)==0x01)                       //Byte ID = 1
  {
  ms_position = data_rx;

  //Update bit 7
  if((ms_position & 0x0020)==0x0020)             //Test for logic 1
    door_position = door_position | 0x0080;      //Set bit 7 to 1
  else
    door_position = door_position & 0xff7f;      //Reset bit 7 to 0

  //Update remaining bits
  ms_position = ((ms_position<<6) & 0x0f00);
  //shift left 6-blank other bits
  door_position = door_position & 0x00ff;        //Blank ms byte
  door_position = door_position | ms_position;   //Update ms byte
  }
else                                             //Byte ID = 0
  {
  ls_position = data_rx;                                   //Update ls_position
                                                           //Shift right 1-blank
  ls_position = ((ls_position >> 1) & 0x007f); //other bits
```

```
if((door_position & 0x0080)==0x0080)
      //Test bit 7 of curr position
  ls_position = ls_position | 0x0080;         //Set bit 7 to logic 1
else
  ls_position = ls_position & 0xff7f;         //Reset bit 7 to 0
door_position = door_position & 0xff00;       //Blank ls byte
door_position = door_position | ls_position; //Update ls byte
}


//Store door position data to EEPROM
EEPROM_WRITE((int) &door_position_EEPROM, door_position);



//format data for DAC and send to DAC on PORT C
DAC_data=(unsigned char)((door_position >> 4)&(0x00FF));


PORTB= DAC_data;
}


//********************************************************************
//end of file
//********************************************************************
```

6.7   SUMMARY

In this chapter, we provided an introduction to the interrupt features available aboard the AT-mega328, the Arduino UNO R3 processing board, the ATmega250, and the Arduino Mega 2560. We also discussed how to program an interrupt for proper operation and provided representative samples for an external interrupt and an internal interrupt.

6.8   REFERENCES

- *Atmel 8–bit AVR Microcontroller with 4/8/16/32K Bytes In–System Programmable Flash, AT-mega48PA, 88PA, 168PA, 328P* data sheet: 8161D–AVR–10/09, Atmel Corporation, 2325 Orchard Parkway, San Jose, CA 95131.

- *Atmel 8–bit AVR Microcontroller with 64/128/256K Bytes In–System Programmable Flash, AT-mega640/V, ATmega1280/V, 2560/V* data sheet: 2549P–AVR–10/2012, Atmel Corporation, 2325 Orchard Parkway, San Jose, CA 95131.

- Barrett S, Pack D (2006) Microcontrollers Fundamentals for Engineers and Scientists. Morgan and Claypool Publishers. DOI: 10.2200/S00025ED1V01Y200605DCS001

- Barrett S and Pack D (2008) Atmel AVR Microcontroller Primer Programming and Interfacing. Morgan and Claypool Publishers. DOI: 10.2200/S00100ED1V01Y200712DCS015

- Barrett S (2010) Embedded Systems Design with the Atmel AVR Microcontroller. Morgan and Claypool Publishers. DOI: 10.2200/S00225ED1V01Y200910DCS025

6.9   CHAPTER PROBLEMS

1. What is the purpose of an interrupt?

2. Describe the flow of events when an interrupt occurs.

3. Describe the interrupt features available with the ATmega328.

4. Describe the built–in interrupt features available with the Arduino Development Environment.

5. What is the interrupt priority? How is it determined?

6. What steps are required by the system designer to properly configure an interrupt?

7. How is the interrupt system turned "ON" and "OFF"?

8. A 10 MHz ceramic resonator is not available. Redo the example of the Timer/Counter0 Overflow interrupt provided with a timebase of 1 MHz and 8 MHz.

9. What is the maximum delay that may be generated with the delay function provided in the text without modification? How could the function be modified for longer delays?

10. In the text, we provided a 24 hour timer (hh:mm:ss:ms) using the Timer/Counter0 Overflow interrupt. What is the accuracy of the timer? How can it be improved?

11. Adapt the 24 hour timer example to generate an active high logic pulse on a microcontroller pin of your choice for three seconds. The pin should go logic high three weeks from now.

12. What are the concerns when using multiple interrupts in a given application?

13. How much time can background processing relative to foreground processing be implemented?

14. What is the advantage of using interrupts over polling techniques?

15. Can the USART transmit and interrupt receive system provided in the chapter be adapted to operate in the Arduino Development Environment? Explain in detail.

CHAPTER 7

Timing Subsystem

Objectives: After reading this chapter, the reader should be able to

- Explain key timing system related terminology.

- Compute the frequency and the period of a periodic signal using a microcontroller.

- Describe functional components of a microcontroller timer system.

- Describe the procedure to capture incoming signal events.

- Describe the procedure to generate time critical output signals.

- Describe the timing related features of the Atmel ATmega328 and ATmega2560.

- Describe the four operating modes of the Atmel ATmega328 and ATmega2560 timing system.

- Describe the register configurations for the ATmega328's Timer 0, Timer 1, and Timer 2.

- Describe the register configurations for the ATmega2560's Timer 0, Timer 1, Timer 2, Timer 3, Timer 4, and Timer 5.

- Program the Arduino UNO R3 and the Arduino Mega 2560 using the built–in timer features of the Arduino Development Environment.

- Program the ATmega328 and the Arduino Mega 2560 timer system for a variety of applications using C.

7.1 OVERVIEW

One of the most important reasons for using microcontrollers is their capability to perform time related tasks. In a simple application, one can program a microcontroller to turn on or turn off an external device at a specific time. In a more involved application, we can use a microcontroller to generate complex digital waveforms with varying pulse widths to control the speed of a DC motor. In this chapter, we review the capabilities of the Atmel ATmega328 and ATmega2560 microcontrollers to perform time related functions. We begin with a review of timing related terminology. We then provide an overview of the general operation of a timing system followed by the timing system features aboard the ATmega328 and the ATmega2560. Next, we present a detailed discussion of each of its timing channels and their different modes of operation. We then review the built–in timing functions of the Arduino Development Environment and conclude the chapter with a variety of examples.

7.2   TIMING RELATED TERMINOLOGY

In this section, we review timing related terminology including frequency, period, and duty cycle.

7.2.1   FREQUENCY

Consider signal $x(t)$ that repeats itself. We call this signal periodic with period T, if it satisfies the following equation:

$$x(t) = x(t + T).$$

To measure the frequency of a periodic signal, we count the number of times a particular event repeats within a one second period. The unit of frequency is the Hertz or cycles per second. For example, a sinusoidal signal with a 60 Hz frequency means that a full cycle of a sinusoid signal repeats itself 60 times each second or every 16.67 ms.

7.2.2   PERIOD

The reciprocal of frequency is the period of a waveform. If an event occurs with a rate of 1 Hz, the period of that event is 1 second. To find the signal period (T), given the signal's frequency (f), we simply need to apply their inverse relationship $f = \frac{1}{T}$. Both the period and frequency of a signal are often used to specify timing constraints of microcontroller–based systems. For example, when your car is on a wintery road and slipping, the engineers who designed your car configured the anti–slippage unit to react within some millisecond period, say 20 milliseconds. The constraint then requires the monitoring system to check slippage at a rate of 50 Hz.

7.2.3   DUTY CYCLE

In many applications, periodic pulses are used as control signals. A good example is the use of a periodic pulse to control a servo motor. To control the direction and sometimes the speed of a motor, a periodic pulse signal with a changing duty cycle over time is used. The periodic pulse signal shown in Figure 7.1(a) is on for 50 percent of the signal period and off for the rest of the period. The pulse shown in (b) is on for only 25 percent of the same period as the signal in (a) and off for 75 percent of the period. The duty cycle is defined as the percentage of the period a signal is on or logic high. Therefore, we refer to the signal in Figure 7.1(a) as a periodic pulse signal with a 50 percent duty cycle and the corresponding signal in (b), a periodic pulse signal with a 25 percent duty cycle.

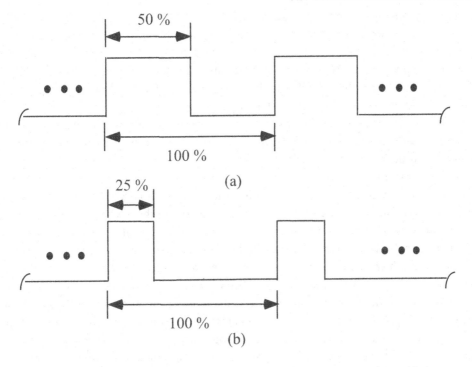

Figure 7.1: Two signals with the same period but different duty cycles. The top figure (a) shows a periodic signal with a 50% duty cycle and the lower figure (b) displays a periodic signal with a 25% duty cycle.

7.3 TIMING SYSTEM OVERVIEW

The heart of the timing system is the time base. The time base's frequency is used to generate a baseline clock signal. For a timer system, the system clock is used to update the contents of a register called the free running counter. The job of the free running counter is to count up (increment) for each rising edge (or a falling edge) of a clock signal. For example, if a clock is operating at a rate of 2 MHz, the free running counter will count up or increment each 0.5 microseconds. All other timer related events reference the contents of the free running counter to perform input and output time related activities: measurement of time periods, capture of timing events, and generation of time related signals.

The ATmega328 and the ATmega2560 may be clocked internally using a user–selectable resistor capacitor (RC) time base or they may be clocked externally. The RC internal time base is selected using programmable fuse bits. You may choose an internal fixed clock operating frequency of 1, 2, 4 or 8 MHz.

To provide for a wider range of frequency selections an external time source may be used. The external time sources, in order of increasing accuracy and stability, are an external RC network, a ceramic resonator, and a crystal oscillator. The system designer chooses the time base frequency and clock source device appropriate for the application at hand. The maximum operating frequency of the ATmega328P with a 5 VDC supply voltage is 20 MHz. The Arduino UNO R3 processing board is equipped with a 16 MHz crystal oscillator time base. The maximum operating frequency of the ATmega2560 with a 5 VDC supply voltage is 16 MHz. The Arduino Mega 2560 processing board is equipped with a 16 MHz crystal oscillator time base.

The timing system may be used to capture the characteristics of an incoming input signal or generate a digital output signal. We provide a brief overview of these capabilities followed by a more detailed treatment in the next section.

For **input** time related activities, all microcontrollers typically have timer hardware components that detect signal logic changes on one or more input pins. Such components rely on the free running counter to capture external event times. We can use these features to measure the period of an incoming signal or the width of an incoming pulse. You can also use the timer input system to measure the pulse width of an aperiodic signal. For example, suppose that the times for the rising edge and the falling edge of an incoming signal are 1.5 sec and 1.6 sec, respectively. We can use these values to easily compute the pulse width of 0.1 second.

For **output** timer functions, a microcontroller uses a comparator, a free running counter, logic switches, and special purpose registers to generate time related signals on one or more output pins. A comparator checks the value of the free running counter for a match with the contents of another special purpose register where a programmer stores a specified time in terms of the free running counter value. The checking process is executed at each clock cycle and when a match occurs, the corresponding hardware system induces a programmed logic change on a programmed output port pin. Using such capability, one can generate a simple logic change at a designated time incident, a pulse with a desired time width, or a pulse width modulated signal to control servo or Direct Current (DC) motors.

From the examples we discussed above, you may have wondered how a microcontroller can be used to compute absolute times from the relative free running counter values, say 1.5 second and 1.6 second. The simple answer is that we can not do so directly. A programmer must use the system clock values and derive the absolute time values. Suppose your microcontroller is clocked by a 2 MHz signal and the system clock uses a 16–bit free running counter. For such a system, each clock period represents 0.5 microsecond and it takes approximately 32.78 milliseconds to count from 0 to 2^{16} (65,536). The timer input system then uses the clock values to compute frequencies, periods, and pulse widths. For example, suppose you want to measure a pulse width of an incoming aperiodic signal. If the rising edge and the falling edge occurred at count values $0010 and $0114, [1] can you find the pulse width when the free running counter is counting at 2 MHz? Let's first convert the two values into their corresponding decimal values, 276 and 16. The pulse width of the signal in the number of counter value is 260. Since we already know how long it takes for the system to increment by one, we can readily compute the pulse width as $260 \times 0.5 \, microseconds \, = \, 130 \, microseconds$.

---

[1]The $ symbol represents that the following value is in a hexadecimal form.

Our calculations do not take into account time increments lasting longer than the rollover time of the counter. When a counter rolls over from its maximum value back to zero, a status flag is set to notify the processor of this event. The rollover events may be counted to correctly determine the overall elapsed time of an event.

To calculate the total elapsed time of an event; the event start time, stop time, and the number of timer overflows (n) that occurred between the start time and stop time must be known. Elapsed time may be calculated using:

$$elapsed\ clock\ ticks = (n \times 2^b) + (stop\ count - start\ count)\ [clock\ ticks]$$

$$elapsed\ time = (elapsed\ clock\ ticks) \times (FRC\ clock\ period)\ [seconds]$$

In this first equation, "n" is the number of Timer Overflow Flag (TOF) events that occur between the start and stop events and "b" is the number of bits in the timer counter. The equation yields the elapsed time in clock ticks. To convert to seconds the number of clock ticks are multiplied by the period of the clock source of the free running counter.

7.4   TIMER SYSTEM APPLICATIONS

In this section, we take a closer look at some important uses of the timer system of a microcontroller to (1) measure an input signal timing event, termed input capture, (2) to count the number of external signal occurrences, (3) to generate timed signals – termed output compare, and, finally, (4) to generate pulse width modulated signals. We first start with a case of measuring the time duration of an incoming signal.

7.4.1   INPUT CAPTURE – MEASURING EXTERNAL TIMING EVENT

In many applications, we are interested in measuring the elapsed time or the frequency of an external event using a microcontroller. Using the hardware and functional units discussed in the previous sections, we now present a procedure to accomplish the task of computing the frequency of an incoming periodic signal. Figure 7.2 shows an incoming periodic signal to our microcontroller.

The first step for input capture is to turn on the timer system. To reduce power consumption a microcontroller usually does not turn on all of its functional systems after reset until they are needed. In addition to a separate timer module, many microcontroller manufacturers allow a programmer to choose the rate of a separate timer clock that governs the overall functions of a timer module.

Once the timer is turned on and the clock rate is selected, a programmer must configure the physical port to which the incoming signal arrives. This step is done using a special input timer port configuration register. The next step is to program the input event to capture. In our current example, we should capture two consecutive rising edges or falling edges of the incoming signal. Again, the programming portion is done by storing an appropriate setup value to a special register.

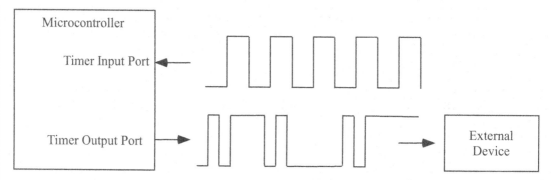

Figure 7.2: Use of the timer input and output systems of a microcontroller. The signal on top is fed into a timer input port. The captured signal is subsequently used to compute the input signal frequency. The signal on the bottom is generated using the timer output system. The signal is used to control an external device.

Now that the input timer system is configured appropriately, you now have two options to accomplish the task. The first one is the use of a polling technique; the microcontroller continuously polls a flag, which holds a logic high signal when a programmed event occurs on the physical pin. Once the microcontroller detects the flag, it needs to clear the flag and record the time when the flag was set using another special register that captures the time of the associated free running counter value. The program needs to continue to wait for the next flag, which indicates the end of one period of the incoming signal. A programmer then needs to record the newly acquired captured time represented in the form of a free running counter value again. The period of the signal can now be computed by computing the time difference between the two captured event times, and, based on the clock speed of the microcontroller's timer system, the programmer can compute the actual time changes and consequently the frequency of the signal.

In many cases, a microcontroller can't afford the time to poll for one event. Such situation introduces the second method: interrupt systems. Most microcontrollers are equipped with built--in interrupt systems with their timer input modules. Instead of continuously polling for a flag, a microcontroller performs other tasks and relies on its interrupt system to detect the programmed event. The task of computing the period and the frequency is the same as the first method, except that the microcontroller will not be tied down to constantly checking the flag, increasing the efficient use of the microcontroller resources. To use interrupt systems, of course, we must pay the price by appropriately configuring the interrupt systems to be triggered when a desired event is detected. Typically, additional registers must be configured, and a special program called an interrupt service routine must be written.

Suppose that for an input capture scenario the two captured times for the two rising edges are $1000 and $5000, respectively. Note that these values are not absolute times but the time hacks captured from the free running counter. The period of the signal is $4000 or 16384 in a decimal form. If we assume that the timer clock runs at 10 MHz, the period of the signal is 1.6384 msec, and the corresponding frequency of the signal is approximately 610.35 Hz.

7.4.2   COUNTING EVENTS

The same capability of measuring the period of a signal can also be used to simply count external events. Suppose we want to count the number of logic state changes of an incoming signal for a given period of time. Again, we can use the polling technique or the interrupt technique to accomplish the task. For both techniques, the initial steps of turning on the timer and configuring a physical input port pin are the same. In this application, however, the programmed event should be any logic state changes instead of looking for a rising or a falling edge as we have done in the previous section.

    If the polling technique is used, at each event detection, the corresponding flag must be cleared and a counter must be updated. If the interrupt technique is used, one must write an interrupt service routine within which the flag is cleared and a counter is updated.

7.4.3   OUTPUT COMPARE – GENERATING TIMING SIGNALS TO INTERFACE EXTERNAL DEVICES

In the previous two sections, we considered two applications of capturing external incoming signals. In this subsection and the next one, we consider how a microcontroller can generate time critical signals for external devices. Suppose in this application, we want to send a signal shown in Figure 7.2 to turn on an external device. The timing signal is arbitrary but the application will show that a timer output system can generate any desired time related signals permitted under the timer clock speed limit of the microcontroller.

    Similar to the use of the timer input system, one must first turn on the timer system and configure a physical pin as a timer output pin using special registers. In addition, one also needs to program the desired external event using another special register associated with the timer output system. To generate the signal shown in Figure 7.2, one must compute the time required between the rising and the falling edges. Suppose that the external device requires a pulse which is 2 milliseconds wide to be activated. To generate the desired pulse, one must first program the logic state for the particular pin to be low and set the time value using a special register with respect to the contents of the free running counter. As was mentioned in Section 7.2, at each clock cycle, the special register contents are compared with the contents of the free running counter and when a match occurs, the programmed logic state appears on the designated hardware pin. Once the rising edge is generated, the program then must reconfigure the event to be a falling edge (logic state low) and change the contents of the special register to be compared with the free running counter. For the particular example in Figure 7.2, let's assume that the main clock runs at 2 MHz, the free running counter is a 16 bit counter, and the name of the special register (16 bit register) where we can put appropriate values is output timer register. To generate the desired pulse, we can put $0000 first to the output timer register, and after the rising edge has been generated, we need to change the program event to

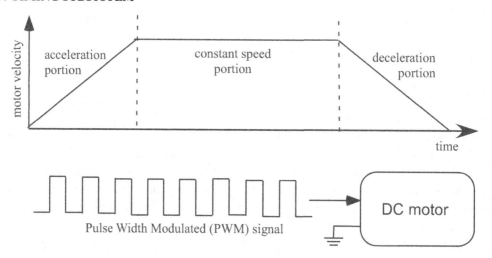

Figure 7.3: The figure shows the speed profile of a DC motor over time when a pulse–width–modulated signal is applied to the motor.

a falling edge and put $0FA0 or 4000 in decimal to the output timer register. As was the case with the input timer system module, we can use output timer system interrupts to generate the desired signals as well.

7.4.4   INDUSTRIAL IMPLEMENTATION CASE STUDY (PWM)

In this section, we discuss a well–known method to control the speed of a DC motor using a pulse width modulated (PWM) signal. The underlying concept is as follows. If we turn on a DC motor and provide the required voltage, the motor will run at its maximum speed. Suppose we turn the motor on and off rapidly, by applying a periodic signal. The motor at some point can not react fast enough to the changes of the voltage values and will run at the speed proportional to the average time the motor was turned on. By changing the duty cycle, we can control the speed of a DC motor as we desire. Suppose again we want to generate a speed profile shown in Figure 7.3. As shown in the figure, we want to accelerate the speed, maintain the speed, and decelerate the speed for a fixed amount of time.

As an example, an elevator control system does not immediately operate the elevator motor at full speed. The elevator motor speed will ramp up gradually from stop to desired speed. As the elevator approaches, the desired floor it will gradually ramp back down to stop.

The first task necessary is again to turn on the timer system, configure a physical port, and program the event to be a rising edge. As a part of the initialization process, we need to put $0000 to the output timer register we discussed in the previous subsection. Once the rising edge is generated, the program then needs to modify the event to a falling edge and change the contents of the output timer register to a value proportional to a desired duty cycle. For example, if we want to start off with 25% duty cycle, we need to input $4000 to the register, provided that we are using a 16 bit free running counter. Once the falling edge is generated, we now need to go back and change the event to be a rising edge and the contents of the output timer counter value back to $0000. If we want to continue to generate a 25% duty cycle signal, then we must repeat the process indefinitely. Note that we are using the time for a free running counter to count from $0000 to $FFFF as one period.

Now suppose we want to increase the duty cycle to 50% over 1 sec and that the clock is running at 2 MHz. This means that the free running counter counts from $0000 to $FFFF every 32.768 milliseconds, and the free running counter will count from $0000 to $FFFF approximately 30.51 times over the period of one second. That is we need to increase the pulse width from $4000 to $8000 in approximately 30 turns, or approximately 546 clock counts every turn. This technique may be used to generate any desired duty cycle.

7.5   OVERVIEW OF THE ATMEL ATMEGA328 AND ATMEGA2560 TIMER SYSTEMS

The Atmel ATmega328 and the ATmega2560 are equipped with a flexible and powerful multiple channel timing system. For the ATmega328, the timer channels are designated Timer 0, Timer 1, and Timer 2. For the ATmega2560, the timer channels are designated Timer 0 through Timer 5. In this section, we review the operation of the timing system in detail. We begin with an overview of the timing system features followed by a detailed discussion of timer channel 0. Space does not permit a complete discussion of the other two types of timing channels; we review their complement of registers and highlight their features not contained in our discussion of timer channel 0. The information provided on timer channel 0 is readily adapted to the other channels.

The features of the timing system are summarized in Figure 7.4. Timer 0 and 2 are 8–bit timers; whereas, Timer 1 for the ATmega 328 and Timers 1, 3–5 for the ATmega2560 are 16–bit timers. Each timing channel is equipped with a prescaler. The prescaler is used to subdivide the main microcontroller clock source (designated $f_{clk\_I/O}$ in upcoming diagrams) down to the clock source for the timing system (clk_{Tn}).

Each timing channel has the capability to generate pulse width modulated signals, generate a periodic signal with a specific frequency, count events, and generate a precision signal using the output compare channels. Additionally, Timer 1 on the ATmega328 and Timer 1, 3–5 on the ATmega2560 are equipped with the Input Capture feature.

All of the timing channels may be configured to operate in one of four operational modes designated : Normal, Clear Timer on Compare Match (CTC), Fast PWM, and Phase Correct PWM. We provide more information on these modes shortly.

ATmega328: Timer 0 ATmega2560: Timer 0	ATmega328: Timer 1 ATmega2560: Timer 1, 3, 4, 5	ATmega328: Timer 2 ATmega2560: Timer 2
Features:	Features:	Features:
- 8-bit timer/counter	- 16-bit timer/counter	- 8-bit timer/counter
- 10-bit clock prescaler	- 10-bit clock prescaler	- 10-bit clock prescaler
- Functions:	- Functions:	- Functions:
-- Pulse width modulation	-- Pulse width modulation	-- Pulse width modulation
-- Frequency generation	-- Frequency generation	-- Frequency generation
-- Event counter	-- Event counter	-- Event counter
-- Output compare -- 2 ch	-- Output compare -- 2 ch	-- Output compare -- 2 ch
- Modes of operation:	-- Input capture	- Modes of operation:
-- Normal	- Modes of operation:	-- Normal
-- Clear timer on compare match (CTC)	-- Normal	-- Clear timer on compare match (CTC)
-- Fast PWM	-- Clear timer on compare match (CTC)	-- Fast PWM
-- Phase correct PWM	-- Fast PWM	-- Phase correct PWM
	-- Phase correct PWM	

Figure 7.4: Atmel timer system overview.

7.6   TIMER 0 SYSTEM

In this section, we discuss the features, overall architecture, modes of operation, registers, and programming of Timer 0. This information may be readily adapted to Timers 1, 3, 4, 5 and Timer 2.

A Timer 0 block diagram is shown in Figure 7.5. The clock source for Timer 0 is provided via an external clock source at the T0 pin of the microcontroller. Timer 0 may also be clocked internally via the microcontroller's main clock ($f_{clk\_I/O}$). This clock frequency may be too rapid for many applications. Therefore, the timing system is equipped with a prescaler to subdivide the main clock frequency down to timer system frequency (clk_{Tn}). The clock source for Timer 0 is selected using the CS0[2:0] bits contained in the Timer/Counter Control Register B (TCCR0B). The TCCR0A register contains the WGM0[1:0] bits and the COM0A[1:0] (and B) bits. Whereas, the TCCR0B register contains the WGM0[2] bit. These bits are used to select the mode of operation for Timer 0 as well as tailor waveform generation for a specific application.

The timer clock source (clk_{Tn}) is fed to the 8–bit Timer/Counter Register (TCNT0). This register is incremented (or decremented) on each clk_{Tn} clock pulse. Timer 0 is also equipped with two 8–bit comparators that constantly compares the numerical content of TCNT0 to the Output

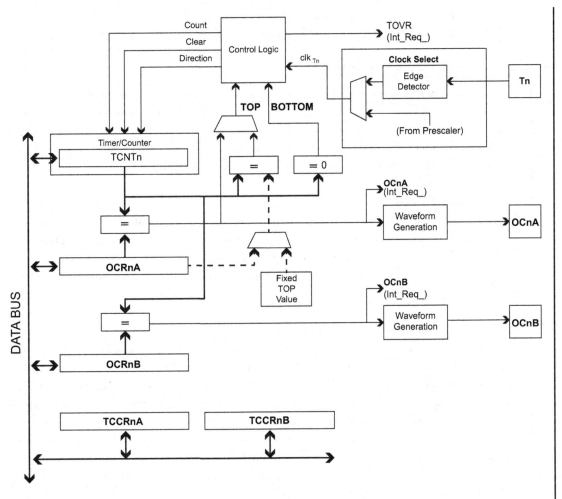

Figure 7.5: Timer 0 block diagram. (Figure used with permission Atmel, Inc.)

Compare Register A (OCR0A) and Output Compare Register B (OCR0B). The compare signal from the 8–bit comparator is fed to the waveform generators. The waveform generators have a number of inputs to perform different operations with the timer system.

The BOTTOM signal for the waveform generation and the control logic, shown in Figure 7.5, is asserted when the timer counter TCNT0 reaches all zeroes (0x00). The MAX signal for the control logic unit is asserted when the counter reaches all ones (0xFF). The TOP signal for the waveform generation is asserted by either reaching the maximum count values of 0xFF on the TCNT0 register or reaching the value set in the Output Compare Register 0 A (OCR0A) or B. The setting for the TOP signal will be determined by the Timer's mode of operation.

Timer 0 also uses certain bits within the Timer/Counter Interrupt Mask Register 0 (TIMSK0) and the Timer/Counter Interrupt Flag Register 0 (TIFR0) to signal interrupt related events.

7.6.1   MODES OF OPERATION

Each of the timer channels may be set for a specific mode of operation: normal, clear timer on compare match (CTC), fast PWM, and phase correct PWM. The system designer chooses the correct mode for the application at hand. The timer modes of operation are summarized in Figure 7.6. A specific mode of operation is selected using the Waveform Generation Mode bits located in Timer/Control Register A (TCCR0A) and Timer/Control Register B (TCCR0B).

7.6.1.1   Normal Mode

In the normal mode, the timer will continually count up from 0x00 (BOTTOM) to 0xFF (TOP). When the TCNT0 returns to zero on each cycle of the counter the Timer/Counter Overflow Flag (TOV0) will be set. The normal mode is useful for generating a periodic "clock tick" that may be used to calculate elapsed real time or provide delays within a system. We provide an example of this application in Section 5.9.

7.6.1.2   Clear Timer on Compare Match (CTC)

In the CTC mode, the TCNT0 timer is reset to zero every time the TCNT0 counter reaches the value set in Output Compare Register A (OCR0A) or B. The Output Compare Flag A (OCF0A) or B is set when this event occurs. The OCF0A or B flag is enabled by asserting the Timer/Counter 0 Output Compare Math Interrupt Enable (OCIE0A) or B flag in the Timer/Counter Interrupt Mask Register 0 (TIMSK0) and when the I–bit in the Status Register is set to one.

The CTC mode is used to generate a precision digital waveform such as a periodic signal or a single pulse. The user must describe the parameters and key features of the waveform in terms of Timer 0 "clock ticks." When a specific key feature is reached within the waveform the next key feature may be set into the OCR0A or B register.

7.6.1.3   Phase Correct PWM Mode

In the Phase Correct PWM Mode, the TCNT0 register counts from 0x00 to 0xFF and back down to 0x00 continually. Every time the TCNT0 value matches the value set in the OCR0A or B register the OCF0A or B flag is set and a change in the PWM signal occurs.

7.6.1.4   Fast PWM

The Fast PWM mode is used to generate a precision PWM signal of a desired frequency and duty cycle. It is called the Fast PWM because its maximum frequency is twice that of the Phase Correct PWM mode. When the TCNT0 register value reaches the value set in the OCR0A or B register it will cause a change in the PWM output as prescribed by the system designer. It continues to count up to the TOP value at which time the Timer/Counter 0 Overflow Flag is set.

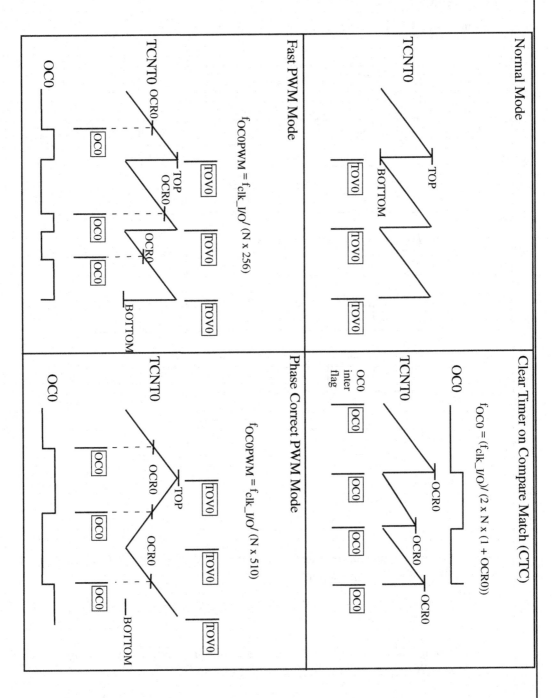

Figure 7.6: Timer 0 modes of operation.

7.6.2   TIMER 0 REGISTERS

A summary of the Timer 0 registers are shown in Figure 7.7.

Timer/Counter Control Register A (TCCR0A)

COM0A1	COM0A0	COM0B1	COM0B0	---	---	WGM01	WGM00
7							0

Timer/Counter Control Register B (TCCR0B)

FOC0A	FOC0B	---	---	WGM02	CS02	CS01	CS00
7							0

Timer/Counter Control Register (TCNT0)

7							0

Output Compare Register A (OCR0A)

7							0

Output Compare Register B (OCR0B)

7							0

Timer/Counter Interrupt Mask Register (TIMSK0)

---	---	---	---	---	OCIE20	OCIE20	TOIE0
7							0

Timer/Counter 2 Interrupt Flag Register (TIFR0)

---	---	---	---	---	OCF0B	OCF20	TOV0
7							0

Figure 7.7: Timer 0 registers.

7.6.2.1 Timer/Counter Control Registers A and B (TCCR0A and TCCR0B)

The TCCR0 register bits are used to:

- Select the operational mode of Timer 0 using the Waveform Mode Generation (WGM0[2:0]) bits,

- Determine the operation of the timer within a specific mode with the Compare Match Output Mode (COM0A[1:0] or COM0B[1:0] or) bits, and

- Select the source of the Timer 0 clock using Clock Select (CS0[2:0]) bits.

The bit settings for the TCCR0 register are summarized in Figure 7.8.

7.6.2.2 Timer/Counter Register(TCNT0)
The TCNT0 is the 8–bit counter for Timer 0.

7.6.2.3 Output Compare Registers A and B (OCR0A and OCR0B)
The OCR0A and B registers holds a user–defined 8–bit value that is continuously compared to the TCNT0 register.

7.6.2.4 Timer/Counter Interrupt Mask Register (TIMSK0)
Timer 0 uses the Timer/Counter 0 Output Compare Match Interrupt Enable A and B (OCIE0A and B) bits and the Timer/Counter 0 Overflow Interrupt Enable (TOIE0) bit. When the OCIE0A or B bit and the I–bit in the Status Register are both set to one, the Timer/Counter 0 Compare Match interrupt is enabled. When the TOIE0 bit and the I–bit in the Status Register are both set to one, the Timer/Counter 0 Overflow interrupt is enabled.

7.6.2.5 Timer/Counter Interrupt Flag Register 0 (TIFR0)
Timer 0 uses the Output Compare Flag A or B (OCF0A and OCF0B) which sets for an output compare match. Timer 0 also uses the Timer/Counter 0 Overflow Flag (TOV0) which sets when Timer/Counter 0 overflows.

7.7 TIMER 1

Timer 1 on the ATmega328 and also Timers 1 and 3–5 on the ATmega2560 are 16–bit timer/counters. These timers share many of the same features of the Timer 0 channel. Due to limited space the shared information will not be repeated. Instead, we concentrate on the enhancements of Timer 1 which include an additional output compare channel and also the capability for input capture. The block diagram for Timer 1 is shown in Figure 7.9.

As discussed earlier in the chapter, the input capture feature is used to capture the characteristics of an input signal including period, frequency, duty cycle, or pulse length. This is accomplished by monitoring for a user–specified edge on the ICP1 microcontroller pin. When the desired edge occurs, the value of the Timer/Counter 1 (TCNT1) register is captured and stored in the Input Capture Register 1 (ICR1).

7.7.1 TIMER 1 REGISTERS

The complement of registers supporting Timer 1 are shown in Figure 7.10. Each register will be discussed in turn.

7.7.1.1 TCCR1A and TCCR1B registers
The TCCR1 register bits are used to:

CS0[2:0]	Clock Source
000	None
001	clk$_{I/O}$
010	clk$_{I/O}$/8
011	clk$_{I/O}$/64
100	clk$_{I/O}$/8clk$_{I/O}$/256
101	clk$_{I/O}$/8clk$_{I/O}$/1024
110	External clock on T0 (falling edge trigger)
111	External clock on T1 (rising edge trigger)

Clock Select

Timer/Counter Control Register B (TCCR0B)

FOC0A	FOC0B	---	---	WGM02	CS02	CS01	CS00

7   0

Timer/Counter Control Register A (TCCR0A)

COM0A1	COM0A0	COM0B1	COM0B0	---	---	WGM01	WGM00

Waveform Generation Mode

Mode	WGM[02:00]	Mode
0	000	Normal
1	001	PWM, Phase Correct
2	010	CTC
3	011	Fast PWM
4	100	Reserved
5	101	PWM, Phase Correct
6	110	Reserved
7	111	Fast PWM

Compare Output Mode, non-PWM Mode

COM0A[1:0]	Description
00	Normal, OC0A disconnected
01	Toggle OC0A on compare match
10	Clear OC0A on compare match
11	Set OC0A on compare match

Compare Output Mode, non-PWM Mode

COM0B[1:0]	Description
00	Normal, OC0B disconnected
01	Toggle OC0B on compare match
10	Clear OC0B on compare match
11	Set OC0B on compare match

Compare Output Mode, Fast PWM Mode

COM0A[1:0]	Description
00	Normal, OC0A disconnected
01	WGM02 = 0: normal operation, OC0A disconnected WGM02 = 1: Toggle OC0A on compare match
10	Clear OC0A on compare match, set OC0A at Bottom (non-inverting mode)
11	Set OC0A on compare match, clear OC0A at Bottom (inverting mode)

Compare Output Mode, Fast PWM Mode

COM0B[1:0]	Description
00	Normal, OC0B disconnected
01	Reserved
10	Clear OC0B on compare match, set OC0B at Bottom (non-inverting mode)
11	Set OC0B on compare match, clear OC0B at Bottom (inverting mode)

Compare Output Mode, Phase Correct PWM

COM0A[1:0]	Description
00	Normal, OC0A disconnected
01	WGM02 = 0: normal operation, OC0A disconnected WGM02 = 1: Toggle OC0A on compare match
10	Clear OC0A on compare match, when upcounting. Set OC0A on compare match when down counting
11	Set OC0A on compare match, when upcounting. Set OC0A on compare match when down counting

Compare Output Mode, Phase Correct PWM

COM0B[1:0]	Description
00	Normal, OC0B disconnected
01	Reserved
10	Clear OC0B on compare match, when upcounting. Set OC0B on compare match when down counting
11	Set OC0B on compare match, when upcounting. Set OC0B on compare match when down counting

Figure 7.8: Timer/Counter Control Registers A and B (TCCR0A and TCCR0B) bit settings.

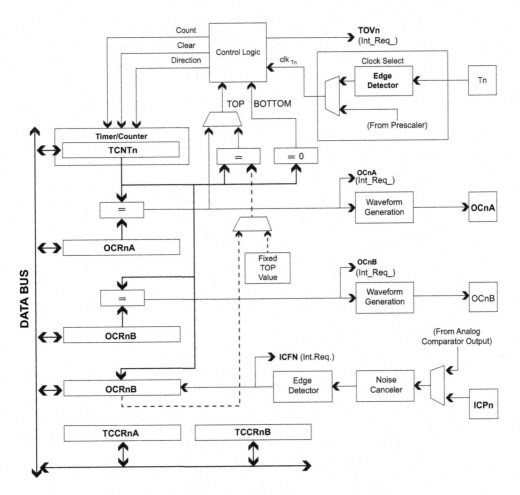

Figure 7.9: Timer 1 block diagram. (Figure used with Permission, Atmel,Inc.)

ATmega328: n=1
ATmega2560: n= 1, 3, 4, 5

Timer/Counter n Control Register A (TCCRnA)

COMnA1	COMnA0	COMnB1	COMnB0	COMnC1	COMnC0	WGMn1	WGMn0
7							0

Timer/Counter n Control Register B (TCCRnB)

ICNCn	ICESn	---	WGMn3	WGMn2	CS12	CS11	CS10
7							0

Timer/Counter n Control Register C (TCCRnC)

FOCnA	FOCnB	FOCnC	---	---	---	---	---
7							0

Timer Counter n (TCNTnH/TCNTnL)

15							8
7							0

Output Compare Register n A (OCRnAH/OCRnAL)

15							8
7							0

Output Compare Register n B (OCRnBH/OCRnBL)

15							8
7							0

Input Capture Register n (ICRnH/ICRnL)

15							8
7							0

Timer/Counter Interrupt Mask Register n (TIMSKn)

7							0

Timer/Counter 1 Interrupt Flag REgister (TIFRn)

7							0

Figure 7.10: Timer 1, 3 through 5 registers.

- Select the operational mode of Timer 1 using the Waveform Mode Generation (WGM1[3:0]) bits,

- Determine the operation of the timer within a specific mode with the Compare Match Output Mode (Channel A: COM1A[1:0] and Channel B: COM1B[1:0]) bits, and

- Select the source of the Timer 1 clock using Clock Select (CS1[2:0]) bits.

The bit settings for the TCCR1A and TCCR1B registers are summarized in Figure 7.11.

7.7.1.2 Timer/Counter Register 1 (TCNT1H/TCNT1L)
The TCNT1 is the 16–bit counter for Timer 1.

7.7.1.3 Output Compare Register 1 (OCR1AH/OCR1AL)
The OCR1A register holds a user–defined 16–bit value that is continuously compared to the TCNT1 register when Channel A is used.

7.7.1.4 OCR1BH/OCR1BL
The OCR1B register holds a user–defined 16–bit value that is continuously compared to the TCNT1 register when Channel B is used.

7.7.1.5 Input Capture Register 1 (ICR1H/ICR1L)
ICR1 is a 16–bit register used to capture the value of the TCNT1 register when a desired edge on ICP1 pin has occurred.

7.7.1.6 Timer/Counter Interrupt Mask Register 1 (TIMSK1)
Timer 1 uses the Timer/Counter 1 Output Compare Match Interrupt Enable (OCIE1A/1B) bits, the Timer/Counter 1 Overflow Interrupt Enable (TOIE1) bit, and the Timer/Counter 1 Input Capture Interrupt Enable (IC1E1) bit. When the OCIE1A/B bit and the I–bit in the Status Register are both set to one, the Timer/Counter 1 Compare Match interrupt is enabled. When the OIE1 bit and the I–bit in the Status Register are both set to one, the Timer/Counter 1 Overflow interrupt is enabled. When the IC1E1 bit and the I–bit in the Status Register are both set to one, the Timer/Counter 1 Input Capture interrupt is enabled.

7.7.1.7 Timer/Counter Interrupt Flag Register (TIFR1)
Timer 1 uses the Output Compare Flag 1 A/B (OCF1A/B) which sets for an output compare A/B match. Timer 1 also uses the Timer/Counter 1 Overflow Flag (TOV1) which sets when Timer/Counter 1 overflows. Timer Channel 1 also uses the Timer/Counter 1 Input Capture Flag (ICF1) which sets for an input capture event.

CS0[2:0]	Clock Source
000	None
001	$clk_{I/O}$
010	$clk_{I/O}/8$
011	$clk_{I/O}/64$
100	$clk_{I/O}/8clk_{I/O}/256$
101	$clk_{I/O}/8clk_{I/O}/1024$
110	External clock on T0 (falling edge trigger)
111	External clock on T1 (rising edge trigger)

Clock Select

Timer/Counter 1 Control Register B (TCCR1B)

ICNC1	ICES1	---	WGM13	WGM12	CS12	CS11	CS10

7   0

Timer/Counter 1 Control Register A (TCCR1A)

COM1A1	COM1A0	COM1B1	COM1B0	---	---	WGM11	WGM10

7   0

Waveform Generation Mode

Mode	WGM[13:12:11:10]	Mode
0	0000	Normal
1	0001	PWM, Phase Correct, 8-bit
2	0010	PWM, Phase Correct, 9-bit
3	0011	PWM, Phase Correct, 10-bit
4	0100	CTC
5	0101	Fast PWM, 8-bit
6	0110	Fast PWM, 9-bit
7	0111	Fast PWM, 10-bit
8	1000	PWM, Phase & Freq Correct
9	1001	PWM, Phase & Freq Correct
10	1010	PWM, Phase Correct
11	1011	PWM, Phase Correct
12	1100	CTC
13	1101	Reserved
14	1110	Fast PWM
15	1111	Fast PWM

Normal, CTC

COMx[1:0]	Description
00	Normal, OC1A/1B disconnected
01	Toggle OC1A/1B on compare match
10	Clear OC1A/1B on compare match
11	Set OC1A/1B on compare match

PWM, Phase Correct, Phase & Freq Correct

COMx[1:0]	Description
00	Normal, OC0 disconnected
01	WGM1[3:0] = 9 or 14: toggle OCnA on compare match, OCnB disconnected WGM1[3:0]= other settings, OC1A/1B disconnected
10	Clear OC0 on compare match when up-counting. Set OC0 on compare match when down counting
11	Set OC0 on compare match when up-counting. Clear OC0 on compare match when down counting.

Fast PWM

COMx[1:0]	Description
00	Normal, OC1A/1B disconnected
01	WGM1[3:0] = 9 or 11, toggle OC1A on compare match OC1B disconnected WGM1[3:0] = other settings, OC1A/1B disconnected
10	Clear OC1A/1B on compare match, set OC1A/1B on Compare Match when down counting
11	Set OC1A/1B on compare match when upcounting. Clear OC1A/1B on Compare Match when upcounting

Figure 7.11: TCCR1A and TCCR1B registers.

7.8   TIMER 2

Timer 2 is another 8–bit timer channel similar to Timer 0. The Timer 2 channel block diagram is provided in Figure 7.12. Its registers are summarized in Figure 7.13.

7.8.0.8 Timer/Counter Control Register A and B (TCCR2A and B)
The TCCR2A and B register bits are used to:

- Select the operational mode of Timer 2 using the Waveform Mode Generation (WGM2[2:0]) bits,

- Determine the operation of the timer within a specific mode with the Compare Match Output Mode (COM2A[1:0] and B) bits, and

- Select the source of the Timer 2 clock using Clock Select (CS2[2:0]) bits.

    The bit settings for the TCCR2A and B registers are summarized in Figure 7.14.

7.8.0.9 Timer/Counter Register(TCNT2)
The TCNT2 is the 8–bit counter for Timer 2.

7.8.0.10 Output Compare Register A and B (OCR2A and B)
The OCR2A and B registers hold a user–defined 8–bit value that is continuously compared to the TCNT2 register.

7.8.0.11 Timer/Counter Interrupt Mask Register 2 (TIMSK2)
Timer 2 uses the Timer/Counter 2 Output Compare Match Interrupt Enable A and B (OCIE2A and B) bits and the Timer/Counter 2 Overflow Interrupt Enable A and B (OIE2A and B) bits. When the OCIE2A or B bit and the I–bit in the Status Register are both set to one, the Timer/Counter 2 Compare Match interrupt is enabled. When the TOIE2 bit and the I–bit in the Status Register are both set to one, the Timer/Counter 2 Overflow interrupt is enabled.

7.8.0.12 Timer/Counter Interrupt Flag Register 2 (TIFR2)
Timer 2 uses the Output Compare Flags 2 A and B (OCF2A and B) which sets for an output compare match. Timer 2 also uses the Timer/Counter 2 Overflow Flag (TOV2) which sets when Timer/Counter 2 overflows.

7.9   PROGRAMMING THE ARDUINO UNO R3 AND MEGA 2560 USING THE BUILT–IN ARDUINO DEVELOPMENT ENVIRONMENT TIMING FEATURES

The Arduino Development Environment has several built–in timing features. These include:

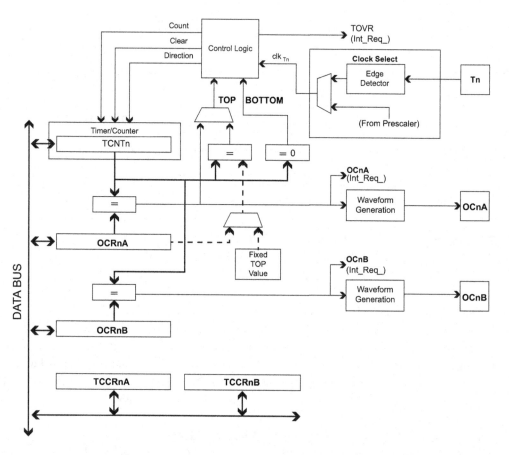

Figure 7.12: Timer 2 block diagram. (Figure used with Permission, Atmel,Inc.)

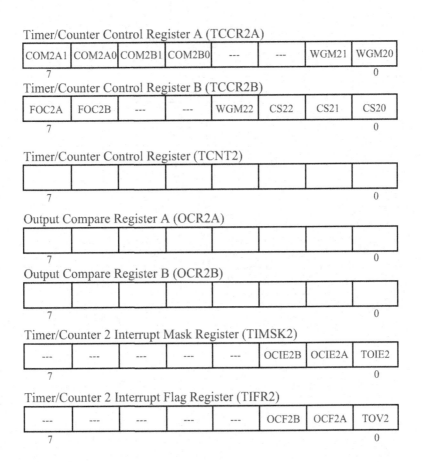

Figure 7.13: Timer 2 registers.

CS2[2:0]	Clock Source
000	None
001	$clk_{I/O}$
010	$clk_{I/O}/8$
011	$clk_{I/O}/32$
100	$clk_{I/O}/64$
101	$clk_{I/O}/128$
110	$clk_{I/O}/256$
111	$clk_{I/O}/1024$

Clock Select

Timer/Counter Control Register B (TCCR2B)

FOC2A	FOC2B	---	---	WGM22	CS22	CS21	CS20

7   0

Timer/Counter Control Register A (TCCR2A)

COM2A1	COM2A0	COM2B1	COM2B0	---	---	WGM21	WGM20

Waveform Generation Mode

Mode	WGM[02:00]	Mode
0	000	Normal
1	001	PWM, Phase Correct
2	010	CTC
3	011	Fast PWM
4	100	Reserved
5	101	PWM, Phase Correct
6	110	Reserved
7	111	Fast PWM

Compare Output Mode, non-PWM Mode

COM2A[1:0]	Description
00	Normal, OC2A disconnected
01	Toggle OC2A on compare match
10	Clear OC2A on compare match
11	Set OC2A on compare match

Compare Output Mode, non-PWM Mode

COM2B[1:0]	Description
00	Normal, OC2B disconnected
01	Toggle OC2B on compare match
10	Clear OC2B on compare match
11	Set OC2B on compare match

Compare Output Mode, Fast PWM Mode

COM2A[1:0]	Description
00	Normal, OC2A disconnected
01	WGM22 = 0: normal operation, OC2A disconnected; WGM22 = 1: Toggle OC2A on compare match
10	Clear OC2A on compare match, set OC2A at Bottom (non-inverting mode)
11	Set OC2A on compare match, clear OC2A at Bottom (inverting mode)

Compare Output Mode, Fast PWM Mode

COM2B[1:0]	Description
00	Normal, OC2B disconnected
01	Reserved
10	Clear OC2B on compare match, set OC2B at Bottom (non-inverting mode)
11	Set OC2B on compare match, clear OC2B at Bottom (inverting mode)

Compare Output Mode, Phase Correct PWM

COM2A[1:0]	Description
00	Normal, OC2A disconnected
01	WGM22 = 0: normal operation, OC2A disconnected; WGM22 = 1: Toggle OC2A on compare match
10	Clear OC2A on compare match, when upcounting. Set OC2A on compare match when down counting
11	Set OC2A on compare match, when upcounting. Set OC2A on compare match when down counting

Compare Output Mode, Phase Correct PWM

COM2B[1:0]	Description
00	Normal, OC2B disconnected
01	Reserved
10	Clear OC2B on compare match, when upcounting. Set OC2B on compare match when down counting
11	Set OC2B on compare match, when upcounting. Set OC2B on compare match when down counting

Figure 7.14: Timer/Counter Control Register A and B (TCCR2A and B) bit settings.

- **delay(unsigned long):** The delay function pauses a sketch for the amount of time specified in milliseconds.

- **delayMicroseconds(unsigned int):** The delayMicroseconds function pauses a sketch for the amount of time specified in microseconds.

- **pulseIn(pin, value):** The pulseIn function measures the length of an incoming digital pulse. If value is specified as HIGH, the function waits for the specified pin to go high and then times until the pin goes low. The pulseIn function returns the length of elapsed time in microseconds as an unsigned long.

- **analogWrite(pin, value):** The analog write function provides a pulse width modulated (PWM) output signal on the specified pin. The PWM frequency is approximately 490 Hz. The duty cycle is specified from 0 (value of 0) to 100 (value of 255) percent.

We have already used the analogWrite function in an earlier chapter to control the motor speed of the Blinky 602A robot.

7.10   PROGRAMMING THE TIMER SYSTEM IN C

In this section, we provide several representative examples of using the timer system for various applications. We will provide examples of using the timer system to generate a prescribed delay, to generate a PWM signal, and to capture an input event.

7.10.1   PRECISION DELAY IN C

In this example, we program the ATmega328 to provide a delay of some number of 6.55 ms interrupts. The Timer 0 overflow is configured to occur every 6.55 ms. The overflow flag is used as a "clock tick" to generate a precision delay. To create the delay the microcontroller is placed in a while loop waiting for the prescribed number of Timer 0 overflows to occur.

```
//Function prototypes
void delay(unsigned int number_of_6_55ms_interrupts);
void init_timer0_ovf_interrupt(void);
void timer0_interrupt_isr(void);


                                      //interrupt handler definition
#pragma interrupt_handler timer0_interrupt_isr:17

//door profile data
```

```
//*********************************************************************
//int_timer0_ovf_interrupt(): The Timer0 overflow interrupt is being
//employed as a time base for a master timer for this project.
//The ceramic resonator operating at 10 MHz is divided by 256.
//The 8-bit Timer0 register (TCNT0) overflows every 256 counts
//or every 6.55 ms.
//*********************************************************************

void init_timer0_ovf_interrupt(void)
{
TCCR0B = 0x04; //divide timer0 timebase by 256, overfl. occurs every 6.55ms
TIMSK0 = 0x01; //enable timer0 overflow interrupt
asm("SEI");    //enable global interrupt
}


//*********************************************************************
//*********************************************************************
//timer0_interrupt_isr:
//Note: Timer overflow 0 is cleared by hardware when executing the
//corresponding interrupt handling vector.
//*********************************************************************


void timer0_interrupt_isr(void)
{
input_delay++;    //input delay processing
}


//*********************************************************************
//*********************************************************************
//delay(unsigned int num_of_6_55ms_interrupts): this generic delay function
//provides the specified delay as the number of 6.55 ms "clock ticks" from
//the Timer0 interrupt.
//Note: this function is only valid when using a 10 MHz crystal or ceramic
//       resonator
//*********************************************************************
```

```
void delay(unsigned int number_of_6_55ms_interrupts)
{
TCNT0 = 0x00;                        //reset timer0
input_delay = 0;
while(input_delay <= number_of_6_55ms_interrupts)
  {
  ;
  }
}
```

//***

7.10.2 PULSE WIDTH MODULATION IN C

The function provided below is used to configure output compare channel B to generate a pulse width modulated signal. An analog voltage provided to ADC Channel 3 is used to set the desired duty cycle from 50 to 100 percent. Note how the PWM ramps up from 0 to the desired speed.

```
//Function Prototypes
void PWM(unsigned int PWM_incr)
{
unsigned int  Open_Speed_int;
float         Open_Speed_float;
int           gate_position_int;


PWM_duty_cycle = 0;
InitADC();                           //Initialize ADC

                                     //Read "Open Speed" volt
Open_Speed_int = ReadADC(0x03);      //setting PA3

                                      //Open Speed Setting unsigned int

                                      //Convert to max duty cycle
  //setting

                                     //0 VDC = 50% = 127,
                                     //5 VDC = 100% =255
Open_Speed_float = ((float)(Open_Speed_int)/(float)(0x0400));
```

```c
                                          //Convert volt to PWM constant
                                          //127-255
Open_Speed_int = (unsigned int)((Open_Speed_float * 127) + 128.0);
                                          //Configure PWM clock
TCCR1A = 0xA1;                            //freq = resonator/510 = 10MHz/510

                                          //freq = 19.607 kHz
TCCR1B = 0x01;                            //no clock source division

PWM_duty_cycle = 0;                       //Initiate PWM duty cycle
                                          //variables

OCR1BH = 0x00;
OCR1BL = (unsigned char)(PWM_duty_cycle);//Set PWM duty cycle CH B to 0%
                                          //Ramp up to Open Speed in 1.6s
OCR1BL = (unsigned char)(PWM_duty_cycle);//Set PWM duty cycle CH B

while (PWM_duty_cycle < Open_Speed_int)
  {
  if(PWM_duty_cycle < Open_Speed_int)      //Increment duty cycle
    PWM_duty_cycle=PWM_duty_cycle + PWM_open_incr;
                                          //Set PWM duty cycle CH B
  OCR1BL = (unsigned char)(PWM_duty_cycle);
  }

//Gate continues to open at specified upper speed (PA3)
:
:
:

//********************************************************************************
```

7.10.3 INPUT CAPTURE MODE IN C

This example was developed by Julie Sandberg, BSEE and Kari Fuller, BSEE at the University of Wyoming as part of their senior design project. In this example, the input capture channel is being used to monitor the heart rate (typically 50–120 beats per minute) of a patient. The microcontroller is set to operate at an internal clock frequency of 1 MHz. Timer/Counter channel 1 is used in this example.

```
//**********************************************************************
//initialize_ICP_interrupt: Initialize Timer/Counter 1 for input capture
//**********************************************************************

void initialize_ICP_interrupt(void)
{
TIMSK=0x20;                         //Allows input capture interrupts
SFIOR=0x04;                         //Internal pull-ups disabled
TCCR1A=0x00;                        //No output comp or waveform
                                    //generation mode
TCCR1B=0x45;                        //Capture on rising edge,
                                    //clock prescalar=1024
TCNT1H=0x00;                        //Initially clear timer/counter 1
TCNT1L=0x00;
asm("SEI");                         //Enable global interrupts
}


//**********************************************************************

void Input_Capture_ISR(void)
{
if(first_edge==0)
  {
  ICR1L=0x00;                       //Clear ICR1 and TCNT1 on first edge
  ICR1H=0x00;
  TCNT1L=0x00;
  TCNT1H=0x00;
  first_edge=1;
  }

else
  {
  ICR1L=TCNT1L;                     //Capture time from TCNT1
  ICR1H=TCNT1H;
  TCNT1L=0x00;
  TCNT1H=0x00;
  first_edge=0;
  }
```

```
heart_rate();                        //Calculate the heart rate
TIFR=0x20;                           //Clear the input capture flag
asm("RETI");                         //Resets the I flag to allow
                                     //global interrupts

}


//*************************************************************************

void heart_rate(void)
{
if(first_edge==0)
  {
  time_pulses_low = ICR1L;           //Read 8 low bits first
  time_pulses_high = ((unsigned int)(ICR1H << 8));
  time_pulses = time_pulses_low | time_pulses_high;
  if(time_pulses!=0)                 //1 counter increment = 1.024 ms
    {                                //Divide by 977 to get seconds/pulse
    HR=60/(time_pulses/977);         //(secs/min)/(secs/beat) =bpm
    }
  else
    {
    HR=0;
    }
  }
else
  {
  HR=0;
  }
}


//*************************************************************************
```

7.11   APPRLICATION 1: SERVO MOTOR CONTROL WITH THE PWM SYSTEM IN C

A servo motor provides an angular displacement from 0 to 180 degrees. Most servo motors provide the angular displacement relative to the pulse length of repetitive pulses sent to the motor as shown in Figure 7.15. A 1 ms pulse provides an angular displacement of 0 degrees while a 2 ms pulse provides a displacement of 180 degrees. Pulse lengths in between these two extremes provide angular

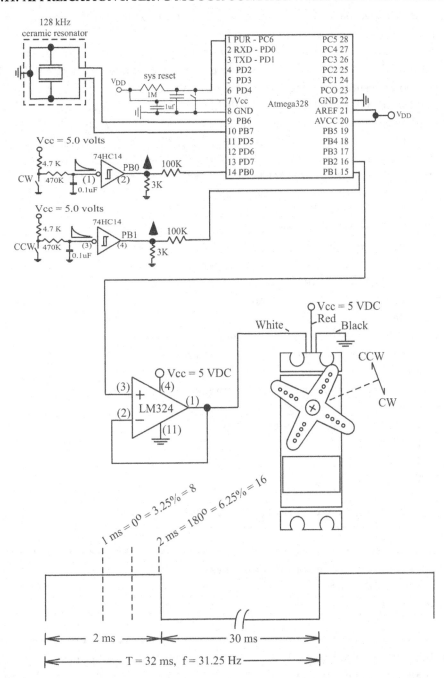

Figure 7.15: Test and interface circuit for a servo motor.

displacements between 0 and 180 degrees. Usually, a 20 to 30 ms low signal is provided between the active pulses.

A test and interface circuit for a servo motor is provided in Figure 7.15. The PB0 and PB1 inputs of the ATmega328 provide for clockwise (CW) and counter–clockwise (CCW) rotation of the servo motor, respectively. The time base for the ATmega328 is provided by a 128 KHz external RC oscillator. Also, the external time base divide–by–eight circuit is active via a fuse setting. Pulse width modulated signals to rotate the servo motor is provided by the ATmega328. A voltage–follower op amp circuit is used as a buffer between the ATmega328 and the servo motor.

The software to support the test and interface circuit is provided below.

```
//****************************************************************************
//target controller: ATMEL ATmega328
//
//ATMEL AVR ATmega328PV Controller Pin Assignments
//Chip Port Function I/O Source/Dest Asserted Notes
//Pin  1 PUR Reset - 1M resistor to Vdd, tact switch to ground,
//        1.0 uF to ground
//Pin  7 Vdd - 1.0 uF to ground
//Pin  8 Gnd
//Pin  9 PB6 ceramic resonator connection
//Pin 10 PB7 ceramic resonator connection
//PORTB:
//Pin 14 PB0 to active high RC debounced switch - CW
//Pin 15 PB1 to active high RC debounced switch - CCW
//Pin 16 PB2 - to servo control input
//Pin 20 AVcc to Vdd
//Pin 21 ARef to Vdd
//Pin 22 AGnd to Ground


//****************************************************************************

//include files***************************************************************
//ATMEL register definitions for ATmega328
#include<iom328pv.h>
#include<macros.h>


//function prototypes*********************************************************
void initialize_ports(void);            //initializes ports
void read_new_input(void);              //used to read input change on PORTB
void init_timer0_ovf_interrupt(void);   //used to initialize timer0 overflow
```

```
//main program*********************************************************
//The main program checks PORTB for user input activity.
//If new activity is found, the program responds.

//global variables
unsigned char   old_PORTB = 0x08;      //present value of PORTB
unsigned char   new_PORTB;             //new values of PORTB
unsigned int    PWM_duty_cycle;

void main(void)
{
initialize_ports();
 //return LED configuration to default


 //external ceramic resonator: 128 KHZ
                                      //fuse set for divide by 8
                                      //configure PWM clock
TCCR1A = 0xA1;
 //freq = oscillator/510 = 128KHz/8/510
                                      //freq = 31.4 Hz
TCCR1B = 0x01;                        //no clock source division

 //duty cycle will vary from 3.1% =
                                      //1 ms = 0 degrees = 8 counts to

 //6.2% = 2 ms = 180 degrees = 16 counts

 //initiate PWM duty cycle variables
PWM_duty_cycle = 12;
OCR1BH = 0x00;
OCR1BL = (unsigned char)(PWM_duty_cycle);

//main activity loop - processor will continually cycle through
//loop for new activity.
//Activity initialized by external signals presented to PORTB[1:0]

while(1)
  {
```

```
  _StackCheck();                          //check for stack overflow
  read_new_input();                       //read input status changes on PORTB
  }
}//end main

//Function definitions
//*********************************************************************
//initialize_ports: provides initial configuration for I/O ports
//*********************************************************************

void initialize_ports(void)
{
//PORTB
DDRB=0xfc;                                //PORTB[7-2] output, PORTB[1:0] input
PORTB=0x00;                               //disable PORTB pull-up resistors

//PORTC
DDRC=0xff;                                //set PORTC[7-0] as output
PORTC=0x00;                               //init low

//PORTD
DDRD=0xff;                                //set PORTD[7-0] as output
PORTD=0x00;                               //initialize low
}


//*********************************************************************
//*********************************************************************

//read_new_input: functions polls PORTB for a change in status. If status
//change has occurred, appropriate function for status change is called
//Pin 1 PB0 to active high RC  debounced switch - CW
//Pin 2 PB1 to active high RC debounced switch  - CCW
//*********************************************************************

void read_new_input(void)
{
new_PORTB = (PINB);
if(new_PORTB != old_PORTB){
  switch(new_PORTB){                      //process change in PORTB input
```

```
    case 0x01:                          //CW
      while(PINB == 0x01)
        {
        PWM_duty_cycle = PWM_duty_cycle + 1;
        if(PWM_duty_cycle > 16) PWM_duty_cycle = 16;
        OCR1BH = 0x00;
        OCR1BL = (unsigned char)(PWM_duty_cycle);
        }
    break;

    case 0x02:                          //CCW
      while(PINB == 0x02)
        {
        PWM_duty_cycle = PWM_duty_cycle - 1;
        if(PWM_duty_cycle < 8) PWM_duty_cycle = 8;
        OCR1BH = 0x00;
        OCR1BL = (unsigned char)(PWM_duty_cycle);
        }
      break;


    default:;                           //all other cases
    }                                   //end switch(new_PORTB)
  }                                     //end if new_PORTB
  old_PORTB=new_PORTB;                  //update PORTB
}


//***********************************************************************
```

7.12   APPLICATION 2: INEXPENSIVE LASER LIGHT SHOW

An inexpensive laser light show may be constructed from two servos. In this example we use two Futaba 180 degree range servos (Parallax 900–00005, available from Jameco #283021) mounted as shown in Figure 7.16. The X and Y control signals are provided by an Arduino processing board. The X and Y control signals are interfaced to the servos via LM324 operational amplifiers. The laser source is provided by an inexpensive laser pointer.

Sample code to drive the servos from an Arduino are provided on the Jameco website (www. jameco.com). Reference Jameco #283021.

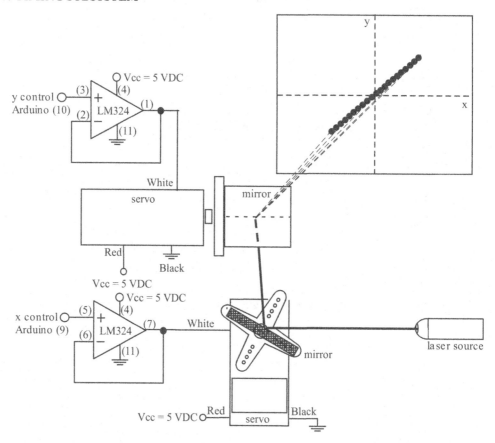

Figure 7.16: Inexpensive laser light show.

```
//*************************************************************
//X-Y ramp
//*************************************************************

#include <Servo.h>     // Use Servo library, included with IDE

Servo myServo_x;       // Create Servo object to control the servo
Servo myServo_y;

void setup() {
  myServo_x.attach(9);   // Servo is connected to digital pin 9
  myServo_y.attach(10);  // Servo is connected to digital pin 10
```

```
}

void loop() {
  int i = 0;
   for(i=0; i<=180; i++)
     {
     myServo_x.write(i);    // Rotate servo counter clockwise
     myServo_y.write(i);    // Rotate servo counter clockwise
     delay(20);             // Wait 2 seconds
     if(i==180)
       delay(5000);
     }
}
//********************************************************************
```

7.13 SUMMARY

In this chapter, we considered a microcontroller timer system, associated terminology for timer related topics, discussed typical functions of a timer subsystem, studied timer hardware operations, and considered some applications where the timer subsystem of a microcontroller can be used. We then took a detailed look at the timer subsystem aboard the ATmega328 and ATmega2560 and reviewed the features, operation, registers, and programming of the three different types of timer channels. We then investigated the built–in timing features of the Arduino Development Environment. We concluded with an example employing a servo motor controlled by the ATmega328 PWM system and an inexpensive laser light show.

7.14 REFERENCES

- Kenneth Short, *Embedded Microprocessor Systems Design: An Introduction Using the INTEL 80C188EB*, Prentice Hall, Upper Saddle River, 1998.

- Frederick Driscoll, Robert Coughlin, and Robert Villanucci, *Data Acquisition and Process Control with the M68HC11 Microcontroller*, Second Edition, Prentice Hall, Upper Saddle River, 2000.

- Todd Morton, *Embedded Microcontrollers*, Prentice Hall, Upper Saddle River, Prentice Hall, 2001.

- *Atmel 8–bit AVR Microcontroller with 16K Bytes In–System Programmable Flash, ATmega328, ATmega328L*, data sheet: 2466L–AVR–06/05, Atmel Corporation, 2325 Orchard Parkway, San Jose, CA 95131.

- *Atmel 8–bit AVR Microcontroller with 16/32/64K Bytes In–System Programmable Flash, AT-mega328P/V, ATmega324P/V, 644P/V* data sheet: 8011I–AVR–05/08, Atmel Corporation, 2325 Orchard Parkway, San Jose, CA 95131.

- *Atmel 8–bit AVR Microcontroller with 64/128/256K Bytes In–System Programmable Flash, AT-mega640/V, ATmega1280/V, 2560/V* data sheet: 2549P–AVR–10/2012, Atmel Corporation, 2325 Orchard Parkway, San Jose, CA 95131.

- Barrett S, Pack D (2006) Microcontrollers Fundamentals for Engineers and Scientists. Morgan and Claypool Publishers. DOI: 10.2200/S00025ED1V01Y200605DCS001

- Barrett S and Pack D (2008) Atmel AVR Microcontroller Primer Programming and Inter-facing. Morgan and Claypool Publishers. DOI: 10.2200/S00100ED1V01Y200712DCS015

- Barrett S (2010) Embedded Systems Design with the Atmel AVR Microcontroller. Morgan and Claypool Publishers. DOI: 10.2200/S00225ED1V01Y200910DCS025

7.15   CHAPTER PROBLEMS

1. Given an 8 bit free running counter and the system clock rate of 24 MHz, find the time required for the counter to count from zero to its maximum value.

2. If we desire to generate periodic signals with periods ranging from 125 nanoseconds to 500 microseconds, what is the minimum frequency of the system clock?

3. Describe how you can compute the period of an incoming signal with varying duty cycles.

4. Describe how one can generate an aperiodic pulse with a pulse width of 2 minutes?

5. Program the output compare system of the ATmega328 to generate a 1 kHz signal with a 10 percent duty cycle.

6. Design a microcontroller system to control a sprinkler controller that performs the following tasks. We assume that your microcontroller runs with 10 MHz clock and it has a 16 bit free running counter. The sprinkler controller system controls two different zones by turning sprinklers within each zone on and off. To turn on the sprinklers of a zone, the controller needs to receive a 152.589 Hz PWM signal from your microcontroller. To turn off the sprinklers of the same zone, the controller needs to receive the PWM signal with a different duty cycle.

    (a) Your microcontroller needs to provide the PWM signal with 10% duty cycle for 10 millisecond to turn on the sprinklers in zone one.

    (b) After 15 minutes, your microcontroller must send the PWM signal with 15% duty cycle for 10 millisecond to turn off the sprinklers in zone one.

(c) After 15 minutes, your microcontroller must send the PWM signal with 20% duty cycle for 10 millisecond to turn on the sprinklers in zone two.

(d) After 15 minutes, your microcontroller must send the PWM signal with 25% duty cycle for 10 millisecond to turn off the sprinklers in zone two.

7. Modify the servo motor example to include a potentiometer connected to PORTA[0]. The servo will deflect 0 degrees for 0 VDC applied to PORTA[0] and 180 degrees for 5 VDC.

8. For the automated cooling fan example, what would be the effect of changing the PWM frequency applied to the fan?

9. Modify the code of the automated cooling fan example to also display the set threshold temperature.

10. Write functions to draw a circle, diamond, and a sine wave with the X–Y laser control system.

CHAPTER 8

Serial Communication Subsystem

Objectives: After reading this chapter, the reader should be able to

- Describe the differences between serial and parallel communication.

- Provide definitions for key serial communications terminology.

- Describe the operation of the Universal Synchronous and Asynchronous Serial Receiver and Transmitter (USART).

- Program the USART for basic transmission and reception using the built–in features of the Arduino Development Environment.

- Program the USART for basic transmission and reception using C.

- Describe the operation of the Serial Peripheral Interface (SPI).

- Program the SPI system using the built–in features of the Arduino Development Environment.

- Program the SPI system using C.

- Describe the purpose of the Two Wire Interface (TWI).

- Program the Arduino UNO R3 and the Arduino Mega 2560 processing board using ISP programming techniques.

8.1 OVERVIEW

Serial communication techniques provide a vital link between an Arduino processing board an certain input devices, output devices, and other microcontrollers. In this chapter, we investigate the serial communication features beginning with a review of serial communication concepts and terminology. We then investigate in turn the following serial communication systems available on the Arduino processing boards: the Universal Synchronous and Asynchronous Serial Receiver and Transmitter (USART), the Serial Peripheral Interface (SPI) and the Two Wire Interface (TWI). We provide guidance on how to program the USART and SPI using built–in Arduino Development Environment features and the C programming language. We conclude the chapter with examples on how to connect an SD card to an Arduino processor, interface a voice chip, and also how to program using In System Programming (ISP) techniques.

8.2   SERIAL COMMUNICATIONS

Microcontrollers must often exchange data with other microcontrollers or peripheral devices. Data may be exchanged by using parallel or serial techniques. With parallel techniques, an entire byte of data is typically sent simultaneously from the transmitting device to the receiver device. While this is efficient from a time point of view, it requires eight separate lines for the data transfer.

In serial transmission, a byte of data is sent a single bit at a time. Once eight bits have been received at the receiver, the data byte is reconstructed. While this is inefficient from a time point of view, it only requires a line (or two) to transmit the data.

The ATmega328 (UNO R3) and the ATmega2560 (Mega 2560) are equipped with a host of different serial communication subsystems including the serial USART, the serial peripheral interface or SPI, and the Two–wire Serial Interface (TWI). What all of these systems have in common is the serial transmission of data. Before discussing the different serial communication features aboard these processors, we review serial communication terminology.

8.3   SERIAL COMMUNICATION TERMINOLOGY

In this section, we review common terminology associated with serial communication.

Asynchronous versus Synchronous Serial Transmission: In serial communications, the transmitting and receiving device must be synchronized to one another and use a common data rate and protocol. Synchronization allows both the transmitter and receiver to be expecting data transmission/reception at the same time. There are two basic methods of maintaining "sync" between the transmitter and receiver: asynchronous and synchronous.

In an asynchronous serial communication system, such as the USART aboard the ATmega328 and the ATmega2560, framing bits are used at the beginning and end of a data byte. These framing bits alert the receiver that an incoming data byte has arrived and also signals the completion of the data byte reception. The data rate for an asynchronous serial system is typically much slower than the synchronous system, but it only requires a single wire between the transmitter and receiver.

A synchronous serial communication system maintains "sync" between the transmitter and receiver by employing a common clock between the two devices. Data bits are sent and received on the edge of the clock. This allows data transfer rates higher than with asynchronous techniques but requires two lines, data and clock, to connect the receiver and transmitter.

Baud rate: Data transmission rates are typically specified as a Baud or bits per second rate. For example, 9600 Baud indicates the data is being transferred at 9600 bits per second.

Full Duplex: Often serial communication systems must both transmit and receive data. To do both transmission and reception, simultaneously, requires separate hardware for transmission and reception. A single duplex system has a single complement of hardware that must be switched from transmission to reception configuration. A full duplex serial communication system has separate hardware for transmission and reception.

Non–return to Zero (NRZ) Coding Format: There are many different coding standards used within serial communications. The important point is the transmitter and receiver must use a common coding standard so data may be interpreted correctly at the receiving end. The Atmel ATmega328 and ATmega2560 use a non–return to zero (NRZ) coding standard. In NRZ coding a logic one is signaled by a logic high during the entire time slot allocated for a single bit; whereas, a logic zero is signaled by a logic low during the entire time slot allocated for a single bit.

The RS–232 Communication Protocol: When serial transmission occurs over a long distance additional techniques may be used to insure data integrity. Over long distances logic levels degrade and may be corrupted by noise. At the receiving end, it is difficult to discern a logic high from a logic low. The RS–232 standard has been around for some time. With the RS–232 standard (EIA–232), a logic one is represented with a –12 VDC level while a logic zero is represented by a +12 VDC level. Chips are commonly available (e.g., MAX232) that convert the 5 and 0 V output levels from a transmitter to RS–232 compatible levels and convert back to 5V and 0 V levels at the receiver. The RS–232 standard also specifies other features for this communication protocol.

Parity: To further enhance data integrity during transmission, parity techniques may be used. Parity is an additional bit (or bits) that may be transmitted with the data byte. The ATmega328 and ATmega2560 employ a single parity bit. With a single parity bit, a single bit error may be detected. Parity may be even or odd. In even parity, the parity bit is set to one or zero such that the number of ones in the data byte including the parity bit is even. In odd parity, the parity bit is set to one or zero such that the number of ones in the data byte including the parity bit is odd. At the receiver, the number of bits within a data byte including the parity bit are counted to insure that parity has not changed, indicating an error, during transmission.

ASCII: The American Standard Code for Information Interchange or ASCII is a standardized, seven bit method of encoding alphanumeric data. It has been in use for many decades, so some of the characters and actions listed in the ASCII table are not in common use today. However, ASCII is still the most common method of encoding alphanumeric data. The ASCII code is provided in Figure 8.1. For example, the capital letter "G" is encoded in ASCII as 0x47. The "0x" symbol indicates the hexadecimal number representation. Unicode is the international counterpart of ASCII. It provides standardized 16–bit encoding format for the written languages of the world. ASCII is a subset of Unicode. The interested reader is referred to the Unicode home page website, www.unicode.org, for additional information on this standardized encoding format.

8.4   SERIAL USART

The serial USART (or Universal Synchronous and Asynchronous Serial Receiver and Transmitter) provide for full duplex (two way) communication between a receiver and transmitter. This is accomplished by equipping the ATmega328 and ATmega2560 with independent hardware for the transmitter and receiver. The ATmega328 is equipped with a USART channel while the ATmega2560

		Most significant digit							
		0x0_	0x1_	0x2_	0x3_	0x4_	0x5_	0x6_	0x7_
	0x_0	NUL	DLE	SP	0	@	P	`	p
	0x_1	SOH	DC1	!	1	A	Q	a	q
	0x_2	STX	DC2	"	2	B	R	b	r
	0x_3	ETX	DC3	#	3	C	S	c	s
	0x_4	EOT	DC4	$	4	D	T	d	t
	0x_5	ENQ	NAK	%	5	E	U	e	u
Least significant digit	0x_6	ACK	SYN	&	6	F	V	f	v
	0x_7	BEL	ETB	'	7	G	W	g	w
	0x_8	BS	CAN	(	8	H	X	h	x
	0x_9	HT	EM	)	9	I	Y	i	y
	0x_A	LF	SUB	*	:	J	Z	j	z
	0x_B	VT	ESC	+	;	K	[	k	{
	0x_C	FF	FS	'	<	L	\	l	|
	0x_D	CR	GS	-	=	M	]	m	}
	0x_E	SO	RS	.	>	N	^	n	~
	0x_F	SI	US	/	?	O	_	o	DEL

Figure 8.1: ASCII Code. The ASCII code is used to encode alphanumeric characters. The "0x" indicates hexadecimal notation in the C programming language.

is equipped with four USART channels. The USART is typically used for asynchronous communication. That is, there is not a common clock between the transmitter and receiver to keep them synchronized with one another. To maintain synchronization between the transmitter and receiver, framing start and stop bits are used at the beginning and end of each data byte in a transmission sequence. The Atmel USART also has synchronous features. Space does not permit a discussion of these USART enhancements.

The ATmega USART is quite flexible. It has the capability to be set to a variety of data transmission or Baud (bits per second) rates. The USART may also be set for data bit widths of 5 to 9 bits with one or two stop bits. Furthermore, the ATmega USART is equipped with a hardware generated parity bit (even or odd) and parity check hardware at the receiver. A single parity bit allows for the detection of a single bit error within a byte of data. The USART may also be configured to operate in a synchronous mode. We now discuss the operation, programming, and application of the USART. Due to space limitations, we cover only the most basic capability of this flexible and powerful serial communication system.

8.4.1   SYSTEM OVERVIEW

The block diagram for the USART is provided in Figure 8.2. The block diagram may appear a bit overwhelming but realize there are four basic pieces to the diagram: the clock generator, the transmission hardware, the receiver hardware, and three control registers (UCSRA, UCSBR, and UCSRC). We discuss each in turn.

8.4.1.1 USART Clock Generator

The USART Clock Generator provides the clock source for the USART system and sets the Baud rate for the USART. The Baud Rate is derived from the overall microcontroller clock source. The overall system clock is divided by the USART Baud rate Registers UBRR[H:L] and several additional dividers to set the Baud rate. For the asynchronous normal mode (U2X bit = 0), the Baud Rate is determined using the following expression:

$$Baud\ rate\ =\ (system\ clock\ frequency)/(16(UBRR\ +\ 1))$$

where UBRR is the contents of the UBRRH and UBRRL registers (0 to 4095). Solving for UBRR yields:

$$UBRR\ =\ ((system\ clock\ generator)/(16\ \times\ Baud\ rate))\ -\ 1$$

8.4.1.2 USART Transmitter

The USART transmitter consists of a Transmit Shift Register. The data to be transmitted is loaded into the Transmit Shift Register via the USART I/O Data Register (UDR). The start and stop framing bits are automatically appended to the data within the Transmit Shift Register. The parity is automatically calculated and appended to the Transmit Shift Register. Data is then shifted out of the Transmit Shift Register via the TxD pin a single bit at a time at the established Baud rate. The USART transmitter is equipped with two status flags: the UDRE and the TXC. The USART Data Register Empty (UDRE) flag sets when the transmit buffer is empty indicating it is ready to receive new data. This bit should be written to a zero when writing the USART Control and Status Register A (UCSRA). The UDRE bit is cleared by writing to the USART I/O Data Register (UDR). The Transmit Complete (TXC) Flag bit is set to logic one when the entire frame in the Transmit Shift Register has been shifted out and there are no new data currently present in the transmit buffer. The TXC bit may be reset by writing a logic one to it.

8.4.1.3 USART Receiver

The USART Receiver is virtually identical to the USART Transmitter except for the direction of the data flow is reversed. Data is received a single bit at a time via the RxD pin at the established Baud Rate. The USART Receiver is equipped with the Receive Complete (RXC) Flag. The RXC flag is logic one when unread data exists in the receive buffer.

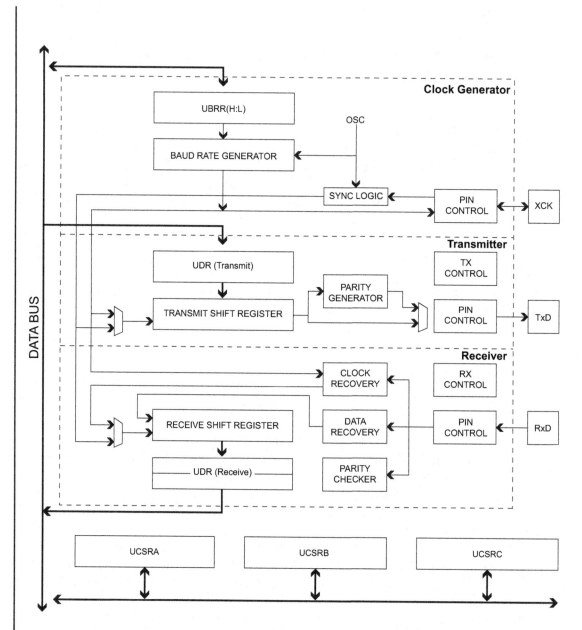

Figure 8.2: Atmel AVR ATmega USART block diagram. (Figure used with permission of Atmel, Incorporated.)

8.4.1.4 USART Registers

In this section, we discuss the register settings for controlling the USART system. We have already discussed the function of the USART I/O Data Register (UDR) and the USART Baud Rate Registers (UBRRH and UBRRL). **Note:** The USART Control and Status Register C (UCSRC) and the USART Baud Rate Register High (UBRRH) are assigned to the same I/O location in the memory map. The URSEL bit (bit 7 of both registers) determine which register is being accessed. The URSEL bit must be one when writing to the UCSRC register and zero when writing to the UBRRH register.

   Note: As previously mentioned, the ATmega328 is equipped with a USART channel while the ATmega2560 is equipped with four USART channels. The registers to configure the ATmega328 is provided in Figure 8.3. The same register configuration is used to individually configure each of the USART channels aboard the ATmega2560. The registers to configure the ATmega2560 is provided in Figure 8.4. Note that each register name has an "n" designator where "n" is the specific USART channel (0, 1, 2, or 3) being configured.

USART Control and Status Register A (UCSRA)   The UCSRA register contains the RXC, TXC, and the UDRE bits. The function of these bits have already been discussed.

USART Control and Status Register B (UCSRB)   The UCSRB register contains the Receiver Enable (RXEN) bit and the Transmitter Enable (TXEN) bit. These bits are the "on/off" switch for the receiver and transmitter, respectively. The UCSRB register also contains the UCSZ2 bit. The UCSZ2 bit in the UCSRB register and the UCSZ[1:0] bits contained in the UCSRC register together set the data character size.

USART Control and Status Register C (UCSRC)   The UCSRC register allows the user to customize the data features to the application at hand. It should be emphasized that both the transmitter and receiver be configured with the same data features for proper data transmission. The UCSRC contains the following bits:

- USART Mode Select (UMSEL) – 0: asynchronous operation, 1: synchronous operation

- USART Parity Mode (UPM[1:0])– 00: no parity, 10: even parity, 11: odd parity

- USART Stop Bit Select (USBS) – 0: 1 stop bit, 1: 2 stop bits

- USART Character Size (data width) (UCSZ[2:0]) – 000: 5–bit, 001: 6–bit, 010: 7–bit, 011: 8–bit, 111: 9–bit

8.5   SYSTEM OPERATION AND PROGRAMMING USING ARDUINO DEVELOPMENT ENVIRONMENT FEATURES

The Arduino Development Environment is equipped with built–in USART communications features.

USART Control and Status Register A (UCSRA)

RXC	TXC	UDRE	FE	DOR	PE	U2X	MPCM
7							0

USART Control and Status Register B (UCSRA)

RXCIE	TXCIE	UDRIE	RXEN	TXEN	UCSZ2	RXB8	TXB8
7							0

USART Control and Status Register C (UCSRC)

URSEL=1	UMSEL	UPM1	UPM0	USBS	UCSZ1	UCSZ0	UCPOL
7							0

USART Data Register - UDR
UDR(Read)

RXB7	RXB6	RXB5	RXB4	RXB3	RXB2	RXB1	RXB0

UDR(Write)

TXB7	TXB6	TXB5	TXB4	TXB3	TXB2	TXB1	TXB0
7							0

USART Baud Rate Registers - UBRRH and UBRRL
UBRRH

URSEL=0	---	---	---	UBRR11	UBRR10	UBRR9	UBRR8

UBRRL

UBRR7	UBRR6	UBRR5	UBRR4	UBRR3	UBRR2	UBRR1	UBRR0
7							0

Figure 8.3: ATmega328 USART Registers.

Arduino UNO R3: The Arduino UNO R3 provides access to the USART transmission (TX) and reception (RX) pins via DIGITAL pins 1 (TX) and 0 (RX). The USART may be connected to a an external USART compatible input device, output device, or microcontroller. The Arduino UNO R3 may also communicate with the host personal computer (PC) via the USB cable.

Arduino Mega 2560: The Arduino Mega 2560 provides access to the following USART channels:

- Channel 1: transmission (TX) on pin 18 and reception (RX) on pin 19.

- Channel 2: transmission (TX) on pin 16 and reception (RX) on pin 17.

- Channel 3: transmission (TX) on pin 14 and reception (RX) on pin 15.

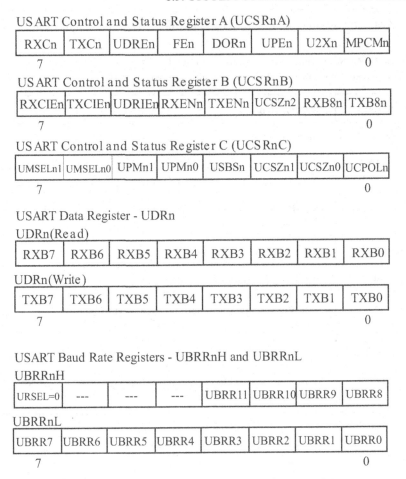

USART Control and Status Register A (UCSRnA)

RXCn	TXCn	UDREn	FEn	DORn	UPEn	U2Xn	MPCMn
7							0

USART Control and Status Register B (UCSRnB)

RXCIEn	TXCIEn	UDRIEn	RXENn	TXENn	UCSZn2	RXB8n	TXB8n
7							0

USART Control and Status Register C (UCSRnC)

UMSELn1	UMSELn0	UPMn1	UPMn0	USBSn	UCSZn1	UCSZn0	UCPOLn
7							0

USART Data Register - UDRn
UDRn(Read)

RXB7	RXB6	RXB5	RXB4	RXB3	RXB2	RXB1	RXB0

UDRn(Write)

TXB7	TXB6	TXB5	TXB4	TXB3	TXB2	TXB1	TXB0
7							0

USART Baud Rate Registers - UBRRnH and UBRRnL
UBRRnH

URSEL=0	---	---	---	UBRR11	UBRR10	UBRR9	UBRR8

UBRRnL

UBRR7	UBRR6	UBRR5	UBRR4	UBRR3	UBRR2	UBRR1	UBRR0
7							0

Figure 8.4: ATmega2560 USART Registers.

The Arduino processor pins are configured for TTL compatible inputs and outputs. That is, logic highs and lows are represented with 5 VDC and 0 VDC signals, respectively. The TX and RX pins are **not** compatible with RS–232 signals. A level shifter such as the MAX232 is required between the Arduino processor and the RS–232 device for communications of this type. When connected to a PC via the USB cable, appropriate level shifting is accomplished via the USB support chip onboard the Arduino processing board.

The Arduino Development Environment commands to provide USART communications is provided in Figure 8.5. A brief description of each command is provided. As before, we will not duplicate the excellent source material and examples provided at the Arduino homepage (www.Arduino.cc).

Arduino Development Environment built-in USART commands [www.Arduino.cc]

Notes:
- **Arduino UNO R3:** omit the "n" from the command
- **Arduino Mega 2560:** insert the USART channel in use (1, 2, or 3) for "n"

Command	Description
Serialn.begin()	Sets Baud rate
Serialn.end()	Disables serial communication. Allows Digital 1(TX) and Digital (0) RX to be used for digital input and output.
Serialn.available()	Determines how many bytes have already been received and stored in the 128 byte buffer.
Serialn.read()	Reads incoming serial data.
Serialn.flush()	Flushes the serial receive buffer of data.
Serialn.print()	Prints data to the serial port as ASCII text. An optional second parameter specifies the format for printing (BYTE, BIN, OCT, DEC, HEX).
Serialn.println()	Prints data to the serial port as ASCII text followed by a carriage return.
Serialn.write()	Writes binary data to the serial port. A single byte, a series of bytes, or an array of bytes may be sent.

Figure 8.5: Arduino Development Environment USART commands.

Example: To illustrate the use of the Arduino Development Environment built–in serial functions using the USART, we will add code to the robot sketch to provide status updates to the host PC. These status updates are a helpful aid during algorithm development. Due to limited space, we will not provide the entire algorithm here (it is five pages long). Instead, we provide a code snapshot that may be used to modify the remaining code.

In the code snapshot, we have included the "Serial.begin(9600)" command in the setup function to set the USART Baud rate at 9600. We have also inserted a " Serial.println" command in the algorithm to provide a status update. These status updates are handy during sketch development. These status updates would not be available while the robot is progressing through the maze since the robot would no longer be connected to the host PC via the USB cable.

```
//***********************************************************************
                                          //analog input pins
#define left_IR_sensor    A0              //analog pin - left IR sensor
```

```
#define center_IR_sensor A1        //analog pin - center IR sensor
#define right_IR_sensor  A2        //analog pin - right IR sensor

                                   //digital output pins
                                   //LED indicators - wall detectors
#define wall_left        3         //digital pin - wall_left
#define wall_center      4         //digital pin - wall_center
#define wall_right       5         //digital pin - wall_right

                                   //LED indicators - turn signals
#define left_turn_signal 2         //digital pin - left_turn_signal
#define right_turn_signal 6        //digital pin - right_turn_signal

                                   //motor outputs
#define left_motor       11        //digital pin - left_motor
#define right_motor      10        //digital pin - right_motor


int left_IR_sensor_value;          //declare var. for left IR sensor
int center_IR_sensor_value;        //declare var. for center IR sensor
int right_IR_sensor_value;         //declare var. for right IR sensor

void setup()
  {
  Serial.begin(9600);              //set USART Baud rate to 9600
                                   //LED indicators - wall detectors
  pinMode(wall_left,    OUTPUT);   //configure pin 1 for digital output
  pinMode(wall_center,  OUTPUT);   //configure pin 2 for digital output
  pinMode(wall_right,   OUTPUT);   //configure pin 3 for digital output

                                   //LED indicators - turn signals
  pinMode(left_turn_signal,OUTPUT);  //configure pin 0 for digital output
  pinMode(right_turn_signal,OUTPUT); //configure pin 4 for digital output

                                   //motor outputs - PWM
  pinMode(left_motor,   OUTPUT);   //configure pin 11 for digital output
  pinMode(right_motor,  OUTPUT);   //configure pin 10 for digital output
  }
```

```
void loop()
  {
                                       //read analog output from IR sensors
    left_IR_sensor_value   = analogRead(left_IR_sensor);
    center_IR_sensor_value = analogRead(center_IR_sensor);
    right_IR_sensor_value  = analogRead(right_IR_sensor);

    //robot action table row 0
    if((left_IR_sensor_value < 512)&&(center_IR_sensor_value < 512)&&
       (right_IR_sensor_value < 512))
      {
      Serial.println(''No walls detected'');   //print status
                                               //wall detection LEDs
      digitalWrite(wall_left,   LOW);          //turn LED off
      digitalWrite(wall_center, LOW);          //turn LED off
      digitalWrite(wall_right,  LOW);          //turn LED off
                                               //motor control
      analogWrite(left_motor,  128);
        //0 (off) to 255 (full speed)
      analogWrite(right_motor, 128);
        //0 (off) to 255 (full speed)
                                               //turn signals
      digitalWrite(left_turn_signal,  LOW);    //turn LED off
      digitalWrite(right_turn_signal, LOW);    //turn LED off
      delay(500);                              //delay 500 ms
      digitalWrite(left_turn_signal,  LOW);    //turn LED off
      digitalWrite(right_turn_signal, LOW);    //turn LED off
      delay(500);                              //delay 500 ms
      digitalWrite(left_turn_signal,  LOW);    //turn LED off
      digitalWrite(right_turn_signal, LOW);    //turn LED off
      delay(500);                              //delay 500 ms
      digitalWrite(left_turn_signal,  LOW);    //turn LED off
      digitalWrite(right_turn_signal, LOW);    //turn LED off
      analogWrite(left_motor, 0);              //turn motor off
      analogWrite(right_motor,0);              //turn motor off
      }

      :
```

8.6   SYSTEM OPERATION AND PROGRAMMING IN C

The basic activities of the USART system consist of initialization, transmission, and reception. These activities are summarized in Figure 8.6. Both the transmitter and receiver must be initialized with the same communication parameters for proper data transmission. The transmission and reception activities are similar except for the direction of data flow. In transmission, we monitor for the UDRE flag to set indicating the data register is empty. We then load the data for transmission into the UDR register. For reception, we monitor for the RXC bit to set indicating there is unread data in the UDR register. We then retrieve the data from the UDR register.

Note: As previously mentioned, the ATmega328 is equipped with a USART channel while the ATmega2560 is equipped with four USART channels. The registers to configure the ATmega328 is provided in Figure 8.3. The same register configuration is used to individually configure each of the USART channels aboard the ATmega2560. The registers to configure the ATmega2560 is provided in Figure 8.4. Note that each register name has an "n" designator where "n" is the specific USART channel (0, 1, 2, or 3) being configured.

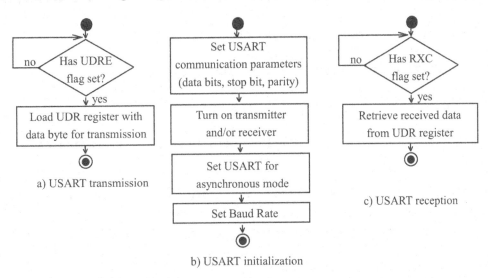

Figure 8.6: USART Activities.

To program the USART, we implement the flow diagrams provided in Figure 8.6. In the sample code provided, we assume the ATmega328 is operating at 10 MHz, and we desire a Baud Rate of 9600, asynchronous operation, no parity, one stop bit, and eight data bits.

To achieve 9600 Baud with an operating frequency of 10 MHz requires that we set the UBRR registers to 64 which is 0x40.

```
//************************************************************************
//USART_init: initializes the USART system
```

```
//***************************************************************************

void USART_init(void)
{
UCSRA = 0x00;                  //control register initialization
UCSRB = 0x08;                  //enable transmitter
UCSRC = 0x86;                  //async, no parity, 1 stop bit, 8 data bits
                               //Baud Rate initialization
UBRRH = 0x00;
UBRRL = 0x40;
}


//***************************************************************************
//USART_transmit: transmits single byte of data
//***************************************************************************

void USART_transmit(unsigned char data)
{
while((UCSRA & 0x20)==0x00)  //wait for UDRE flag
  {
  ;
  }
UDR = data;                    //load data to UDR for transmission
}


//***************************************************************************
//USART_receive: receives single byte of data
//***************************************************************************

unsigned char USART_receive(void)
{
while((UCSRA & 0x80)==0x00)  //wait for RXC flag
  {
  ;
  }
data = UDR;                    //retrieve data from UDR
return data;
}
```

//***

8.6.1   SERIAL PERIPHERAL INTERFACE–SPI

The ATmega Serial Peripheral Interface or SPI also provides for two–way serial communication between a transmitter and a receiver. In the SPI system, the transmitter and receiver share a common clock source. This requires an additional clock line between the transmitter and receiver but allows for higher data transmission rates as compared to the USART. The SPI system allows for fast and efficient data exchange between microcontrollers or peripheral devices. There are many SPI compatible external systems available to extend the features of the microcontroller. For example, a liquid crystal display or a digital–to–analog converter could be added to the microcontroller using the SPI system.

8.6.1.1   SPI Operation

The SPI may be viewed as a synchronous 16–bit shift register with an 8–bit half residing in the transmitter and the other 8–bit half residing in the receiver as shown in Figure 8.7. The transmitter is designated the master since it is providing the synchronizing clock source between the transmitter and the receiver. The receiver is designated as the slave. A slave is chosen for reception by taking its Slave Select ($\overline{SS}$) line low. When the $\overline{SS}$ line is taken low, the slave's shifting capability is enabled.

SPI transmission is initiated by loading a data byte into the master configured SPI Data Register (SPDR). At that time, the SPI clock generator provides clock pulses to the master and also to the slave via the SCK pin. A single bit is shifted out of the master designated shift register on the Master Out Slave In (MOSI) microcontroller pin on every SCK pulse. The data is received at the MOSI pin of the slave designated device. At the same time, a single bit is shifted out of the Master In Slave Out (MISO) pin of the slave device and into the MISO pin of the master device. After eight master SCK clock pulses, a byte of data has been exchanged between the master and slave designated SPI devices. Completion of data transmission in the master and data reception in the slave is signaled by the SPI Interrupt Flag (SPIF) in both devices. The SPIF flag is located in the SPI Status Register (SPSR) of each device. At that time, another data byte may be transmitted.

8.6.1.2   Registers

The registers for the SPI system are provided in Figure 8.8. We will discuss each one in turn.

SPI Control Register (SPCR)   The SPI Control Register (SPCR) contains the "on/off" switch for the SPI system. It also provides the flexibility for the SPI to be connected to a wide variety of devices with different data formats. It is important that both the SPI master and slave devices be configured for compatible data formats for proper data transmission. The SPCR contains the following bits:

- SPI Enable (SPE) is the "on/off" switch for the SPI system. A logic one turns the system on and logic zero turns it off.

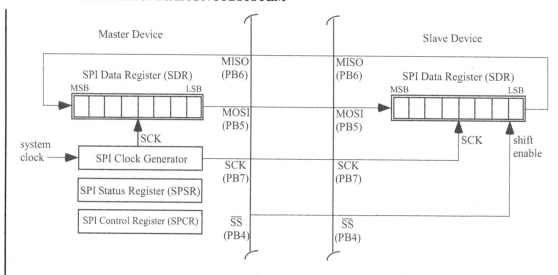

Figure 8.7: SPI Overview.

SPI Control Register - SPCR

SPIE	SPE	DORD	MSTR	CPOL	CPHA	SPR1	SPR0
7							0

SPI Status Register - SPSR

SPIF	WCOL	---	---	---	---	---	SPI2X
7							0

SPI Data Register - SPDR

MSB							LSB
7							0

Figure 8.8: SPI Registers

- Data Order (DORD) allows the direction of shift from master to slave to be controlled. When the DORD bit is set to one, the least significant bit (LSB) of the SPI Data Register (SPDR) is transmitted first. When the DORD bit is set to zero the Most Significant Bit (MSB) of the SPDR is transmitted first.

- The Master/Slave Select (MSTR) bit determines if the SPI system will serve as a master (logic one) or slave (logic zero).

- The Clock Polarity (CPOL) bit allows determines the idle condition of the SCK pin. When CPOL is one, SCK will idle logic high; whereas, when CPOL is zero, SCK will idle logic zero.

- The Clock Phase (CPHA) determines if the data bit will be sampled on the leading (0) or trailing (1) edge of the SCK.

- The SPI SCK is derived from the microcontroller's system clock source. The system clock is divided down to form the SPI SCK. The SPI Clock Rate Select bits SPR[1:0] and the Double SPI Speed Bit (SPI2X) are used to set the division factor. The following divisions may be selected using SPI2X, SPR1, SPR0:

  - 000: SCK = system clock/4
  - 001: SCK = system clock/16
  - 010: SCK = system clock/64
  - 011: SCK = system clock/1284
  - 100: SCK = system clock/2
  - 101: SCK = system clock/8
  - 110: SCK = system clock/32
  - 111: SCK = system clock/64

SPI Status Register (SPSR)   The SPSR contains the SPI Interrupt Flag (SPIF). The flag sets when eight data bits have been transferred from the master to the slave. The SPIF bit is cleared by first reading the SPSR after the SPIF flag has been set and then reading the SPI Data Register (SPDR). The SPSR also contains the SPI2X bit used to set the SCK frequency.

SPI Data Register (SPDR)   As previously mentioned, writing a data byte to the SPDR initiates SPI transmission.

8.7 SPI PROGRAMMING IN THE ARDUINO DEVELOPMENT ENVIRONMENT

The Arduino Development Environment provides the "shiftOut" command to provide ISP style serial communications [www.Arduino.cc]. The shiftOut command requires four parameters when called:

- **dataPin:** the Arduino UNO R3 DIGITAL pin to be used for serial output.

- **clockPin:** the Arduino UNO R3 DIGITAL pin to be used for the clock.

- **bitOrder:** indicates whether the data byte will be sent most significant bit first (MSBFIRST) or least significant bit first (LSBFIRST).

- **value:** the data byte that will be shifted out.

To use the shiftOut command, the appropriate pins are declared as output using the pinMode command in the setup() function. The shiftOut command is then called at the appropriate place within the loop() function using the following syntax:

```
shiftOut(dataPin, clockPin, LSBFIRST, value);
```

As a result of the this command, the value specified will be serially shifted out of the data pin specified, least significant bit first, at the clock rate provided at the clock pin.

8.8 SPI PROGRAMMING IN C

To program the SPI system in C, the system must first be initialized with the desired data format. Data transmission may then commence. Functions for initialization, transmission and reception are provided below. In this specific example, we divide the clock oscillator frequency by 128 to set the SCK clock frequency.

```
//*****************************************************************************
//spi_init: initializes spi system
//*****************************************************************************

void spi_init(unsigned char control)
{
DDRB = 0xA0;              //Set SCK (PB7), MOSI (PB5) for output,
                         //others to input
SPCR = 0x53;             //Configure SPI Control Register (SPCR)
//SPIE:0,SPE:1,DORD:0,MSTR:1,CPOL:0,CPHA:0,SPR:1,SPR0:1
}


//*****************************************************************************
//spi_write: Used by SPI master to transmit a data byte
//*****************************************************************************

void spi_write(unsigned char byte)
{
SPDR = byte;
while (!(SPSR & 0x80));
}


//*****************************************************************************
//spi_read: Used by SPI slave to receive data byte
```

```
//***********************************************************************

unsigned char spi_read(void)
{
while (!(SPSR & 0x80));

return SPDR;
}


//***********************************************************************
```

8.9   TWO–WIRE SERIAL INTERFACE–TWI

The TWI subsystem [1] allows the system designer to network a number of related devices (microcontrollers, transducers, displays, memory storage, etc.) together into a system using a two wire interconnecting scheme. The TWI allows a maximum of 128 devices to be connected together that reside in a small, circuit board size area. Each device has its own unique address and may both transmit and receive over the two wire bus at frequencies up to 400 kHz. This allows the device to freely exchange information with other devices in the network within a small area.

The TWI system consists of a two wire 100k bps (bit per second) bus. The 100k bps bus speed is termed the standard mode but the bus may also operate at higher data rates. There are multiple TWI compatible peripheral components (e.g. LCD displays, sensors, etc.) [I2C].

A large number of devices (termed nodes) may be connected to the TWI bus. The TWI system uses a standard protocol to allow the nodes to send and receive data from the other devices. All nodes on the bus are assigned a unique 7–bit address. The eighth bit of the address register is used to specify the operation to be performed (read or write). Additional devices may be added to the TWI based system as it evolves [I2C].

The basic TWI bus architecture is shown in Figure 8.9. The two wire bus consists of the serial clock line (SCL) and the serial data line (SDA). These lines are pulled up to logic high by the SCL and the SDA pull up resistors. Nodes connected to the bus can drive either of the bus lines to ground (logic 0). Devices within an TWI bus configuration must share a common ground [I2C].

8.9.1   ARDUINO DEVELOPMENT ENVIRONMENT

The Arduino Development Environment is equipped with the Wire Library. The Wire Library provides support for the TWI system. It allows communication between the Arduino processing board and TWI/I2C compatible devices.

The TWI may be accessed on the following Arduino pins:

- Arduino UNO R3: TWI SDA–pin A4, TWI SCL–Pin A5

---

[1]The TWI system is also referred to as I2C or IIC.

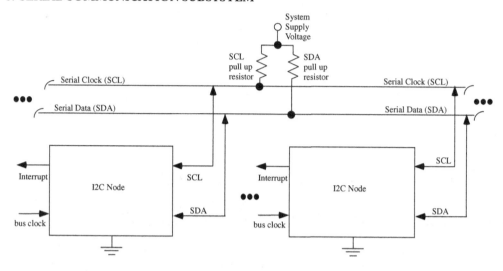

Figure 8.9: TWI configuration.

- Arduino Mega 2560: TWI SDA–pin 20, TWI SCL–Pin 21

Full documentation for this library is provided at the Arduino website (www.arduino.cc).

8.10 APPLICATION 1: USART COMMUNICATION WITH LCD

In Chapter 4 we discussed interfacing a parallel configured Liquid Crystal Display (LCD) to a processor. Typically a parallel connected LCD requires ten lines (8 data, 2 control). If processor pins are limited, a serial configured LCD may be employed. A serial configured LCD requires a single line for communication between the processor and LCD and a common ground. A serial connected LCD typically costs three times as much as a similarly featured parallel LCD.

In this application, we connect a Scott Edwards Electronics (www.seetron.com) ILM–216L integrated serial LCD module (2 by 16 character) to the Arduino UNO R3. The interface between the LCD and the Arduino board is illustrated in Figure 8.10. **Note:** The ILM–216L requires an inverted ASCII input. The ASCII character value may be inverted in software or hardware. In this example we use a 74HC04 hex inverter to invert the serial ASCII stream for the Arduino UNO R3.

A simple sketch to repeatedly print the letter to the LCD is provided below:

```
//************************************************************
void setup()
{
Serial.begin(2400);
delay(1000);
}
```

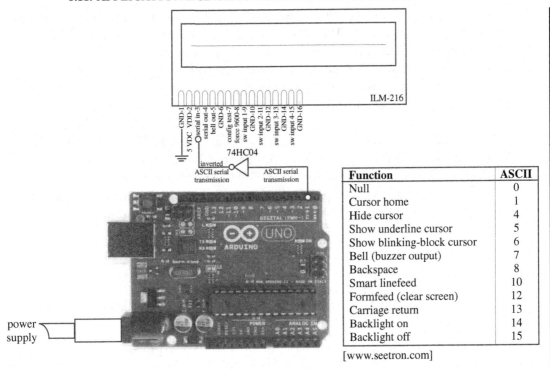

Function	ASCII
Null	0
Cursor home	1
Hide cursor	4
Show underline cursor	5
Show blinking-block cursor	6
Bell (buzzer output)	7
Backspace	8
Smart linefeed	10
Formfeed (clear screen)	12
Carriage return	13
Backlight on	14
Backlight off	15

[www.seetron.com]

Figure 8.10: Serial LCD display. The ILM–216L requires an inverted ASCII input. The ASCII character value may be inverted in software or hardware. In this example we use a 74HC04 hex inverter to invert the serial ASCII stream for the Arduino UNO R3. (UNO R3 illustration used with permission of the Arduino Team (CC BY–NC–SA) www.arduino.cc).

```
void loop()
{
Serial.print("G");
delay(500);
}
//*************************************************************
```

8.11   APPLICATION 2: SD/MMC CARD MODULE EXTENSION VIA THE USART

The Secure Digital/Multi Media Card (SD/MMC) provides a "hard drive" capability to the Arduino UNO R3. That is, it provides a large capacity storage media to log and retrieve data. SD/MMC cards have become a common method of storing data in commercial industry. The card is formatted using the File Allocation Table (FAT) 16 standard. This standard has been around for some time.

In this example, we show how to connect a Comfile Technology SD/MMC SD–COM5 card to the Arduino Microcontroller and the associated Arduino Development Environment commands required to interact with the card [www.comfiletech.com]. The commands will be passed to the SD/MMC card via the serial USART functions of the Arduino Development Environment. We also provide the commands for communicating with the SD/MMC via the C programming language.

The interface circuit between the Arduino UNO R3 and the SD/MMC card is provided in Figure 8.11. The TX and RX pins (DIGITAL 1 and 0) of the Arduino UNO R3 are connected to the RXD and TXD pins (19 and 20) of the SD/MMC card breakout board. Power and ground are also provided to the SD/MMC card as shown in the figure. Also, the reset pin of the SD/MMC breakout board (pin 15) is pulled up to the 5 VDC supply via a 10K resistor.

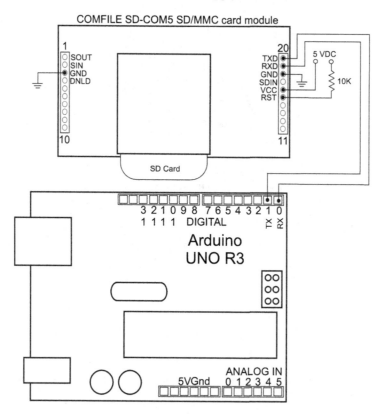

Figure 8.11: Arduino UNO R3 and SD/MMC card interface circuit [Comfile Technology].

Figure 8.12 provides a summary of commands to communicate with the SD/MMC card. The command format is shown at the top of the figure. The commands are issued from the Arduino

Command format: Command [Filename] [Option] [Data] [CR] [LF]	
In C: printf("fputs test.txt /w Hello World \r\n");	In Arduino Development Environment: serial.print("fputs test.txt /w Hello World \r\n");
Command	**Brief Description**
mode[Option][CR][LF]	Select mode of operation: /t terminal or /m MCU mode
init [CR][LF]	Initializes SD/MMC card
cd [Change Directory][CR][LF]	Change director
dir [CR][LF]	List directory
fsize [Filename][CR][LF]	Display file size
dsize[CR][LF]	Display SD/MMC disk space
ftime [Filename][CR][LF]	File creation and last modified time
md[Directory][CR][LF]	Make directory
rd[Directory][CR][LF]	Remove directory
del[Filename][CR][LF]	Delete file
fcreate[Filename][CR][LF]	Create a new file
rename[Source Filename][Destination Filename][CR][LF]	Rename the file
fopen[Filename][/Option][CR][LF]	Open the file: /r Read or /w Write or /a Append
fclose[CR][LF]	Close file
fputc[Filename][/Option][1 Byte Data][CR][LF]	Write a byte to file
fputs[Filename][/Option][String][CR][LF]	Write a string to file (limited 256 characters)
fwrite[/# of bytes to write][CR][LF]	Write up to 512 bytes to file
fgetc[/# of bytes to read][CR][LF]	Read up to 256 bytes from file
fgets[CR][LF]	Read one line of string
fread[Filename][CR][LF]	Read all data in file
reset[CR][LF]	Reset card
baud[Baud rate][CR][LF]	Set Baud rate
card[CR][LF]	Card status

Figure 8.12: SD/MMC commands [COMFILE Technology].

UNO R3 using the built–in "serial.print" command of the Arduino Development Environment. For example, to send the phrase "Hello World" to the SD/MMC the fputs command is used. The format of the command includes the file onboard the SD/MMC where the command should be stored and the file option. In this example we have used the "w" option to indicate a write to the file. The phrase for storage is then provided followed by a carriage return (\r) and a line feed (\n).

Before data can be written to the file, some preparatory steps are required:

- Set the Baud rate for communication.

- Set the SD/MMC for MCU (microcontroller) mode. This mode provides simplified responses back to the Arduino UNO R3.

- Initialize the SD/MMC card.

- Create a file.

Commands are provided for each of these actions in Figure 8.12. We will provide a complete sketch to communicate with the SD/MMC in the Applications section of the next chapter.

Once data has been written to an SD/MMC card, it may be removed from the card socket in the breakout board and read via a PC using a universal card reader. Universal card readers are readily available for under $20. This would make the SD/MMC useful for a remote data logging application. Once the data has been collected, the card may be accessed via the PC, the data pulled into a spreadsheet application such as MS Office Excel and analyzed.

8.12 APPLICATION 3: EQUIPPING AN ARDUINO PROCESSOR WITH A VOICE CHIP

A fun addition to an Arduino–based project is voice capability via a speech synthesis chip. For speech synthesis we use the SP0–512 text to speech chip (www.speechchips.com). The SP0–512 accepts an USART compatible serial text stream. The text stream is converted to phoneme codes used to generate an audio output. The chip requires a 9600 Baud bit stream with no parity, 8 data bits and a stop bit. Full documentation on commands, tones and features on the chip are available at www. speechchips.com.

The support circuit for the SP0–512 is provided in Figure 8.13. The Arduino UNO R3 issues ASCII text strings to the SP0–512 speech synthesizer. When the "Serial.println" command is used, a carriage return (CR) character is appended to the ASCII string. The SPO–512 uses the CR to begin processing the string. The UNO R3 provides TTL compatible logic levels (0 and 5 VDC); whereas, the SPO–512 operates a 3.3 VDC logic levels. Therefore, a 5 VDC–to–3.3 VDC level shifter is required between the UNO R3 and the SP0–512. The DAC output from the SP0–512 is fed to an audio amplifier. The audio amplifier employs an LM386N–3 (www.speechchips.com).

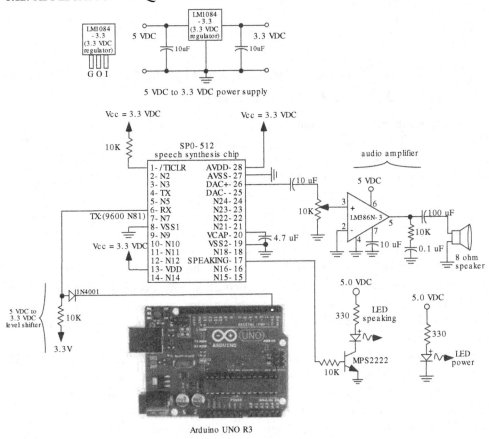

Figure 8.13: Speech synthesis support circuit [www.speechchips.com]. (UNO R3 illustration used with permission of the Arduino Team (CC BY–NC–SA) www.arduino.cc).

```
//***********************************************************
void setup()
{
Serial.begin(9600);                //set Baud rate 9600 bps
delay(1000);                       //delay 1s
}

void loop()
{
Serial.println("[v14][S4][E2] My name is HAL?");
delay(3000);                       //delay 3s
```

```
}
//***********************************************************
```

8.13   APPLICATION 4: PROGRAMMING THE ARDUINO UNO R3 ATMEGA328 VIA THE ISP

An alternate method of programming the Arduino UNO R3 processing board is via In–System Programming (ISP) techniques. We highly recommend that you use the Arduino Development Environment for programming the Arduino UNO R3. The ISP programming techniques are used to program features of the ATmega328P hosted onboard the Arduino UNO R3 that are not currently supported within the Arduino Development Environment.

Programming the ATmega328 requires several hardware and software tools. We briefly mention required components here. Please refer to the manufacturer's documentation for additional details at www.atmel.com.

Software Tools: Throughout the text, we use the ImageCraft ICC AVR compiler. This is a broadly used, user–friendly compiler. There are other excellent compilers available. The compiler is used to translate the source file (filename.c) into machine language for loading into the ATmega328 hosted onboard the Arduino UNO R3. We use Atmel's AVR Studio to load the machine code into the ATmega328.

Hardware Tools: We use Atmel's STK500 AVR Flash MCU Starter Kit (STK500) for programming the ATmega328. The STK500 provides the interface hardware between the host PC and the ATmega328 for machine code loading. The STK500 is equipped with a complement of DIP sockets which allows for programming all of the microcontrollers in the Atmel AVR line. The STK500 also allows for In–System Programming (ISP) [Atmel]. In this example, we use the ISP programming features of the STK500.

8.13.1   PROGRAMMING PROCEDURE

In this section, we provide a step–by–step procedure to program the ATmega328 hosted onboard the Arduino UNO R3 using the STK500 AVR Flash MCU Starter Kit. Please refer to Figure 8.14.

1. Load AVR Studio (free download from www.atmel.com).

2. Ensure that the STK500 is powered down.

3. Connect the STK500 as shown in Figure 8.14. Note: For ISP programming, the 6–wire ribbon cable is connected from the ISP6PIN header pin on the STK500 to the 6–pin header pin on the Arduino UNO R3. Note the position of the red guide wire in the diagram.

4. Power up the STK500.

5. Start up AVR Studio on your PC.

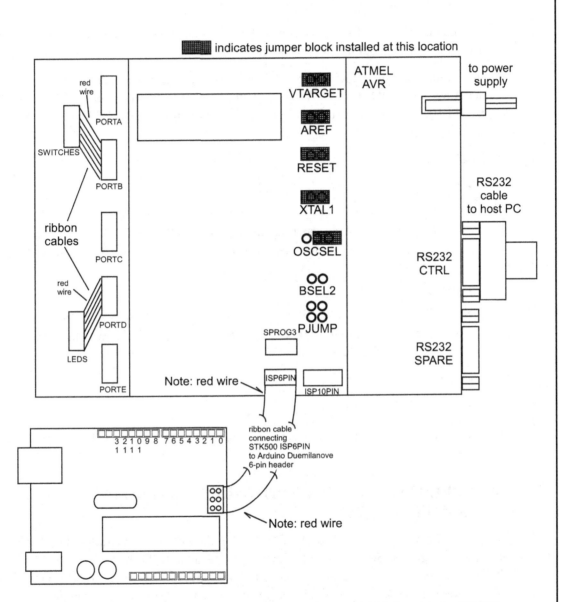

Figure 8.14: Programming the ATmega328 onboard the Arduino UNO R3 with the STK500.

6. Pop up window "Welcome to AVR Studio" should appear. Close this window by clicking on the "Cancel button."

7. Click on the "AVR icon." It looks like the silhouette of an integrated circuit. It is on the second line of the toolbar about half way across the screen.

8. This should bring up a STK500 pop up window with eight tabs (Main, Program, Fuses, Lockbits, Advanced, HW Settings, HW Info). At the bottom of the Main tab window, verify that the STK500 was autodetected. Troubleshoot as necessary to ensure STK500 was autodetected by AVR Studio.

9. Set all tab settings:

- Main:
    - Device and Signature Bytes: ATmega328P
    - Programming Mode and Target Setting: ISP Mode
    - Depress "Read Signature" to insure the STK500 is communicating with the Arduino UNO R3
- Program:
    - Flash: Input HEX file, Browse and find machine code file: <yourfilename.hex>
    - EEPROM: Input HEX file, Browse and find machine code file: <yourfilename.EEP>

10. Programming step:

- Program Tab: click program

11. Power down the STK500. Disconnect the STK500 from the Arduino UNO R3 processing board.

8.14 APPLICATION 5: TMS1803 3–BIT LED DRIVE CONTROLLER

Titan Micro Electronics manufactures the TMS1803 3–bit LED drive controller. The TMS1803 is illustrated in Figure 8.15a). It is used to control three separate LEDs (red, green, and blue). The TMS1803 has a serial data input, a serial data output, and three PWM outputs to independently set the illumination of each LED. Each LEDs' intensity is individually set with a value from 0 (low) to 256 (high). A 24-bit data stream is shifted into the TMS1803. When a new 24–bit data stream is shifted in the current 24–bit control value is shifted out. A series of TMS1803s may be connected as shown in Figure 8.15b)[Titan Micro].

RadioShack(www.RadioShack.com) distributes a 1m tricolor LED strip (#2760249). The strip is easily interfaced to the Arduino UNO R3:

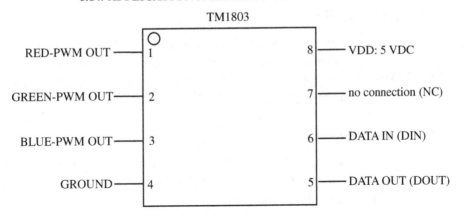

a) Titan Micro Electronics TM1803 3-bit LED drive controller

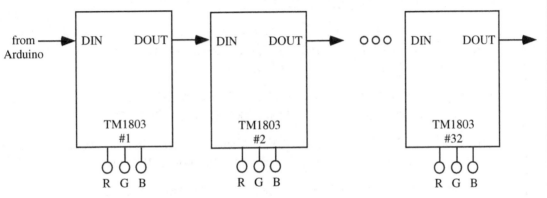

b) serially connected string of 32 TM1803 3-bit LED drive controllers

Figure 8.15: TMS1803 3–bit LED drive controller [Titan Micro Electronics].

- Tricolor +12V (red wire) to Arduino UNO R3 Vin

- Tricolor DIN (green wire) to Arduino UNO R3 A0

- Tricolor GND (black wire) to Arduino UNO R3 GND

The tricolor LED strip can be programmed for a number of special effects. They may also be connected together to form longer LED strings. A sample program, titled "strip_1m.pde," may be downloaded from the RadioShack website (www.RadioShack.com).

8.15 SUMMARY

In this chapter, we have discussed the differences between parallel and serial communications and key serial communication related terminology. We then in turn discussed the operation of USART, SPI and TWI serial communication systems. We also provided basic code examples to communicate with the USART and SPI systems.

8.16 REFERENCES

- *Atmel 8–bit AVR Microcontroller with 4/8/16/32K Bytes In–System Programmable Flash, ATmega48PA/88PA/168PA/328P* data sheet: 8161D–AVR–10/09, Atmel Corporation, 2325 Orchard Parkway, San Jose, CA 95131.

- *Atmel 8–bit AVR Microcontroller with 4/8/16/32K Bytes In–System Programmable Flash, ATmega48PA, 88PA, 168PA, 328P* data sheet: 8171D–AVR–05/11, Atmel Corporation, 2325 Orchard Parkway, San Jose, CA 95131.

- *Atmel 8–bit AVR Microcontroller with 64/128/256K Bytes In–System Programmable Flash, ATmega640/V, ATmega1280/V, 2560/V* data sheet: 2549P–AVR–10/2012, Atmel Corporation, 2325 Orchard Parkway, San Jose, CA 95131.

- Barrett S, Pack D (2006) Microcontrollers Fundamentals for Engineers and Scientists. Morgan and Claypool Publishers. DOI: 10.2200/S00025ED1V01Y200605DCS001

- Barrett S and Pack D (2008) Atmel AVR Microcontroller Primer Programming and Interfacing. Morgan and Claypool Publishers. DOI: 10.2200/S00100ED1V01Y200712DCS015

- Barrett S (2010) Embedded Systems Design with the Atmel AVR Microcontroller. Morgan and Claypool Publishers. DOI: 10.2200/S00225ED1V01Y200910DCS025

- Serial SD/MMC Card Module User Manual, Comfile Technology, Inc., www.comfiletech.com.

- *The I2C–Bus Specification.* Version 2.1, Philips Semiconductor, January 2000.

8.17 CHAPTER PROBLEMS

1. Summarize the differences between parallel and serial conversion.

2. Summarize the differences between the USART, SPI, and TWI methods of serial communication.

3. Draw a block diagram of the USART system, label all key registers, and all keys USART flags.

4. Draw a block diagram of the SPI system, label all key registers, and all keys USART flags.

5. If an ATmega328 microcontroller is operating at 12 MHz what is the maximum transmission rate for the USART and the SPI?

6. What is the ASCII encoded value for "Arduino"?

7. Draw the schematic of a system consisting of two ATmega328s that will exchange data via the SPI system.

8. Write the code to implement the system described in the question above.

9. In Chapter 5, Analog to Digital Conversion, a design was provided for a KNH Instrumentation Array using two large LED arrays. Re-accomplish the design employing two 1m Tricolor LED Strips (RadioShack #2760249). What advantage does this design have over the one provided in Chapter 5?

CHAPTER 9

Extended Examples

Objectives: After reading this chapter, the reader should be able to

- Construct an Arduino based system using concepts provided throughout the book.

9.1 OVERVIEW

This chapter provides a series of extended examples to illustrate concepts provided throughout the book.

9.2 EXTENDED EXAMPLE 1: AUTOMATED FAN COOLING SYSTEM

In this section, we describe an embedded system application to control the temperature of a room or some device. The system is illustrated in Figure 9.1. An LM34 temperature sensor (PORTC[0]) is used to monitor the instantaneous temperature of the room or device of interest. The current temperature is displayed on the Liquid Crystal Display (LCD).

We send a 1 KHz PWM signal to a cooling fan (M) whose duty cycle is set from 50% to 90% using the potentiometer connected to PORTC[2]. The PWM signal should last until the temperature of the LM34 cools to a value as set by another potentiometer (PORTC[1]). When the temperature of the LM34 falls below the set level, the cooling fan is shut off. If the temperature falls while the fan is active, the PWM signal should gently return to zero, and wait for further temperature changes.

Provided below is the embedded code for the system. This solution was developed by Geoff Luke, UW MSEE, as a laboratory assignment for an Industrial Control class.

```
//****************************************************************************
//Geoff Luke
//EE 5880 - Industrial Controls
//PWM Fan Control
//Last Updated: April 10, 2010
//****************************************************************************
//Description: This program reads the voltage from an LM34 temperature
//sensor then sends the corresponding temperature to an LCD.
//If the sensed temperature is greater than the temperature designated
//by a potentiometer, then a PWM signal is turned on to trigger a fan
```

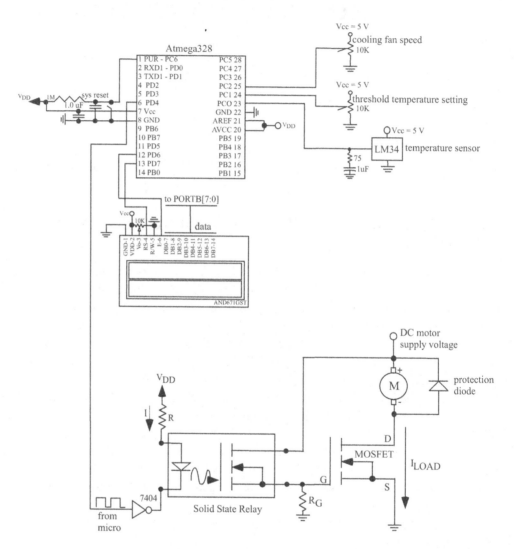

Figure 9.1: Automated fan cooling system.

```
//with duty cycle designated by another potentiometer.
//
//Ports:
//   PORTB[7:0]: data output to LCD
//   PORTD[7:6]: LCD control pins
//   PORTC[2:0]:
//   PORTC[0]: LM34 temperature sensor
//   PORTC[1]: threshold temperature
//   PORTC[2]: fan speed
//   PORTD[4]  : PWM channel B output
//
//****************************************************************************

//include files************************************************************
#include<iom328pv.h>

//function prototypes******************************************************
void initializePorts();
void initializeADC();
unsigned int readADC(unsigned char);
void LCD_init();
void putChar(unsigned char);
void putcommand(unsigned char);
void voltageToLCD(unsigned int);
void temperatureToLCD(unsigned int);
void PWM(unsigned int);
void delay_5ms();

int main(void)
{
unsigned int tempVoltage, tempThreshold;

initializePorts();
initializeADC();
LCD_init();

while(1)
  {
  tempVoltage = readADC(0);
```

```
   temperatureToLCD(tempVoltage);
   tempThreshold = readADC(1);
   if(tempVoltage > tempThreshold)
     {
     PWM(1);
     while(tempVoltage > tempThreshold)
       {
       tempVoltage = readADC(0);
       temperatureToLCD(tempVoltage);
       tempThreshold = readADC(1);
       }
     OCR1BL = 0x00;
     }
   }
return 0;
}


//***************************************************************************

void initializePorts()
{
DDRD = 0xFF;
DDRC = 0xFF;
DDRB = 0xFF;
}


//***************************************************************************

void initializeADC()
{
//select channel 0
ADMUX = 0;

//enable ADC and set module enable ADC and
//set module prescalar to 8
ADCSRA = 0xC3;

//Wait until conversion is ready
while(!(ADCSRA & 0x10));
```

```
//Clear conversion ready flag

ADCSRA |= 0x10;
}

//***********************************************************************

unsigned int readADC(unsigned char channel)
{
unsigned int binary_weighted_voltage, binary_weighted_voltage_low;
unsigned int binary_weighted_voltage_high; //weighted binary

ADMUX = channel;                       //Select channel
ADCSRA |= 0x43;                        //Start conversion

                                       //Set ADC module prescalar to 8
                                       //critical accurate ADC results
while (!(ADCSRA & 0x10));              //Check if conversion is ready
ADCSRA |= 0x10;                        //Clear conv rdy flag - set the bit

binary_weighted_voltage_low = ADCL;
 //Read 8 low bits first (important)
   //Read 2 high bits, multiply by 256
binary_weighted_voltage_high = ((unsigned int)(ADCH << 8));
binary_weighted_voltage = binary_weighted_voltage_low +
                      binary_weighted_voltage_high;

return binary_weighted_voltage;        //ADCH:ADCL
}

//***********************************************************************
//LCD_Init: initialization for an LCD connected in the following manner:
//LCD: AND671GST 1x16 character display
//LCD configured as two 8 character lines in a 1x16 array
//LCD data bus (pin 14-pin7) ATMEL ATmega16: PORTB
//LCD RS (pin 4) ATMEL ATmega16: PORTD[7]
//LCD E (pin 6) ATMEL ATmega16: PORTD[6]
//***********************************************************************
```

```
void LCD_init(void)
{
delay_5ms();
delay_5ms();
delay_5ms();

                                   // output command string to
                                   //initialize LCD
putcommand(0x38);                  //function set 8-bit
delay_5ms();
putcommand(0x38);                  //function set 8-bit
delay_5ms();
putcommand(0x38);                  //function set 8-bit
putcommand(0x38);                  //one line, 5x7 char
putcommand(0x0E);                  //display on
putcommand(0x01);                  //display clear-1.64 ms
putcommand(0x06);                  //entry mode set
putcommand(0x00);                  //clear display, cursor at home
putcommand(0x00);                  //clear display, cursor at home
}



//************************************************************************
//putchar:prints specified ASCII character to LCD
//************************************************************************

void putChar(unsigned char c)
{
DDRB = 0xff;        //set PORTB as output
DDRD = DDRD|0xC0;   //make PORTD[7:6] output
PORTB = c;
PORTD = PORTD|0x80; //RS=1
PORTD = PORTD|0x40; //E=1
PORTD = PORTD&0xbf; //E=0
delay_5ms();
}


//************************************************************************
```

```
//performs specified LCD related command
//************************************************************************

void putcommand(unsigned char d)
{
DDRB = 0xff;          //set PORTB as output
DDRD = DDRD|0xC0;     //make PORTD[7:6] output
PORTD = PORTD&0x7f;   //RS=0
PORTB = d;
PORTD = PORTD|0x40;   //E=1
PORTD = PORTD&0xbf;   //E=0
delay();
}



//************************************************************************
//delay
//************************************************************************

void delay(void)
{
unsigned int i;

for(i=0; i<2500; i++)
  {
  asm("nop");
  }
}

//************************************************************************
void voltageToLCD(unsigned int ADCValue)

{
float voltage;
unsigned int ones, tenths, hundredths;

voltage = (float)ADCValue*5.0/1024.0;

ones = (unsigned int)voltage;
```

```
tenths = (unsigned int)((voltage-(float)ones)*10);
hundredths = (unsigned int)(((voltage-(float)ones)*10-(float)tenths)*10);

putcommand(0x80);

putChar((unsigned char)(ones)+48);
putChar('.');
putChar((unsigned char)(tenths)+48);
putChar((unsigned char)(hundredths)+48);
putChar('V');
putcommand(0xC0);
}

//*****************************************************************************

void temperatureToLCD(unsigned int ADCValue)

{
float voltage,temperature;
unsigned int tens, ones, tenths;

voltage = (float)ADCValue*5.0/1024.0;
temperature = voltage*100;

tens = (unsigned int)(temperature/10);
ones = (unsigned int)(temperature-(float)tens*10);
tenths = (unsigned int)(((temperature-(float)tens*10)-(float)ones)*10);

putcommand(0x80);
putChar((unsigned char)(tens)+48);
putChar((unsigned char)(ones)+48);
putChar('.');
putChar((unsigned char)(tenths)+48);
putChar('F');
}

//*****************************************************************************

void PWM(unsigned int PWM_incr)
```

```
{

unsigned int fan_Speed_int;
float fan_Speed_float;
int PWM_duty_cycle;

fan_Speed_int = readADC(0x02); //fan Speed Setting

//unsigned int convert to max duty cycle setting:
//   0 VDC =  50% = 127,
//   5 VDC = 100% = 255

fan_Speed_float = ((float)(fan_Speed_int)/(float)(0x0400));

//convert volt to PWM constant 127-255
fan_Speed_int = (unsigned int)((fan_Speed_float * 127) + 128.0);

//Configure PWM clock

TCCR1A = 0xA1;                       //freq = resonator/510 = 4 MHz/510
                 //freq = 19.607 kHz
TCCR1B = 0x02;                             //clock source
                                           //division of 8: 980 Hz


   //Initiate PWM duty cycle variables
PWM_duty_cycle = 0;
OCR1BH = 0x00;
OCR1BL = (unsigned char)(PWM_duty_cycle);//set PWM duty cycle Ch B to 0%
                                         //Ramp up to fan Speed in 1.6s
OCR1BL = (unsigned char)(PWM_duty_cycle);//set PWM duty cycle Ch B

while (PWM_duty_cycle < fan_Speed_int)
  {
  if(PWM_duty_cycle < fan_Speed_int)     //increment duty cycle
  PWM_duty_cycle=PWM_duty_cycle + PWM_incr;
                                         //set PWM duty cycle Ch B
  OCR1BL = (unsigned char)(PWM_duty_cycle);
  }
}
```

```
//*****************************************************************************
```

9.3   EXTENDED EXAMPLE 2: FINE ART LIGHTING SYSTEM

In Chapter 2 and 5, we investigated an illumination system for a painting. The painting was to be illuminated via high intensity white LEDs. The LEDs could be mounted in front of or behind the painting as the artist desired. In Chapter 5, we equipped the lighting system with an IR sensor to detect the presence of someone viewing the piece. We also wanted to adjust the intensity of the lighting based on how the close viewer was to the art piece. In this example, we enhance the lighting system with three IR sensors. The sensors will determine the location of viewers and illuminate portions of the painting accordingly. The analogWrite function is used to set the intensity of the LEDs using PWM techniques.

The Arduino Development Environment sketch to sense how away the viewer is and issue a proportional intensity control signal to illuminate the LED is provided below. The analogRead function is used to obtain the signal from the IR sensor. The analogWrite function is used to issue a proportional signal.

```
//*****************************************************************************
                                        //analog input pins
#define left_IR_sensor    A0            //analog pin - left IR sensor
#define center_IR_sensor  A1            //analog pin - center IR sensor
#define right_IR_sensor   A2            //analog pin - right IR sensor

                                        //digital output pins
                                        //LED indicators - wall detectors
#define col_0_control     0             //digital pin - column 0 control
#define col_1_control     1             //digital pin - column 1 control
#define col_2_control     2             //digital pin - column 2 control

int left_IR_sensor_value;               //declare var. for left IR sensor
int center_IR_sensor_value;             //declare var. for center IR sensor

int right_IR_sensor_value;              //declare var. for right IR sensor

void setup()
  {
                                        //LED indicators - wall detectors
  pinMode(col_0_control, OUTPUT);       //configure pin 0 for digital output
  pinMode(col_1_control, OUTPUT);       //configure pin 1 for digital output
  pinMode(col_2_control, OUTPUT);       //configure pin 2 for digital output
```

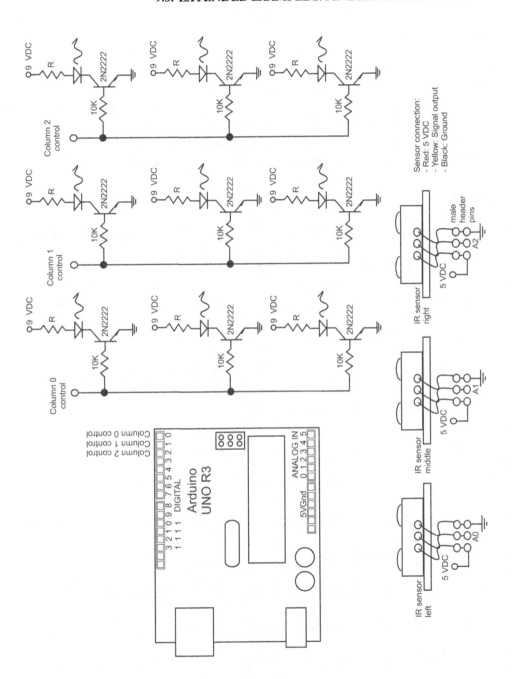

Figure 9.2: IR sensor interface.

```
  }

void loop()
  {
                                    //read analog output from IR sensors
  left_IR_sensor_value   = analogRead(left_IR_sensor);
  center_IR_sensor_value = analogRead(center_IR_sensor);
  right_IR_sensor_value  = analogRead(right_IR_sensor);

  if(left_IR_sensor_value < 128)
    {
    analogWrite(col_0_control, 31);   //0 (off) to 255 (full speed)
    }
  else if(left_IR_sensor_value < 256)
    {
    analogWrite(col_0_control, 63);   //0 (off) to 255 (full speed)
    }
  else if(left_IR_sensor_value < 384)
    {
    analogWrite(col_0_control, 95);   //0 (off) to 255 (full speed)
    }
  else if(left_IR_sensor_value < 512)
    {
    analogWrite(col_0_control, 127);  //0 (off) to 255 (full speed)
    }
  else if(left_IR_sensor_value < 640)
    {
    analogWrite(col_0_control, 159);  //0 (off) to 255 (full speed)
    }
  else if(left_IR_sensor_value < 768)
    {
    analogWrite(col_0_control, 191);  //0 (off) to 255 (full speed)
    }
  else if(left_IR_sensor_value < 896)
    {
    analogWrite(col_0_control, 223);  //0 (off) to 255 (full speed)
    }
  else
```

```
      {
      analogWrite(col_0_control, 255);  //0 (off) to 255 (full speed)
      }

if(center_IR_sensor_value < 128)
      {
      analogWrite(col_1_control, 31);  //0 (off) to 255 (full speed)
      }
   else if(center_IR_sensor_value < 256)
      {
      analogWrite(col_1_control, 63);  //0 (off) to 255 (full speed)
      }
   else if(center_IR_sensor_value < 384)
      {
      analogWrite(col_1_control, 95);  //0 (off) to 255 (full speed)
      }
   else if(center_IR_sensor_value < 512)
      {
      analogWrite(col_1_control, 127); //0 (off) to 255 (full speed)
      }
   else if(center_IR_sensor_value < 640)
      {
      analogWrite(col_1_control, 159); //0 (off) to 255 (full speed)
      }
   else if(center_IR_sensor_value < 768)
      {
      analogWrite(col_1_control, 191); //0 (off) to 255 (full speed)
      }
   else if(center_IR_sensor_value < 896)
      {
      analogWrite(col_1_control, 223); //0 (off) to 255 (full speed)
      }
   else
      {
      analogWrite(col_1_control, 255); //0 (off) to 255 (full speed)
      }

if(right_IR_sensor_value < 128)
      {
```

```
      analogWrite(col_2_control, 31);   //0 (off) to 255 (full speed)
      }
   else if(right_IR_sensor_value < 256)
      {
      analogWrite(col_2_control, 63);   //0 (off) to 255 (full speed)
      }
   else if(right_IR_sensor_value < 384)
      {
      analogWrite(col_2_control, 95);   //0 (off) to 255 (full speed)
      }
   else if(right_IR_sensor_value < 512)
      {
      analogWrite(col_2_control, 127); //0 (off) to 255 (full speed)
      }
   else if(right_IR_sensor_value < 640)
      {
      analogWrite(col_2_control, 159); //0 (off) to 255 (full speed)
      }
   else if(right_IR_sensor_value < 768)
      {
      analogWrite(col_2_control, 191);  //0 (off) to 255 (full speed)
      }
   else if(right_IR_sensor_value < 896)
      {
      analogWrite(col_2_control, 223);  //0 (off) to 255 (full speed)
      }
   else
      {
      analogWrite(col_2_control, 255);  //0 (off) to 255 (full speed)
      }

delay(500);                              //delay 500 ms
}


//********************************************************************
```

9.4   EXTENDED EXAMPLE 3: FLIGHT SIMULATOR PANEL

We close the chapter with an extended example of developing a flight simulator panel. This panel was actually designed and fabricated for a middle school to allow students to react to space mission like status while using a program that allowed them to travel about the planets. An Atmel ATmega328 microcontroller is used because its capabilities best fit the requirements for the project.

The panel face is shown in Figure 9.3. It consists of a joystick that is connected to a host computer for the flight simulator software. Below the joystick is a two line liquid crystal display equipped with a backlight LED (Hantronix HDM16216L–7, Jameco# 658988). Below the LCD is a buzzer to alert students to changing status. There are also a complement of other status indicators. From left–to–right is the Trip Duration potentiometer. At the beginning of the flight episode students are prompted to set this from 0 to 60 minutes to communicate to the microcontroller the length of the flight episode. This data is used to calculate different flight increments. There are also a series of simulated circuit breakers: system reset (SYS Reset), oxygen (O_2 CB), auxiliary fuel (AUX FUEL CB), and the main power circuit breakers (MAIN PWR CB). These are not actual circuit breakers but normally open (NO) single pole single throw (SPST) momentary pushbutton switches that allow the students to interact with the microcontroller. There are also a series of LEDs that form a Y pattern on the panel face. They are also used to indicate status changes.

To interface the flight simulator panel to the microcontroller a number of different techniques previously discussed in the book were employed. The interface diagram is shown in Figure 9.4. Pin 1 is a reset for the microcontroller. When the switch is depressed, pin 1 is taken low and resets the microcontroller. Port D of the microcontroller (pins 2–6, 11–13) forms the data connection for the LCD. Pin 9 is used to turn the buzzer on and off. This pin is routed through a transistor interface described earlier in the text. Port B[5:0] (pins 14–19) is used to control the LEDs on the front panel. Each individual LED is also supported with a transistor interface circuit. Conveniently, these small NPN signal transistors come in four to a 14 pin DIP package (MPQ2222). Port C [0] (pin 23) is used as an analog input pin. It is connected to the trip duration potentiometer. Port C pins [3:1] (pins 24 to 26) are used to connect the NO SPST tact switches to the microcontroller. Port C pins [4:5] (pins 27, 28) are used for the LCD control signals.

The software flowchart is shown in Figure 9.5. After startup the students are prompted via the LCD display to set the trip duration and then press the main power circuit breaker. This potentiometer setting is then used to calculate four different trip increments. Countdown followed by blastoff then commences. At four different trip time increments, students are presented with status that they must respond to. Real clock time is kept using the TCNT0 timer overflow configured as a 65.5 ms "clock tick." The overall time base for the microcontroller was its internal 1 MHz clock that may be selected during programming with the STK500.

Provided below is the code listing for the flight simulator panel.

```
//****************************************************************
//file name: flight_sim.c
//author: Steve Barrett, Ph.D., P.E.
//last revised: April 10, 2010
//function: Controls Flight Simulator Control Panel for Larimer County
//          School District #1
//
```

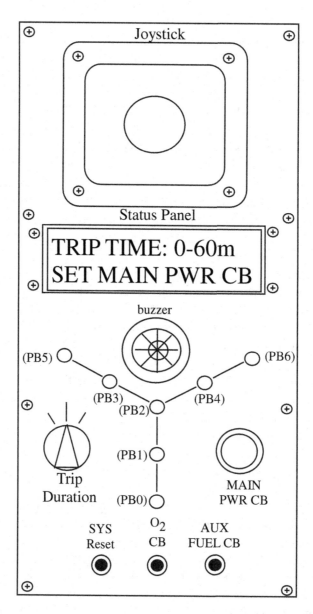

Figure 9.3: Flight Simulator Panel.

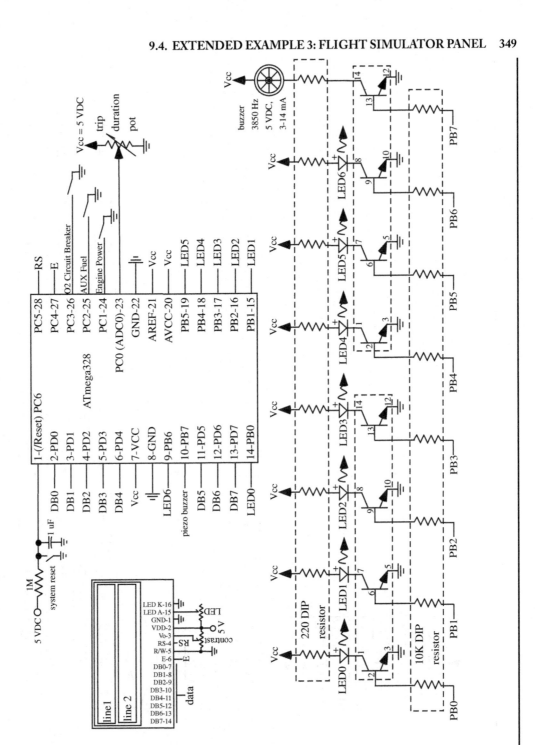

Figure 9.4: Interface diagram for the flight simulator panel.

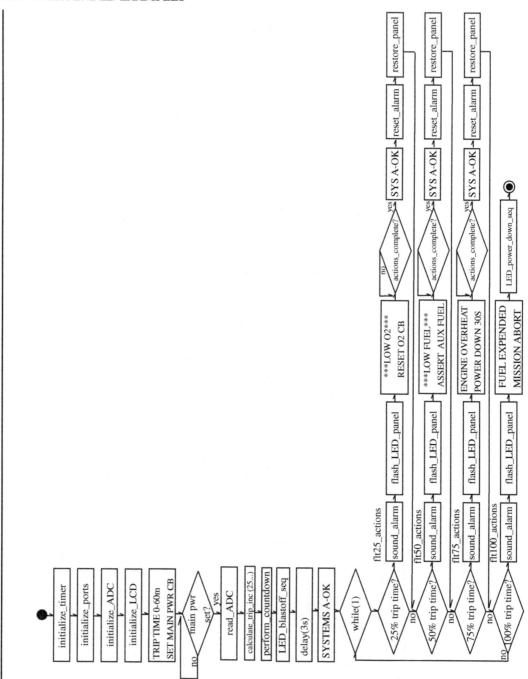

Figure 9.5: Software flow for the flight simulator panel.

```
//ATMEL AVR ATmega328
//Chip Port Function I/O Source/Dest Asserted Notes
//****************************************************************************
//Pin 1: /Reset
//Pin 2: PD0 to DB0 LCD
//Pin 3: PD1 to DB1 LCD
//Pin 4: PD2 to DB2 LCD
//Pin 5: PD3 to DB3 LCD
//Pin 6: PD4 to DB4 LCD
//Pin 7: Vcc
//Pin 8: Gnd
//Pin 9: PB6 to LED6
//Pin 10: PB7 to piezo buzzer
//Pin 11: PD5 to DB6 LCD
//Pin 12: PD6 to DB6 LCD
//Pin 13: PD7 to DB7 LCD
//Pin 14: PB0 to LED0
//Pin 15: PB1 to LED1
//Pin 16: PB2 to LED2
//Pin 17: PB3 to LED3
//Pin 18: PB4 to LED4
//Pin 19: PB5 to LED5
//Pin 20: AVCC to Vcc
//Pin 21: AREF to Vcc
//Pin 22 Gnd
//Pin 23 ADC0 to trip duration potentiometer
//Pin 24 PC1 Engine Power Switch
//Pin 25 PC2 AUX Fuel circuit breaker
//Pin 26 PC3 02 circuit breaker
//Pin 27 PC4 to LCD Enable (E)
//Pin 28 PC5 to LCD RS
//

//include files**************************************************************

//ATMEL register definitions for ATmega8
#include<iom328v.h>
```

```
//function prototypes*********************************************************
void delay(unsigned int number_of_65_5ms_interrupts);
void init_timer0_ovf_interrupt(void);
void InitADC(void);                        //initialize ADC
void initialize_ports(void);               //initializes ports
void power_on_reset(void);
     //returns system to startup state
unsigned int  ReadADC(unsigned char chan);//read value from ADC results
void clear_LCD(void);                      //clears LCD display
void LCD_Init(void);                       //initialize AND671GST LCD
void putchar(unsigned char c);             //send character to LCD
void putcommand(unsigned char c);          //send command to LCD
unsigned int  ReadADC(unsigned char chan);//read value from ADC results
void timer0_interrupt_isr(void);
void flt25_actions(void);
void flt50_actions(void);
void flt75_actions(void);
void flt100_actions(void);
void sound_alarm(void);
void turn_off_LEDs(void);
void reset_alarm(void);
void restore_panel(void);
void LED_blastoff_sequence(void);
void LED_power_down_sequence(void);
void monitor_main_power_CB(void);
void monitor_O2_CB_reset(void);
void monitor_aux_fuel_CB(void);
void perform_countdown(void);
void print_LOWO2(void);
void print_LOW_FUEL(void);
void print_fuel_expended(void);
void print_OVERHEAT(void);
void print_trip_dur(void);
void flash_LED_panel(void);
void clear_LCD(void);
void calculate_trip_int(void);
void systems_A_OK(void);

//program constants
```

```
#define TRUE     1
#define FALSE    0
#define OPEN     1
#define CLOSE    0
#define YES      1
#define NO       0
#define SAFE     1
#define UNSAFE   0
#define ON       1
#define OFF      0


//interrupt handler definition
#pragma interrupt_handler timer0_interrupt_isr:17


//main program***********************************************************


//global variables
unsigned int flt_25, flt_50, flt_75, flt_100;
unsigned int action25_done=NO, action50_done=NO;
unsigned int action75_done=NO, action100_done=NO;
unsigned int achieved25=NO, achieved50=NO;
unsigned int achieved75=NO, achieved100=NO;
unsigned int flt_timer=0;
unsigned int trip_duration_volt;
unsigned char PORTC_pullup_mask = 0x0e;
unsigned int flash_timer;
unsigned int PORTB_LEDs;
unsigned int flash_panel=NO;
unsigned int delay_timer;
unsigned int troubleshooting = 1;
void convert_display_voltage_LCD(int trip_duration_volt);
void convert_int_to_string_display_LCD(unsigned int total_integer_value);


void main(void)
{
init_timer0_ovf_interrupt();
//initialize Timer0 to serve as elapsed
initialize_ports();                      //initialize ports
InitADC();                               //initialize ADC
```

```
LCD_Init();                               //initialize LCD
print_trip_dur();
//prompt user to enter trip duration
monitor_main_power_CB();
clear_LCD();

trip_duration_volt = ReadADC(0x00);    //Read trip duration ADC0
if(troubleshooting)
   {                                      //display voltage LCD
   convert_display_voltage_LCD(trip_duration_volt);
   delay(46);
   }
calculate_trip_int();
if(troubleshooting)
   {
   convert_int_to_string_display_LCD(flt_25);
   delay(46);
   }
perform_countdown();
LED_blastoff_sequence();
sound_alarm();
delay(46);
reset_alarm();
systems_A_OK();

while(1)
   {
   if(flt_timer > flt_25)  achieved25  = YES;
   if(flt_timer > flt_50)  achieved50  = YES;
   if(flt_timer > flt_75)  achieved75  = YES;
   if(flt_timer > flt_100) achieved100 = YES;

     if((achieved25==YES)&&(action25_done==NO))
         //25% flight complete
      {
      flt25_actions();
      action25_done=YES;
   systems_A_OK();
      }
```

```
   if((achieved50==YES)&&(action50_done==NO))
       //50% flight complete
    {
    flt50_actions();
    action50_done=YES;
 systems_A_OK();
    }

   if((achieved75==YES)&&(action75_done==NO))
       //75% flight complete
    {
    flt75_actions();
    action75_done=YES;
 systems_A_OK();
    }

   if((achieved100==YES)&&(action100_done==NO))
     //100% flight complete
    {
    flt100_actions();
    action100_done=YES;
    }
   }//end while
}//end main

//function definitions**************************************************

//*********************************************************************
//initialize_ports: provides initial configuration for I/O ports
//
//Note: when the RSTDISBL fuse is unprogrammed, the RESET circuitry is
//      connected to the pin, and the pin can not be used as an I/O pin.
//*********************************************************************

void initialize_ports(void)
{
DDRB = 0xff;              //PORTB[7:0] as output
PORTB= 0x00;              //initialize low
```

```
DDRC = 0xb0;                     //set PORTC as output  OR00_IIII 1011_0000
PORTC= PORTC_pullup_mask;        //initialize pullups PORTC[3:1]
DDRD = 0xff;                     //set PORTD as output
PORTD =0x00;                     //initialize low
}


//***************************************************************************
//delay(unsigned int num_of_65_5ms_interrupts): this generic delay function
//provides the specified delay as the number of 65.5 ms "clock ticks"
//from the Timer0 interrupt.
//Note: this function is only valid when using a 1 MHz crystal or ceramic
//      resonator
//***************************************************************************

void delay(unsigned int number_of_65_5ms_interrupts)
{
TCNT0 = 0x00;                            //reset timer0
delay_timer = 0;
while(delay_timer <= number_of_65_5ms_interrupts)
  {
  ;
  }
}


//***************************************************************************
//InitADC: initialize ADC converter
//***************************************************************************

void InitADC( void)
{
ADMUX = 0;                       //Select channel 0
ADCSRA = 0xC3;                   //Enable ADC & start 1st dummy conversion
                                 //Set ADC module prescalar to 8
  //critical for accurate ADC results
while (!(ADCSRA & 0x10));         //Check if conversation is ready
ADCSRA |= 0x10;                   //Clear conv rdy flag - set the bit
}


//***************************************************************************
```

```
//ReadADC: read analog voltage from ADC-the desired channel for conversion
//is passed in as an unsigned character variable. The result is returned
//as a left justified, 10 bit binary result.
//The ADC prescalar must be set to 8 to slow down the ADC clock at higher
//external clock frequencies (10 MHz) to obtain accurate results.
//*********************************************************************

unsigned int ReadADC(unsigned char channel)
{
unsigned int binary_weighted_voltage, binary_weighted_voltage_low;
                                     //weighted binary voltage
unsigned int binary_weighted_voltage_high;

ADMUX = channel;                     //Select channel
ADCSRA |= 0x43;                      //Start conversion
                                     //Set ADC module prescalar to 8
   //critical for accurate ADC results
while (!(ADCSRA & 0x10));            //Check if conversion is ready
ADCSRA |= 0x10;                      //Clear Conv rdy flag - set the bit
binary_weighted_voltage_low = ADCL;  //Read 8 low bits first (important)
                                     //Read 2 high bits, multiply by 256
binary_weighted_voltage_high = ((unsigned int)(ADCH << 8));
binary_weighted_voltage=binary_weighted_voltage_low |
                    binary_weighted_voltage_high;
return binary_weighted_voltage;      //ADCH:ADCL
}


//*********************************************************************
//int_timer0_ovf_interrupt(): The Timer0 overflow interrupt is being
//employed as a time base for a master timer for this project.
//The internal time base is set to operate at 1 MHz and then
//is divided by 256.  The 8-bit Timer0 register (TCNT0) overflows
//every 256 counts or every 65.5 ms.
//*********************************************************************

void init_timer0_ovf_interrupt(void)
{
TCCR0 = 0x04; //divide timer0 timebase by 256, overfl. occurs every 65.5ms
TIMSK = 0x01; //enable timer0 overflow interrupt
```

```
asm("SEI");    //enable global interrupt
}


//*************************************************************************
//LCD_Init: initialization for an LCD connected in the following manner:
//LCD: AND671GST 1x16 character display
//LCD configured as two 8 character lines in a 1x16 array
//LCD data bus (pin 14-pin7) ATMEL 8: PORTD
//LCD RS (pin 28)  ATMEL 8: PORTC[5]
//LCD E  (pin 27)  ATMEL 8: PORTC[4]
//*************************************************************************

void LCD_Init(void)
{
delay(1);
delay(1);
delay(1);
                   // output command string to initialize LCD
putcommand(0x38);  //function set 8-bit
delay(1);
putcommand(0x38);  //function set 8-bit
putcommand(0x38);  //function set 8-bit
putcommand(0x38);  //one line, 5x7 char
putcommand(0x0C);  //display on
putcommand(0x01);  //display clear-1.64 ms
putcommand(0x06);  //entry mode set
putcommand(0x00);  //clear display, cursor at home
putcommand(0x00);  //clear display, cursor at home
}


//*************************************************************************
//putchar:prints specified ASCII character to LCD
//*************************************************************************

void putchar(unsigned char c)
{
DDRD  = 0xff;                          //set PORTD as output
DDRC  = DDRC|0x30;                     //make PORTC[5:4] output
PORTD = c;
```

```
PORTC = (PORTC|0x20)|PORTC_pullup_mask;     //RS=1
PORTC = (PORTC|0x10)|PORTC_pullup_mask;;    //E=1
PORTC = (PORTC&0xef)|PORTC_pullup_mask;;    //E=0
delay(1);
}

//************************************************************************
//putcommand: performs specified LCD related command
//************************************************************************

void putcommand(unsigned char d)
{
DDRD  = 0xff;                                //set PORTD as output
DDRC  = DDRC|0xC0;                           //make PORTA[5:4] output
PORTC = (PORTC&0xdf)|PORTC_pullup_mask;      //RS=0
PORTD = d;
PORTC = (PORTC|0x10)|PORTC_pullup_mask;      //E=1
PORTC = (PORTC&0xef)|PORTC_pullup_mask;      //E=0
delay(1);
}

//************************************************************************
//clear_LCD: clears LCD
//************************************************************************

void clear_LCD(void)
{
putcommand(0x01);
}

//************************************************************************
//*void calculate_trip_int(void)
//************************************************************************

void calculate_trip_int(void)
{
unsigned int      trip_duration_sec;
unsigned int      trip_duration_int;
```

```
trip_duration_sec=(unsigned int)(((double)(trip_duration_volt)/1024.0)*
                 60.0*60.0);
trip_duration_int = (unsigned int)((double)(trip_duration_sec)/0.0655);
flt_25 = (unsigned int)((double)(trip_duration_int) * 0.25);
flt_50 = (unsigned int)((double)(trip_duration_int) * 0.50);
flt_75 = (unsigned int)((double)(trip_duration_int) * 0.75);
flt_100 = trip_duration_int;
}


//****************************************************************************
//void timer0_interrupt_isr(void)
//****************************************************************************

void timer0_interrupt_isr(void)
{
delay_timer++;
flt_timer++;                       //increment flight timer

if(flash_panel==YES)
  {
  if(flash_timer <= 8)
    {
    flash_timer++;
}
  else
    {
flash_timer = 0;
if(PORTB_LEDs == OFF)
  {
  PORTB = 0xff;
  PORTB_LEDs = ON;
  }
else
  {
  PORTB = 0x00;
  PORTB_LEDs = OFF;
  }
 }
  }
```

```
else
  {
  flash_timer = 0;
  }
}

//***************************************************************************
//void flt25_actions(void)
//***************************************************************************

void flt25_actions(void)
{
sound_alarm();
flash_LED_panel();
print_LOWO2();
monitor_O2_CB_reset();
reset_alarm();
restore_panel();
action25_done = YES;
}

//***************************************************************************
//void flt50_actions(void)
//***************************************************************************

void flt50_actions(void)
{
sound_alarm();
flash_LED_panel();
print_LOW_FUEL();
monitor_aux_fuel_CB();
reset_alarm();
restore_panel();
action50_done = YES;
}

//***************************************************************************
//void flt75_actions(void)
//***************************************************************************
```

```
void flt75_actions(void)
{
sound_alarm();
flash_LED_panel();
print_OVERHEAT();
delay(458);                              //delay 30s
monitor_main_power_CB();
reset_alarm();
restore_panel();
action75_done = YES;
}


//****************************************************************************
//void flt100_actions(void)
//****************************************************************************

void flt100_actions(void)
{
sound_alarm();
flash_LED_panel();
print_fuel_expended();
turn_off_LEDs();
action100_done = YES;
}



//****************************************************************************
//void sound_alarm(void)
//****************************************************************************

void sound_alarm(void)
{
PORTB = PORTB | 0x80;
}


//****************************************************************************
//void turn_off_LEDs(void)
//****************************************************************************
```

```
void turn_off_LEDs(void)
{
PORTB = PORTB & 0x80;
}



//*****************************************************************************
//void reset_alarm(void)
//*****************************************************************************

void reset_alarm(void)
{
PORTB = PORTB & 0x7F;
}


//*****************************************************************************
//void restore_panel(void)
//*****************************************************************************

void restore_panel(void)
{
flash_panel = NO;
PORTB = PORTB | 0x7F;
}


//*****************************************************************************
//void LED_blastoff_sequence(void)
//*****************************************************************************

void LED_blastoff_sequence(void)
{
PORTB = 0x00;           //0000_0000
delay(15);              //delay 1s
PORTB = 0x01;           //0000_0001
delay(15);              //delay 1s
PORTB = 0x03;           //0000_0011
delay(15);              //delay 1s
PORTB = 0x07;           //0000_0111
```

```
delay(15);              //delay 1s
PORTB = 0x1F;           //0001_1111
delay(15);              //delay 1s
PORTB = 0x7F;           //0111_1111
delay(15);              //delay 1s
}


//*************************************************************************
//void LED_power_down_sequence(void)
//*************************************************************************

void LED_power_down_sequence(void)
{
PORTB = 0x7F;           //0111_1111
delay(15);              //delay 1s
PORTB = 0x1F;           //0001_1111
delay(15);              //delay 1s
PORTB = 0x07;           //0000_0111
delay(15);              //delay 1s
PORTB = 0x03;           //0000_0011
delay(15);              //delay 1s
PORTB = 0x01;           //0000_0001
delay(15);              //delay 1s
PORTB = 0x00;           //0000_0000
delay(15);              //delay 1s
}


//*************************************************************************
//void monitor_main_power_CB(void)
//*************************************************************************

void monitor_main_power_CB(void)
{
  while((PINC & 0x02) == 0x02)
  {
  ;     //wait for PC1 to be exerted low
  }
}
```

```
//******************************************************************
//void monitor_O2_CB_reset(void)
//******************************************************************

void monitor_O2_CB_reset(void)
{
  while((PINC & 0x08) == 0x08)
    {
    ;    //wait for PC3 to be exerted low
    }
}


//******************************************************************
//void monitor_aux_fuel_CB(void)
//******************************************************************

void monitor_aux_fuel_CB(void)
{
  while((PINC & 0x04) == 0x04)
    {
    ;    //wait for PC2 to be exerted low
    }
}


//******************************************************************
//void perform_countdown(void)
//******************************************************************

void perform_countdown(void)
{
clear_LCD();
putcommand(0x01);               //cursor home
putcommand(0x80);               //DD RAM location 1 - line 1
putchar('1'); putchar ('0');    //print 10
delay(15);                      //delay 1s

putcommand(0x01);               //cursor home
putcommand(0x80);               //DD RAM location 1 - line 1
putchar('9');                   //print 9
```

```
delay(15);                      //delay 1s

putcommand(0x01);               //cursor home
putcommand(0x80);               //DD RAM location 1 - line 1
putchar('8');                   //print 8
delay(15);                      //delay 1s

putcommand(0x01);               //cursor home
putcommand(0x80);               //DD RAM location 1 - line 1
putchar('7');                   //print 7
delay(15);                      //delay 1s

putcommand(0x01);               //cursor home
putcommand(0x80);               //DD RAM location 1 - line 1
putchar('6');                   //print 6
delay(15);                      //delay 1s

putcommand(0x01);               //cursor home
putcommand(0x80);               //DD RAM location 1 - line 1
putchar('5');                   //print 5
delay(15);                      //delay 1s

putcommand(0x01);               //cursor home
putcommand(0x80);               //DD RAM location 1 - line 1
putchar('4');                   //print 4
delay(15);                      //delay 1s

putcommand(0x01);               //cursor home
putcommand(0x80);               //DD RAM location 1 - line 1
putchar('3');                   //print 3
delay(15);                      //delay 1s

putcommand(0x01);               //cursor home

putcommand(0x80);               //DD RAM location 1 - line 1
putchar('2');                   //print 2
delay(15);                      //delay 1s

putcommand(0x01);               //cursor home
```

```
putcommand(0x80);              //DD RAM location 1 - line 1
putchar('1');                  //print 1
delay(15);                     //delay 1s

putcommand(0x01);              //cursor home
putcommand(0x80);              //DD RAM location 1 - line 1
putchar('0');                  //print 0
delay(15);                     //delay 1s


//BLASTOFF!
putcommand(0x01);              //cursor home
putcommand(0x80);              //DD RAM location 1 - line 1
putchar('B'); putchar('L'); putchar('A'); putchar('S'); putchar('T');
putchar('0'); putchar('F'); putchar('F'); putchar('!');

}

//****************************************************************************
//void print_LOWO2(void)
//****************************************************************************

void print_LOWO2(void)
{
clear_LCD();
putcommand(0x01);              //cursor home
putcommand(0x80);              //DD RAM location 1 - line 1
putchar('L'); putchar('0'); putchar('W'); putchar(' '); putchar('0');
putchar('2');

putcommand(0xC0);//DD RAM location 1 - line 2
putchar('R'); putchar('E'); putchar('S'); putchar('E'); putchar('T');
putchar(' '); putchar('0'); putchar('2'); putchar(' '); putchar('C');
putchar('B');
}

//****************************************************************************
//void print_LOW_FUEL(void)
//****************************************************************************
```

```
void print_LOW_FUEL(void)
{
clear_LCD();
putcommand(0x01);                //cursor home
putcommand(0x80);                //DD RAM location 1 - line 1
putchar('L'); putchar('O'); putchar('W'); putchar(' '); putchar('F');
putchar('U'); putchar('E'); putchar('L');

putcommand(0xC0);//DD RAM location 1 - line 2
putchar('A'); putchar('S'); putchar('S'); putchar('E'); putchar('R');
putchar('T'); putchar(' '); putchar('A'); putchar('U'); putchar('X');
putchar('F'); putchar('U'); putchar('E'); putchar('L'); putchar('C');
putchar('B');
}


//***************************************************************************
//void print_fuel_expended(void)
//***************************************************************************

void print_fuel_expended(void)
{
clear_LCD();
putcommand(0x01);                //cursor home
putcommand(0x80);                //DD RAM location 1 - line 1
putchar('F'); putchar('U'); putchar('E'); putchar('L'); putchar(' ');
putchar('E'); putchar('X'); putchar('P'); putchar('E'); putchar('N');
putchar('D'); putchar('E'); putchar('D');

putcommand(0xC0);//DD RAM location 1 - line 2
putchar('S'); putchar('H'); putchar('U'); putchar('T'); putchar('T');
putchar('I'); putchar('N'); putchar('G'); putchar(' '); putchar('D');
putchar('O'); putchar('W'); putchar('N'); putchar('.'); putchar('.');
putchar('.');
}


//***************************************************************************
//void print_trip_dur(void);
//***************************************************************************
```

```
void print_trip_dur(void)
{
clear_LCD();
putcommand(0x01);               //cursor home
putcommand(0x80);               //DD RAM location 1 - line 1
putchar('T'); putchar('R'); putchar('I'); putchar('P');
putchar('T'); putchar('I'); putchar('M'); putchar('E'); putchar(':');
putchar('0'); putchar('-'); putchar('6'); putchar('0');

putcommand(0xC0);//DD RAM location 1 - line 2
putchar('S'); putchar('E'); putchar('T'); putchar(' '); putchar('M');
putchar('A'); putchar('I'); putchar('N'); putchar(' '); putchar('P');
putchar('W'); putchar('R'); putchar(' '); putchar('C'); putchar('B');
}


//*****************************************************************************
//void print_OVERHEAT(void)
//*****************************************************************************

void print_OVERHEAT(void)
{
clear_LCD();
putcommand(0x01);               //cursor home
putcommand(0x80);               //DD RAM location 1 - line 1
putchar('E'); putchar('N'); putchar('G'); putchar('I'); putchar('N');
putchar('E'); putchar(' '); putchar('O'); putchar('V'); putchar('E');
putchar('R'); putchar('H'); putchar('E'); putchar('A'); putchar('T');

putcommand(0xC0);//DD RAM location 1 - line 2
putchar('R'); putchar('E'); putchar('S'); putchar('E'); putchar('T');
putchar(' '); putchar('E'); putchar('N'); putchar('G'); putchar(' ');
putchar('C'); putchar('B'); putchar(' '); putchar('3'); putchar('0');
putchar('S');
}


//*****************************************************************************
//void systems_A_OK(void)
```

```
//***************************************************************************

void systems_A_OK(void)
{
clear_LCD();
putcommand(0x01);              //cursor home
putcommand(0x80);              //DD RAM location 1 - line 1
putchar('S'); putchar('Y'); putchar('S'); putchar('T'); putchar('E');
putchar('M'); putchar('S'); putchar(' '); putchar('A'); putchar('-');
putchar('O'); putchar('K'); putchar('!'); putchar('!'); putchar('!');
}


//***************************************************************************
//void flash_LED_panel(void)
//***************************************************************************

void flash_LED_panel(void)
{
flash_panel = YES;
flash_timer = 0;
PORTB = 0x00;
PORTB_LEDs = OFF;
}


//***************************************************************************
//convert_display_voltage_LCD: converts binary weighted voltage to ASCII
//representation and prints result to LCD screen
//***************************************************************************

void convert_display_voltage_LCD(int binary_voltage)
{
float actual_voltage;                    //voltage between 0 and 5 volts
int    all_integer_voltage;
  //integer representation of voltage
                                          //int representation of voltage
int    hundreths_place, tens_place, ones_place;
                                          //char representation of voltage
char   hundreths_place_char, tens_place_char, ones_place_char;
```

```
                                        // display analog voltage on LCD
putcommand(0xC0);                       //LCD cursor to line 2
                                        //scale float voltage 0..5V
actual_voltage = ((float)(binary_voltage)/(float)(0x3FF))*5.0;
                                        //voltage represented 0 to 500
all_integer_voltage=actual_voltage * 100;//represent as all integer
hundreths_place = all_integer_voltage/100;//isolate first digit
hundreths_place_char = (char)(hundreths_place + 48); //convert to ascii
putchar(hundreths_place_char);          //display first digit
putchar('.');                           //print decimal point to LCD
                                        //isolate tens place
tens_place = (int)((all_integer_voltage - (hundreths_place*100))/10);
tens_place_char=(char)(tens_place+48);  //convert to ASCII
putchar(tens_place_char);               //print to LCD
                                        //isolate ones place
ones_place = (int)((all_integer_voltage - (hundreths_place*100))%10);
ones_place_char=(char)(ones_place+48);  //convert to ASCII
putchar(ones_place_char);               //print to LCD
putchar('V');                           //print unit V
}


//*************************************************************************
//convert_int_to_string_display_LCD: converts 16 bit to unsigned integer
//values range from 0 to 65,535
//prints result to LCD screen
//*************************************************************************

void convert_int_to_string_display_LCD(unsigned int total_integer_value)
{
int    ten_thousandths_place, thousandths_place;
int    hundreths_place, tens_place, ones_place;
char   ten_thousandths_place_char, thousandths_place_char;
char   hundreths_place_char, tens_place_char, ones_place_char;

putcommand(0xC0);                       //LCD cursor to line 2
                                        //10,000th place
ten_thousandths_place = total_integer_value/10000;
ten_thousandths_place_char = (char)(ten_thousandths_place+48);
putchar(ten_thousandths_place_char);
```

```
                                               //1,000th place
thousandths_place = (int)((total_integer_value -
                          (ten_thousandths_place*10000))/1000);
thousandths_place_char = (char)(thousandths_place+48);
putchar(thousandths_place_char);
                                            //100th place
hundreths_place = (int)((total_integer_value -
                        (ten_thousandths_place*10000)-
(thousandths_place*1000))/100);
hundreths_place_char = (char)(hundreths_place + 48);
putchar(hundreths_place_char);
                                            //10th place
tens_place = (int)((total_integer_value -(ten_thousandths_place*10000)-
   (thousandths_place*1000)-(hundreths_place*100))/10);
tens_place_char=(char)(tens_place+48);   //convert to ASCII
putchar(tens_place_char);                //print to LCD
                                         //isolate ones place
ones_place = (int)((total_integer_value -(ten_thousandths_place*10000)-
   (thousandths_place*1000)-(hundreths_place*100))%10);
ones_place_char=(char)(ones_place+48);   //convert to ASCII
putchar(ones_place_char);                //print to LCD
}



//*********************************************************************
//end of file: flight_sim.c
//*********************************************************************
```

9.5   EXTENDED EXAMPLE 4: SUBMERSIBLE ROBOT

The area of submersible robots is fascinating and challenging. To design a robot is quite complex. To add the additional requirement of waterproofing key components adds an additional level of challenge. (Water and electricity do not mix!) In this section we develop a control system for a remotely operated vehicle, an ROV. By definition an ROV is equipped with a tether umbilical cable that provides power and control signals from a surface support platform. An Autonomous Underwater Vehicle (AUV) carries its own power and control equipment and does not require surface support.

We limit our discussion to the development of an Arduino based control system. Details on the construction and waterproofing of an ROV are provided in the excellent and fascinating "Build Your Own Underwater Robot and Other Wet Projects" by Harry Bohm and Vickie Jensen. We

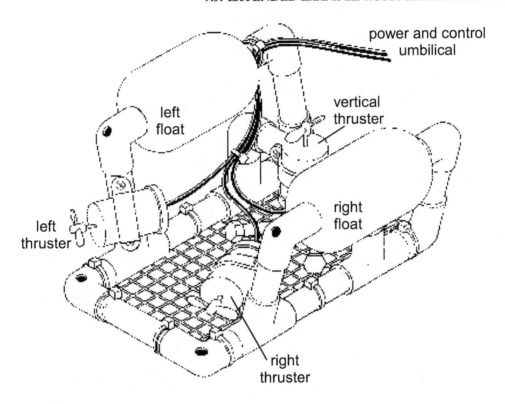

Figure 9.6: SeaPerch ROV. (Adapted and used with permission of Bohm and Jensen, West Coast Words Publishing.)

develop a control system for the SeaPerch style of ROV as shown in Figure 9.6. There is a national competition for students based on the SeaPerch ROV. The goal of the program is to stimulate interest in the next generation of marine related engineering specialties [seaperch].

9.5.1 REQUIREMENTS

The requirements for this system include:

- Develop a control system to allow a three motor ROV to move forward, left (port) and right (starboard).

- The ROV will be pushed down to a shallow depth via a vertical thruster and return to surface based on its own, slightly positive buoyancy.

- ROV movement will be under joystick control.

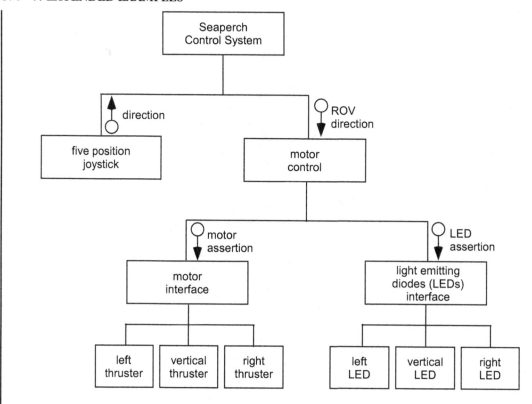

Figure 9.7: SeaPerch ROV structure chart.

- All power and control circuitry will be maintained in a surface support platform.

9.5.2 STRUCTURE CHART

The Sea Perch structure chart is provided in Figure 9.7. As can be seen in the figure, the SeaPerch control system will accept input from the five position joystick (left, right, select, up and down). We will use the joystick shield available from SparkFun Electronics and illustrated in Figure 9.8. The joystick schematic and connections to the Arduino UNO R3 are provided in Figure 9.9.

In response to user joystick input, the SeaPerch control algorithm will issue a control command indicating desired ROV direction. In response to this desired direction command, the motor control algorithm will issue control signals to assert the appropriate motors and LEDs.

9.5.3 CIRCUIT DIAGRAM

The circuit diagram for the SeaPerch control system is provided in Figure 9.9. The Arduino compatible joystick shield is used to select desired ROV direction. There are three LED interface circuits connected to digital pins D7, D8 and D9. The prime mover for the ROV will be two waterproofed motors or two bilge pumps. Details on motor waterproofing may be found in "Build Your Own Underwater Robot and Other Wet Projects." The motors are driven by the pulse width modulation channels (d10 and D11) via power FETs as shown in Figure9.9. Both the LED and the motor interfaces were discussed earlier in the chapter.

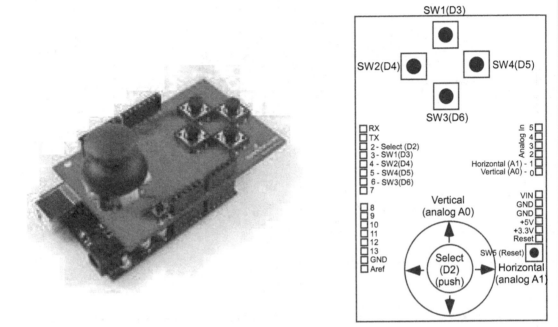

Figure 9.8: Joystick shield. (Used with permission of SparkFun Electronics.)(CC BY–NC–SA)

9.5.4   UML ACTIVITY DIAGRAM

The SeaPerch control system UML activity diagram is provided in Figure 9.10. After initializing ports the control algorithm is placed in a continuous loop awaiting user input. In response to user input, the algorithm determines desired direction of ROV travel and asserts appropriate control signals for the LED and motors.

9.5.5   MICROCONTROLLER CODE

In this example we use switch 2 (SW2–D4) to control the left thruster (motor or bilge pump) and switch 4 (SW4-D5) to control the right thruster motor. This is a simple on/off control. That is, when a specific switch is asserted the motor will be on and the corresponding LED will illuminate.

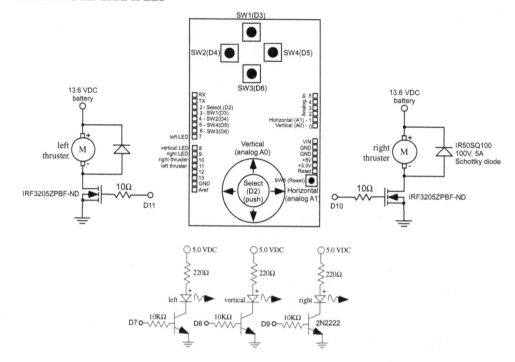

Figure 9.9: SeaPerch ROV interface control.

```
//**************************************************************************
                              //digital input pins
#define left_switch    4      //digital pin left switch
#define right_switch   5      //digital pin right switch


                              //digital output pins
                              //LED indicators
#define left_LED       7      //digital pin left LED
#define right_LED      9      //digital pin right LED


                              //motor outputs
#define left_motor     11     //digital pin left_motor
#define right_motor    10     //digital pin right_motor

int  left_sw_value, right_sw_value;

void setup()
```

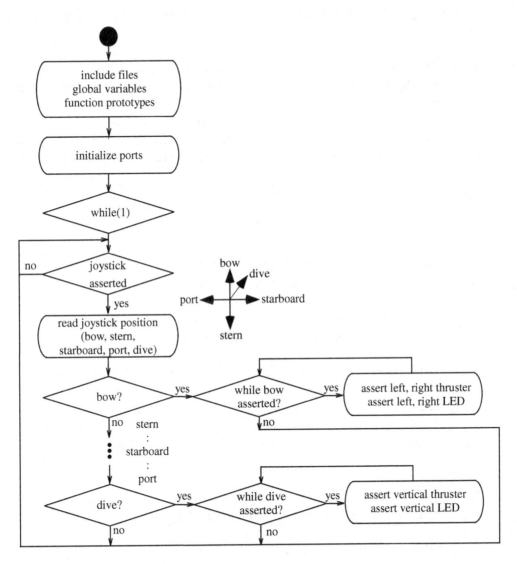

Figure 9.10: SeaPerch ROV UML activity diagram.

```
  {
                                        //switch inputs
  pinMode(left_switch, INPUT);          //configure pin 4 for digital input
  pinMode(right_switch,INPUT);          //configure pin 5 for digital input

                                        //LED indicators
  pinMode(left_LED, OUTPUT);            //configure pin 7 for digital output
  pinMode(right_LED, OUTPUT);           //configure pin 9 for digital output

                                        //motor outputs PWM
  pinMode(left_motor,  OUTPUT);         //config pin 11 for digital output
  pinMode(right_motor, OUTPUT);         //config pin 10 for digital output
  }

void loop()
  {
  //read input switches
  left_sw_value  = digitalRead(left_switch);
  right_sw_value = digitalRead(right_switch);

  //left switch asserted - turn left by asserting right motor
  if((left_sw_value==HIGH)&&(right_sw_value==LOW))
     {
     digitalWrite(left_LED, LOW);           //turn left LED off
     digitalWrite(left_motor, LOW);         //turn left motor off
     digitalWrite(right_LED, HIGH);         //turn right LED on
     digitalWrite(right_motor, HIGH);       //turn right motor on
     delay(500);                            //delay 500 ms
     }

  //right switch asserted - turn right by asserting left motor
  if((left_sw_value==LOW)&&(right_sw_value==HIGH))
     {
     digitalWrite(left_LED, HIGH);          //turn left LED on
     digitalWrite(left_motor, HIGH);        //turn left motor on
     digitalWrite(right_LED, LOW);          //turn right LED off
     digitalWrite(right_motor, LOW);        //turn right motor off
     delay(500);                            //delay 500 ms
     }
```

```
//both switches asserted - move ROV forward
if((left_sw_value==HIGH)&&(right_sw_value==HIGH))
    {
    digitalWrite(left_LED, HIGH);           //turn left LED on
    digitalWrite(left_motor, HIGH);         //turn left motor on
    digitalWrite(right_LED, HIGH);          //turn right LED on
    digitalWrite(right_motor, HIGH);        //turn right motor on
    delay(500);                             //delay 500 ms
    }
}
//*****************************************************************
```

9.5.6   PROJECT EXTENSIONS

The control system provided above has a set of very basic features. Here are some possible extensions for the system:

- Provide adjustable speed control for each motor. This advanced feature requires several modifications including a proportional joystick and pulse width modulation signals to the motor. The joystick features currently in use provides a simple on and off signal. A proportional joystick provides an X (horizontal) and Y (vertical) output DC signal proportional to the joystick deflection on each axis. These signals may be fed to the analog–to–digital converter aboard the Arduino UNO R3. The X and Y signals captured from the joystick may be used to provide a pulse width modulated signal with a duty cycle proportional to deflection. The Arduino shield joystick illustrated in 9.8 is equipped with an analog joystick. The X and Y channel of the joystick is also already connected to the Arduino Analog In channels A0 (vertical) and A1 (horizontal) through the shield printed circuit board connections. Equipping the ROV with this modification is given as an assignment at the end of the chapter.

- Provide a powered dive and surface thruster. To provide for a powered dive and surface capability, the ROV must be equipped with a vertical thruster equipped with an H–bridge to allow for motor forward and reversal. This modification is given as an assignment at the end of the chapter.

- Left and right thruster reverse. Currently the left and right thrusters may only be powered in one direction. To provide additional maneuverability, the left and right thrusters could be equipped with an H–bridge to allow bi–directional motor control. This modification is given as an assignment at the end of the chapter.

- Proportional speed control with bi–directional motor control. Both of these advanced features may be provided by driving the H–bridge circuit with PWM signals. This modification is given as an assignment at the end of the chapter.

9.6 EXTENDED EXAMPLE 5: WEATHER STATION

In this project we design a weather station to sense wind direction and ambient temperature. The sensed values will be displayed on an LCD Fahrenheit. The wind direction will also be displayed on LEDs arranged in a circular pattern.

9.6.1 REQUIREMENTS

The requirements for this system include:

- Design a weather station to sense wind direction and ambient temperature.

- Sensed wind direction and temperature will be displayed on an LCD.

- Sensed temperature will be displayed in the Fahrenheit temperature scale.

- Wind direction will be displayed on LEDs arranged in a circular pattern.

9.6.2 STRUCTURE CHART

To begin the design process a structure chart is used to partition the system into definable pieces. We employ a top–down design/bottom–up implementation approach. The structure chart for the weather station is provided in Figure 9.11. The main microcontroller subsystem needed for this project is the ADC system to convert the analog voltage from the LM34 temperature sensor and weather vane into digital signals, and the wind direction display. This display consists of a 74154, 4–to–16 decoder and 16 individual LEDs to display wind direction. The system is partitioned until the lowest level of the structure chart contains "doable" pieces of hardware components or software functions. Data flown is shown on the structure chart as directed arrows.

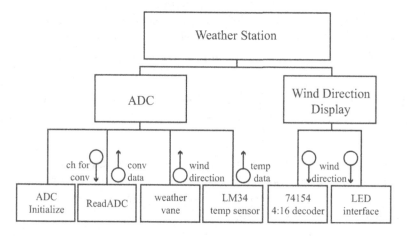

Figure 9.11: Weather station structure chart.

9.6.3 CIRCUIT DIAGRAM

The circuit diagram for the weather station is provided in Figure 9.12. The weather station is equipped with two input sensors: the LM34 to measure temperature and the weather vane to measure wind direction. Both of the sensors provide an analog output that is fed to Analog Input 0 (LM34) and Analog Input 1 (weather vane) of the Arduino UNO R3. The LM34 provides 10 mV output per degree Fahrenheit. The weather vane provides 0 to 5 VDC for 360 degrees of vane rotation. The weather vane must be oriented to a known direction with the output voltage at this direction noted. We assume that 0 VDC corresponds to North and the voltage increases as the vane rotates clockwise to the East. The vane output voltage continues to increase until North is again reached at 5 VDC and then rolls over back to - volts. All other directions are derived from this reference point.

An LCD is connected to digital pins [7:0] (PORTD [7:0] of the ATmega328) for data and digital pins 8 and 9 for the enable and command/data control lines (PORTB[1:0] of the ATmega328). There are 16 different LEDs to drive for the wind speed indicator. Rather than use 16 microcontroller pins, the binary value of the LED for illumination is provided to the 74154 4–to–16 decoder. The decoder provides a "one cold" output as determined by the binary code provided by PORTC[5:2] of the ATmega328 (Analog In [5:2] on the Arduino UNO R3). For example, when A_{16} is provided to the 74154 input, output $/Y10$ is asserted low, while all other outputs remain at logic high. The 74154 is from the standard TTL family. It has sufficient current sink capability ($I_{OL} = 16\,mA$) to meet the current requirements of an LED ($V_f = 1.5\,VDC$, $I_f = 15\,mA$).

9.6.4 UML ACTIVITY DIAGRAMS

The UML activity diagram for the main program is provided in Figure 9.13. After initializing the subsystems, the program enters a continuous loop where temperature and wind direction is sensed and displayed on the LCD and the LED display. The system then enters a delay to set how often the temperature and wind direction parameters are updated. We have you construct the individual UML activity diagrams for each function as an end of chapter exercise.

9.6.5 MICROCONTROLLER CODE

```
//include file*****************************************************************
#include <iom328pv.h>

//function prototypes**********************************************************
void initialize_ports(void);
void initialize_ADC(void);
void temperature_to_LCD(unsigned int ADCValue);
unsigned int readADC(unsigned char);
void LCD_init(void);
void putChar(unsigned char);
void putcommand(unsigned char);
```

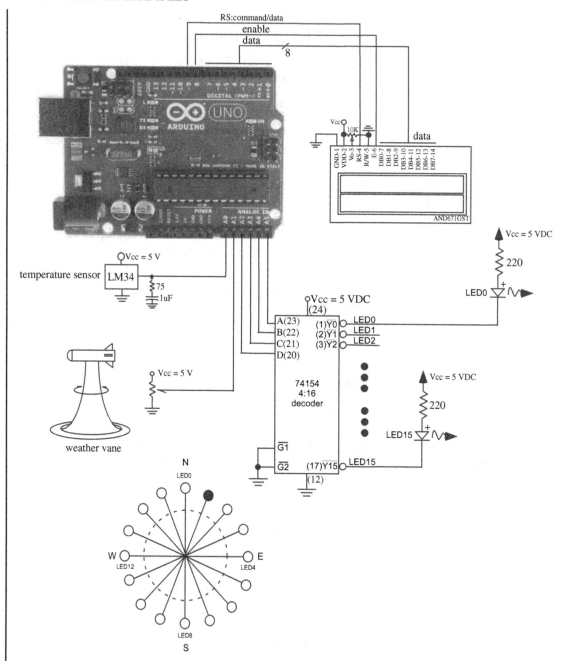

Figure 9.12: Circuit diagram for weather station. (UNO R3 illustration used with permission of the Arduino Team (CC BY–NC–SA) www.arduino.cc).

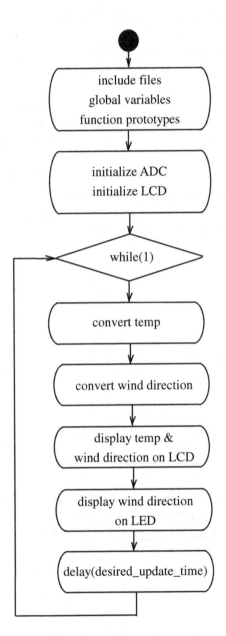

Figure 9.13: Weather station UML activity diagram.

```
void display_data(void);
void convert_wind_direction(unsigned int);
void delay(unsigned int number_of_6_55ms_interrupts);
void init_timer0_ovf_interrupt(void);
void timer0_interrupt_isr(void);


                                           //interrupt handler definition
#pragma interrupt_handler timer0_interrupt_isr:17
                                           //door profile data

//Global variables*****************************************************
unsigned int temperature, wind_direction;
unsigned int binary_weighted_voltage_low, binary_weighted_voltage_high;
unsigned char dir_tx_data;

void main(void)
{
initialize_ports();
initialize_ADC();
LCD_init();
init_timer0_ovf_interrupt();

while(1)
  {
  //temperature data: read \---> display \---> transmit
  temperature = readADC(0x00);          //Read Temp - LM34
  temperature_to_LCD (temperature);     //Convert and display temp on LCD

  //wind direction data: read \---> display \---> transmit
  wind_direction = readADC(0x01);         //Read wind direction
  convert_wind_direction(wind_direction); //Convert wind direction

  //delay 15 minutes
  delay(13740):
  }
}


//*******************************************************************************
```

```
void initialize_ports()
{
DDRB = 0x03;      //PORTB[1:0] as outputs
DDRC = 0x3c;      //PORTC[5:2] as outputs
DDRD = 0xFF;      //PORTD[7:0] as outputs
}

//**********************************************************************

void initialize_ADC()
{
ADMUX = 0;                        //select channel 0
                                  //enable ADC and set module enable ADC
ADCSRA = 0xC3;                    //set module prescalar to 8
while(!(ADCSRA & 0x10));          //Wait until conversion is ready
ADCSRA |= 0x10;                   //Clear conversion ready flag
}

//**********************************************************************

unsigned int readADC(unsigned char channel)
{
unsigned int binary_weighted_voltage, binary_weighted_voltage_low;
unsigned int binary_weighted_voltage_high;//weighted binary voltage

ADMUX = channel;                  //Select channel
ADCSRA |= 0x43;                   //Start conversion
                                  //Set ADC module prescalar
                                  //to 8 critical for
                                  //accurate ADC results
while (!(ADCSRA & 0x10));         //Check if conversion is ready
ADCSRA |= 0x10;                   //Clear conv rdy flag set the bit
binary_weighted_voltage_low = ADCL;   //Read 8 low bits first (important)
                                      //Read 2 high bits, multiply by 256
binary_weighted_voltage_high = ((unsigned int)(ADCH << 8));
binary_weighted_voltage=binary_weighted_voltage_low+binary_weighted_voltage_high;
return binary_weighted_voltage;   //ADCH:ADCL
}
```

```
//*************************************************************************
//LCD_Init: initialization for an LCD connected in the following manner:
//LCD: AND671GST 1x16 character display
//LCD configured as two 8 character lines in a 1x16 array
//LCD data: bus PORTD[7:0]
//LCD RS:    PORTB[1]
//LCD E:     PORTB[0]
//*************************************************************************

void LCD_init(void)
{
delay(1);
delay(1);
delay(1);
                                        //output command string to
                                        //initialize LCD
putcommand(0x38);                       //function set 8-bit
delay(1);
putcommand(0x38);                       //function set 8-bit
delay(1);
putcommand(0x38);                       //function set 8-bit
putcommand(0x38);                       //one line, 5x7 char
putcommand(0x0E);                       //display on
putcommand(0x01);                       //display clear-1.64 ms
putcommand(0x06);                       //entry mode set
putcommand(0x00);                       //clear display, cursor at home
putcommand(0x00);                       //clear display, cursor at home
}



//*************************************************************************

void putChar(unsigned char c)
{
DDRD = 0xff;                            //set PORTD as output
DDRB = DDRB|0x03;                       //make PORTB[1:0] output
PORTD = c;
PORTB = PORTB|0x02;                     //RS=1
PORTB = PORTB|0x01;                     //E=1
```

```
PORTB = PORTB&0xfe;                          //E=0
delay(1);
}


//************************************************************************

void putcommand(unsigned char d)
{
DDRD = 0xff;                                 //set PORTD as output
DDRB = DDRB|0x03;                            //make PORTB[1:0] output
PORTB = PORTB & 0xfd;                        //RS=0
PORTD = d;
PORTB = PORTB|0x01;                          //E=1
PORTB = PORTB&0xfe;                          //E=0
delay(1);
}


//************************************************************************

void temperature_to_LCD(unsigned int ADCValue)
{
float voltage,temperature;
unsigned int tens, ones, tenths;

voltage = (float)ADCValue*5.0/1024.0;

temperature = voltage*100;


tens = (unsigned int)(temperature/10);
ones = (unsigned int)(temperature-(float)tens*10);
tenths = (unsigned int)(((temperature-(float)tens*10)-(float)ones)*10);

putcommand(0x01);                            //cursor home
putcommand(0x80);                            //DD RAM location 1-line 1
putChar((unsigned char)(tens)+48);
putChar((unsigned char)(ones)+48);
putChar('.');
putChar((unsigned char)(tenths)+48);
```

```
putChar('F');
}


//*************************************************************************

void convert_wind_direction(unsigned int wind_dir_int)
{
float wind_dir_float;
                                           //convert wind direction to float
wind_dir_float = ((float)wind_dir_int)/1024.0) * 5;

//N: LED0
if((wind_dir_float <= 0.15625)||(wind_dir_float > 4.84375))
   {
   putcommand(0x01);                       //cursor to home
   putcommand(0xc0);                       //DD RAM location 1-line 2
   putchar('N');                           //LCD displays: N
   PORTC = 0x00;                           //illuminate LED 0


   }

//NNE: LED1
if((wind_dir_float > 0.15625)||(wind_dir_float <= 0.46875))
   {
   putcommand(0x01);                       //cursor to home
   putcommand(0xc0);                       //DD RAM location 1-line 2
   putchar('N');                           //LCD displays: NNE
   putchar('N');
   putchar('E');
   PORTC = 0x04;                           //illuminate LED 1
   }

//NE: LED2
if((wind_dir_float > 0.46875)||(wind_dir_float <= 0.78125))
   {
   putcommand(0x01);                       //cursor to home
   putcommand(0xc0);                       //DD RAM location 1-line 2
   putchar('N');                           //LCD displays: NE
   putchar('E');
```

```
    PORTC = 0x08;                          //illuminate LED 2
    }


//ENE: LED3
if((wind_dir_float > 0.78125)||(wind_dir_float <= 1.09375))
    {
    putcommand(0x01);                      //cursor to home
    putcommand(0xc0);                      //DD RAM location 1-line 2
    putchar('E');                          //LCD displays: NNE
    putchar('N');
    putchar('E');
    PORTC = 0x0c;                          //illuminate LED 3


    }


//E: LED4
if((wind_dir_float > 1.09375)||(wind_dir_float <= 1.40625))
    {
    putcommand(0x01);                      //cursor to home
    putcommand(0xc0);                      //DD RAM location 1-line 2
    putchar('E');                          //LCD displays: E
    PORTC = 0x10;                          //illuminate LED 4
    }


//ESE: LED5
if((wind_dir_float > 1.40625)||(wind_dir_float <= 1.71875))
    {
    putcommand(0x01);                      //cursor to home
    putcommand(0xc0);                      //DD RAM location 1-line 2
    putchar('E');                          //LCD displays: ESE
    putchar('S');
    putchar('E');
    PORTC = 0x14;                          //illuminate LED 5


    }


//SE: LED6
if((wind_dir_float > 1.71875)||(wind_dir_float <= 2.03125))
    {
```

```
    putcommand(0x01);                      //cursor to home
    putcommand(0xc0);                      //DD RAM location 1-line 2
    putchar('S');                          //LCD displays: SE
    putchar('E');
    PORTC = 0x10;                          //illuminate LED 6
    }

//SSE: LED7
if((wind_dir_float > 2.03125)||(wind_dir_float <= 2.34875))
   {
   putcommand(0x01);                       //cursor to home
   putcommand(0xc0);                       //DD RAM location 1-line 2
   putchar('S');                           //LCD displays: SSE
   putchar('S');
   putchar('E');
   PORTC = 0x1c;                           //illuminate LED 7
   }

//S: LED8
if((wind_dir_float > 2.34875)||(wind_dir_float <= 2.65625))
   {
   putcommand(0x01);                       //cursor to home
   putcommand(0xc0);                       //DD RAM location 1-line 2
   putchar('S');                           //LCD displays: S

   PORTC = 0x20;                           //illuminate LED 8

   }

//SSW: LED9
if((wind_dir_float > 2.65625)||(wind_dir_float <= 2.96875))
   {
   putcommand(0x01);                       //cursor to home
   putcommand(0xc0);                       //DD RAM location 1-line 2
   putchar('S');                           //LCD displays: SSW
   putchar('S');
   putchar('W');
   PORTC = 0x24;                           //illuminate LED 9
   }
```

```
//SW: LED10 (A)
if((wind_dir_float > 2.96875)||(wind_dir_float <= 3.28125))
   {
   putcommand(0x01);                    //cursor to home
   putcommand(0xc0);                    //DD RAM location 1-line 2
   putchar('S');                        //LCD displays: SW
   putchar('W');
   PORTC = 0x28;                        //illuminate LED 10 (A)
   }



//WSW: LED11 (B)
if((wind_dir_float > 3.28125)||(wind_dir_float <= 3.59375))
   {
   putcommand(0x01);                    //cursor to home
   putcommand(0xc0);                    //DD RAM location 1-line 2
   putchar('W');                        //LCD displays: WSW
   putchar('S');
   putchar('W');
   PORTC = 0x2c;                        //illuminate LED 11 (B)
   }

//W: LED12 (C)
if((wind_dir_float > 3.59375)||(wind_dir_float <= 3.90625))
   {
   putcommand(0x01);                    //cursor to home
   putcommand(0xc0);                    //DD RAM location 1-line 2
   putchar('W');                        //LCD displays: W
   PORTC = 0x30;                        //illuminate LED 12 (C)
   }

//WNW: LED13 (D)
if((wind_dir_float > 3.90625)||(wind_dir_float <= 4.21875))
   {
   putcommand(0x01);                    //cursor to home
   putcommand(0xc0);                    //DD RAM location 1-line 2
   putchar('W');                        //LCD displays: WNW
   putchar('N');
```

```
    putchar('W');
    PORTC = 0x34;                            //illuminate LED 13 (D)
    }


//NW: LED14 (E)
if((wind_dir_float > 4.21875)||(wind_dir_float <= 4.53125))
    {
    putcommand(0x01);                        //cursor to home
    putcommand(0xc0);                        //DD RAM location 1-line 2
    putchar('N');                            //LCD displays: NW
    putchar('W');
    PORTC = 0x38;                            //illuminate LED 14 (E)
    }


//NNW: LED15(F)
if((wind_dir_float > 4.53125)||(wind_dir_float < 4.84375))
    {
    putcommand(0x01);                        //cursor to home
    putcommand(0xc0);                        //DD RAM location 1-line 2
    putchar('N');                            //LCD displays: NNW
    putchar('N');
    putchar('W');
    PORTC = 0x3c;                            //illuminate LED 15 (F)


    }
}


//*************************************************************************
//delay(unsigned int num_of_65_5ms_interrupts): this generic delay
//function provides the specified delay as the number of 65.5 ms clock
//ticks from the Timer0 interrupt.
//Note: this function is only valid when using a 1 MHz crystal or ceramic
//       resonator.  Function time constants need to be adjusted for other
//       values of the timebase.
//*************************************************************************


void delay(unsigned int number_of_65_5ms_interrupts)
{
TCNT0 = 0x00;                                //reset timer0
```

```
delay_timer = 0;
while(delay_timer <= number_of_65_5ms_interrupts)
  {
  ;
  }
}


//*********************************************************************
//int_timer0_ovf_interrupt(): The Timer0 overflow interrupt is being
//employed as a time base for a master timer for this project.
//The internal time base is set to operate at 1 MHz and then
//is divided by 256.  The 8-bit Timer0 register (TCNT0) overflows
//every 256 counts or every 65.5 ms.
//*********************************************************************

void init_timer0_ovf_interrupt(void)
{
TCCR0 = 0x04; //div timer0 timebase by 256, overfl. occurs every 65.5ms
TIMSK = 0x01; //enable timer0 overflow interrupt
asm("SEI");   //enable global interrupt
}


//*********************************************************************
//timer0_interrupt_isr:
//Note: Timer overflow 0 is cleared by hardware when executing the
//corresponding interrupt handling vector.
//*********************************************************************

void timer0_interrupt_isr(void)
{
input_delay++;                          //input delay processing
}


//*********************************************************************
```

9.7   AUTONOMOUS MAZE NAVIGATING ROBOTS

In the next two examples we investigate two different autonomous navigating robot designs. Before delving into these designs, it would be helpful to review the fundamentals of robot steering and motor control. Figure 9.14 illustrates the fundamental concepts. Robot steering is dependent upon the number of powered wheels and whether the wheels are equipped with unidirectional or bidirectional control. Additional robot steering configurations are possible.

Recall from a previous chapter that an H–bridge is typically required for bidirectional control of a DC motor. An H–bridge configured for controlling a single motor is provided in Figure 9.15 a). The H–bridge has two inputs: one for controlling motor direction and one for controlling motor speed via a pulse width modulated signal. Provided in Figure 9.15b) is a motor control system for a two motor robot.

9.8   EXTENDED EXAMPLE 6: BLINKY 602A ROBOT–REVISITED

Graymark (www.graymarkint.com) manufacturers many low–cost, excellent robot platforms. In this example we modify the Blinky 602A robot discussed earlier in the book to be controlled by the ATmega328. In this example we program the control algorithm in C. The Blinky 602A kit contains the hardware and mechanical parts to construct a line following robot. The processing electronics for the robot consists of analog circuitry. The robot is controlled by two 3 VDC motors which independently drive a left and right wheel. A third non–powered drag wheel provides tripod stability for the robot.

In this project we equip the Blinky 602A robot platform with three Sharp GP12D IR sensors as shown in Figure 9.16. The robot is placed in a maze with reflective walls. The project goal is for the robot to detect wall placement and navigate through the maze. It is important to note the robot is not provided any information about the maze. The control algorithm for the robot is hosted on the ATmega328 aboard the Arduino UNO R3.

9.8.1   REQUIREMENTS

The requirements for this project are simple, the robot must autonomously navigate through the maze without touching maze walls.

9.8.2   CIRCUIT DIAGRAM

The circuit diagram for the robot is provided in Figure 9.17. The three IR sensors (left, middle, and right) are mounted on the leading edge of the robot to detect maze walls. The output from the sensor is fed to three ADC channels (PORTC[2:0]). The robot motors will be driven by PWM channels A and B, OC1A and OC1B (PORTB[2:1]). The microcontroller is interfaced to the motors via a transistor with enough drive capability to handle the maximum current requirements of the motor. Since the microcontroller is powered at 5 VDC and the motors are rated at 3 VDC, two 1N4001 diodes are placed in series with the motor. This reduces the supply voltage to the motor to be approximately 3 VDC. The robot will be powered by a 9 VDC battery which is fed to a 5 VDC voltage regulator. Alternatively, a 9 VDC power supply rated at several amps may be used in place of the 9 VDC battery. The supply may be connected to the robot via a flexible umbilical cable.

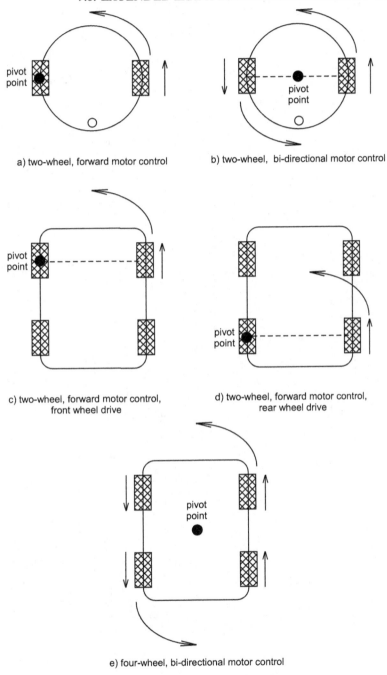

a) two-wheel, forward motor control

b) two-wheel, bi-directional motor control

c) two-wheel, forward motor control, front wheel drive

d) two-wheel, forward motor control, rear wheel drive

e) four-wheel, bi-directional motor control

Figure 9.14: Robot control configurations.

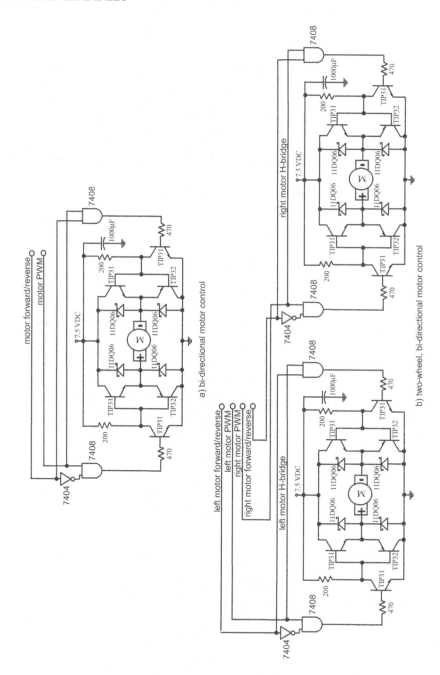

Figure 9.15: Robot motor control.

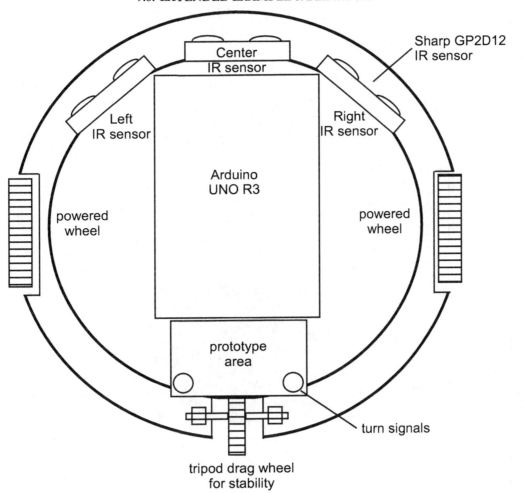

Figure 9.16: Robot layout.

9.8.3   STRUCTURE CHART

The structure chart for the robot project is provided in Figure 9.18.

9.8.4   UML ACTIVITY DIAGRAMS

The UML activity diagram for the robot is provided in Figure 9.19.

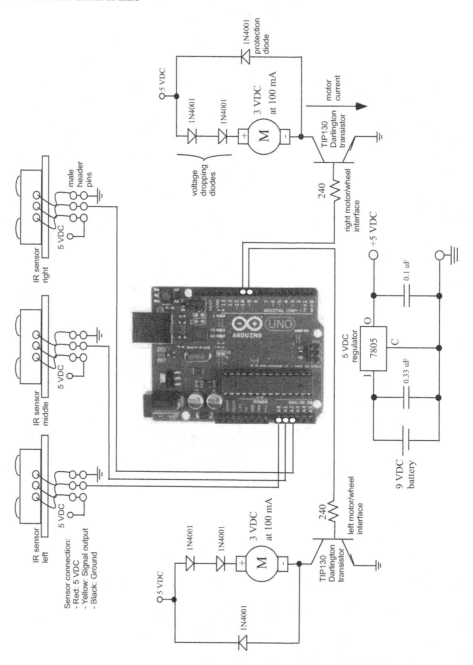

Figure 9.17: Robot circuit diagram. (UNO R3 illustration used with permission of the Arduino Team (CC BY–NC–SA) www.arduino.cc).

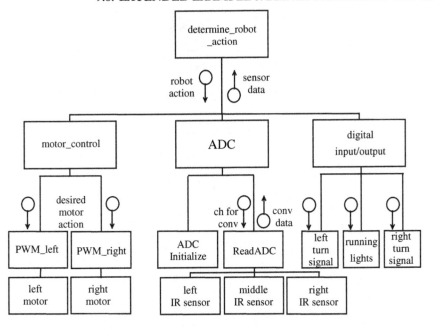

Figure 9.18: Robot structure diagram.

9.8.5   MICROCONTROLLER CODE

Provided below is the basic framework for the code. As illustrated in the Robot UML activity diagram, the control algorithm initializes various ATmega328 subsystems (ports, ADC, and PWM), senses wall locations, and issues motor control signals to avoid walls.

It is helpful to characterize the infrared sensor response to the maze walls. This allows a threshold to be determined indicating the presence of a wall. In this example we assume that a threshold of 2.5 VDC has been experimentally determined.

It is important to note that the amount of robot turn is determined by the PWM duty cycle (motor speed) and the length of time the turn is executed. For motors without optical tachometers, the appropriate values for duty cycle and motor on time must be experimentally determined. In the example functions provided the motor PWM and on time are fixed. Functions where the motor duty cycle and on time are passed to the function as local variables are left as a homework assignment.

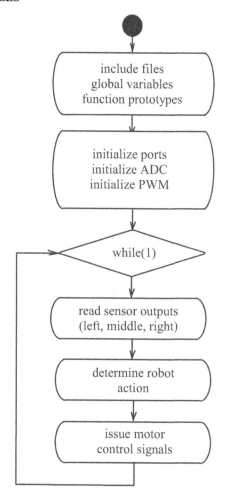

Figure 9.19: Robot UML activity diagram.

```
//*****************************************************************************
#include<iom328v.h>                        //ATmega328 include files
                                           //function prototypes
void init_ADC(void);
unsigned int Read_ADC(unsigned char channel);
void PWM(unsigned char Duty_Cycle_Left, unsigned char Duty_Cycle_Right);
void ADC_values(void);
void PWM_forward(void);
void PWM_left(void);
```

```c
void PWM_right(void);
void delay(unsigned int number_of_8_192ms_interrupts);
void init_timer2_ovf_interrupt(void);
void timer2_interrupt_isr(void);
void initialize_ports(void);

//interrupt handler definition
#pragma interrupt_handler timer2_interrupt_isr:12

//Global variables
float left_IR_voltage = 0.0;
float right_IR_voltage = 0.0;
float center_IR_voltage = 0.0;
unsigned int input_delay;

void main(void)
{
initialize_ports();                    //initialize ports
init_timer2_ovf_interrupt();           //initialize interrupts
init_ADC();                            //initialize ADC

ADC_values();
determine_robot_action();
}

//********************************************************************
// void initialize_ports(void)
//    1: output, 0: input
//********************************************************************

void initialize_ports(void)
{
DDRB = 0xFF;
DDRC = 0xF8;                           //PORTC[2:0] input
DDRD = 0xFF;
}


//********************************************************************
//void determine_robot_action(void)
```

```
//In this example we assume that a threshold of 2.5 VDC has been
//experimentally determined.
//*********************************************************************
void determine_robot_action(void)
{
                                    //wall on left and front, turn right
if((left_IR_voltage >= 2.5)&&(center_IR_voltage >= 2.5 VDC)
  &&(right_IR_voltage < 2.5))
  {
  PWM_right();
  }
else if                           //provide other cases here


  :
  :
  :
}


//*********************************************************************
//void ATD_values(void)
//
//  PORTA[0] - Left IR Sensor
//  PORTA[1] - Center IR Sensor
//  PORTA[2] - Right IR Sensor
//*********************************************************************

void ATD_values(void)
{
left_IR_Voltage   = (Read_ADC(0)*5.0)/1024.0;
center_IR_Voltage = (Read_ADC(1)*5.0)/1024.0;
right_IR_Voltage  = (Read_ADC(2)*5.0)/1024.0;
}


//*********************************************************************
//void PWM_forward(void): the PWM is configured to make the motors go
//forward.
//Implementation notes:
//  - The left motor is controlled by PWM channel OC1B
//  - The right motor is controlled by PWM channel OC1A
```

```
//  - To go forward the same PWM duty cycle is applied to both the left
//    and right motors.
//  - The length of the delay controls the amount of time the motors are
//    powered.
//*********************************************************************

void PWM_forward(void)
{

TCCR1A = 0xA1;                          //freq = resonator/510 = 10 MHz/510
                                        //freq = 19.607 kHz
TCCR1B = 0x01;                          //no clock source division
                                        //Initiate PWM duty cycle variables
                                        //Set PWM for left and right motors
                                        //to 50%
OCR1BH = 0x00;                          //PWM duty cycle CH B left motor
OCR1BL = (unsigned char)(128);
OCR1AH = 0x00;                          //PWM duty cycle CH B right motor
OCR1AL = (unsigned char)(128);
delay(122);                             //delay 1s
OCR1BL = (unsigned char)(0);            //motors off
OCR1AL = (unsigned char)(0);
}


//*********************************************************************
//void PWM_left(void) //Implementation notes:
//  - The left motor is controlled by PWM channel OC1B
//  - The right motor is controlled by PWM channel OC1A
//  - To go left the left motor is stopped and the right motor is
//    provided a PWM signal.  The robot will pivot about the left motor.
//  - The length of the delay controls the amount of time the motors are
//    powered.
//*********************************************************************

void PWM_left(void)
{
TCCR1A = 0xA1;                          //freq = resonator/510 = 10 MHz/510
                                        //freq = 19.607 kHz
TCCR1B = 0x01;                          //no clock source division
```

```
                                        //Initiate PWM duty cycle variables
                                        //Set PWM for left motor at 0%
                                        //and the right motor to 50%
OCR1BH = 0x00;                          //PWM duty cycle CH B left motor
OCR1BL = (unsigned char)(0);
OCR1AH = 0x00;                          //PWM duty cycle CH B right motor
OCR1AL = (unsigned char)(128);
delay(122);                             //delay 1s
OCR1BL = (unsigned char)(0);            //motors off
OCR1AL = (unsigned char)(0);
}


//************************************************************************
// void PWM_right(void)
//  - The left motor is controlled by PWM channel OC1B
//  - The right motor is controlled by PWM channel OC1A
//  - To go right the right motor is stopped and the left motor is
//    provided a PWM signal.  The robot will pivot about the right motor.
//  - The length of the delay controls the amount of time the motors are
//    powered.
//************************************************************************

void PWM_right(void)
{
TCCR1A = 0xA1;                          //freq = resonator/510 = 10 MHz/510
                                        //freq = 19.607 kHz
TCCR1B = 0x01;                          //no clock source division
                                        //Initiate PWM duty cycle variables
                                        //Set PWM for left motor to 50%
                                        //and right motor to 0%
OCR1BH = 0x00;                          //PWM duty cycle CH B left motor
OCR1BL = (unsigned char)(128);
OCR1AH = 0x00;                          //PWM duty cycle CH B right motor
OCR1AL = (unsigned char)(0);
delay(122);                             //delay 1s
OCR1BL = (unsigned char)(0);            //motors off
OCR1AL = (unsigned char)(0);
}
```

```c
//***********************************************************************
void Init_ADC(void)
{
ADMUX = 0;                       //Select channel 0
ADCSRA = 0xC3;                   //Enable ADC & start 1st dummy conversion
                                 //Set ADC module prescalar to 8
                                 //critical for accurate ADC results
while (!(ADCSRA & 0x10));        //Check if conversation is ready
ADCSRA |= 0x10; //Clear conv rdy flag - set the bit
}


//***********************************************************************
unsigned int Read_ADC(unsigned char channel)
{
unsigned int binary_weighted_voltage = 0x00;
unsigned intbinary_weighted_voltage_low = 0x00;
unsigned int binary_weighted_voltage_high = 0x00;

ADMUX = channel;                     //Select channel
ADCSRA |= 0x43;                      //Start conversion
                                     //Set ADC module prescalar to 8
          //critical for accurate ADC results
while (!(ADCSRA & 0x10));            //Check if conversion is ready
ADCSRA |= 0x10;                      //Clear Conv rdy flag - set the bit


binary_weighted_voltage_low = ADCL; //Read 8 low bits first
                                     //Read 2 high bits, multiply by 256
                                     //Shift to the left 8 times to get
                                     //the upper "ADC" result
                                     //into the correct position to be
                                     //ORed with the Lower result.

binary_weighted_voltage_high = ((unsigned int)(ADCH << 8));
                                     //Cast to unsigned int
                                     //OR the two results together
                                     //to form the 10 bit result
binary_weighted_voltage = binary_weighted_voltage_low
                    | binary_weighted_voltage_high;
```

```
return binary_weighted_voltage;        //ADCH:ADCL
}


//*******************************************************************
void delay(unsigned int number_of_8_192ms_interrupts)
{
TCNT2 = 0x00;                           //reset timer2
input_delay = 0;                        //reset timer2 overflow counter
while(input_delay <= number_of_8_192ms_interrupts)
  {
  ;                                     //wait specified number of interrupts
  }
}


//*********************************************************************
void init_timer2_ovf_interrupt(void)
{
TCCR2A = 0x00;    //Do nothing with this register, not needed for counting
TCCR2B = 0x06;    //divide timer2 timebase by 256, overflow every 8.192ms
TIMSK2 = 0x01;    //enable timer2 overflow interrupt
asm("SEI");       //enable global interrupt
}


//*******************************************************************


void timer2_interrupt_isr(void)
{
input_delay++;                                  //increment overflow counter
}


//*********************************************************************
```

The design provided is very basic. As end of chapter homework assignments, extend the design to include the following:

- Modify the PWM turning commands such that the PWM duty cycle and the length of time the motors are on are sent in as variables to the function.

- Equip the motor with another IR sensor that looks down to the maze floor for "land mines." A land mine consists of a paper strip placed in the maze floor that obstructs a portion of the maze. If a land mine is detected, the robot must deactivate it by rotating three times and flashing a large LED while rotating.

- Develop a function for reversing the robot.

9.9 EXTENDED EXAMPLE 7: MOUNTAIN MAZE NAVIGATING ROBOT

In this project we extend the maze navigating project to a three–dimensional mountain pass. Also, we use a robot equipped with four motorized wheels. Each of the wheels is equipped with an H–bridge to allow bidirectional motor control. In this example we will only control two wheels. We leave the development of a 4WD robot as an end of chapter homework assignment.

9.9.1 DESCRIPTION

For this project a DF Robot 4WD mobile platform kit was used (DFROBOT ROB0003, Jameco #2124285). The robot kit is equipped with four powered wheels. As in the Blinky 602A project, we equipped the DF Robot with three Sharp GP12D IR sensors as shown in Figure 9.20. The robot will be placed in a three dimensional maze with reflective walls modeled after a mountain pass. The goal of the project is for the robot to detect wall placement and navigate through the maze. The robot will not be provided any information about the maze. The control algorithm for the robot is hosted on the ATmega328.

9.9.2 REQUIREMENTS

The requirements for this project are simple, the robot must autonomously navigate through the maze without touching maze walls.

9.9.3 CIRCUIT DIAGRAM

The circuit diagram for the robot is provided in Figure 9.21. The three IR sensors (left, middle, and right) will be mounted on the leading edge of the robot to detect maze walls. The output from the sensor is fed to three ADC channels (PORTC[2:0]). The robot motors will be driven by PWM channels A and B, OC1A and OC1B, PORTB[2:1] via an H–bridge. The robot is powered by a 7.5 VDC battery pack (5 AA batteries) which is fed to a 5 VDC voltage regulator. Alternatively, the robot may be powered by a 7.5 VDC power supply rated at several amps. In this case the power is delivered to the robot by a flexible umbilical cable.

9.9.4 STRUCTURE CHART

The structure chart for the robot project is provided in Figure 9.22.

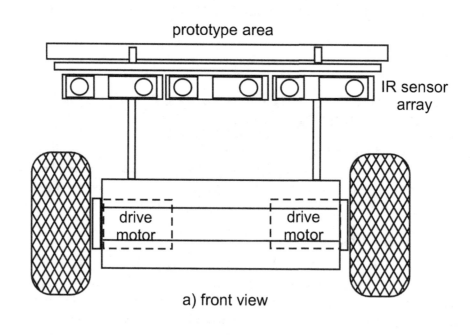

a) front view

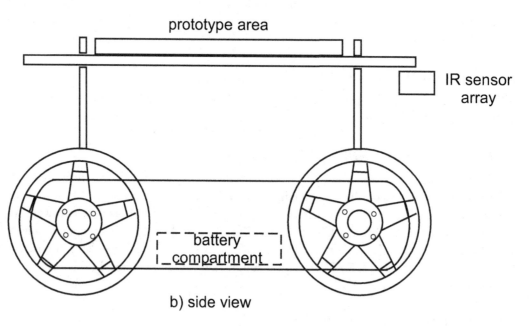

b) side view

Figure 9.20: Robot layout.

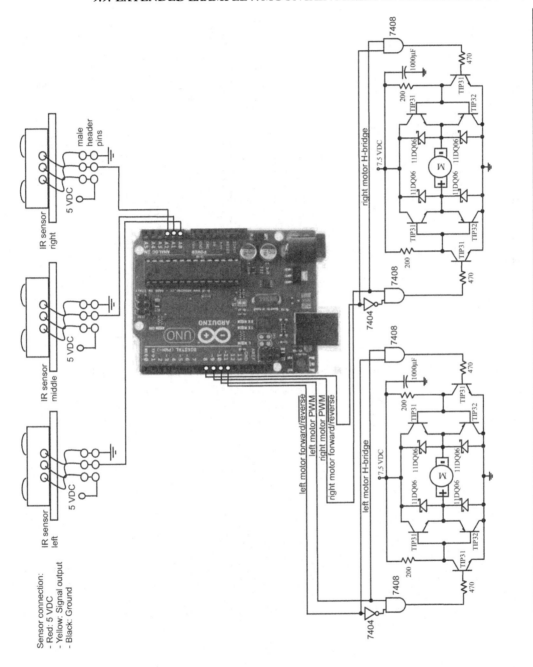

Figure 9.21: Robot circuit diagram. (UNO R3 illustration used with permission of the Arduino Team (CC BY–NC–SA) `www.arduino.cc`).

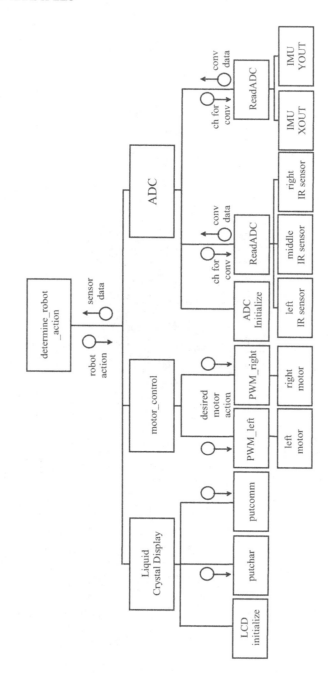

Figure 9.22: Robot structure diagram.

9.9.5   UML ACTIVITY DIAGRAMS

The UML activity diagram for the robot is provided in Figure 9.23.

9.9.6   MICROCONTROLLER CODE

The code for the robot may be adapted from that for the Blinky602A robot. Since the motors are equipped with an H–bridge, slight modifications are required to the robot turning code. These modifications are provided below.

```
//*********************************************************************
//void PWM_forward(void): the PWM is configured to make the motors go
//forward.
//Implementation notes:
//  - The left motor is controlled by PWM channel OC1B
//  - The right motor is controlled by PWM channel OC1A
//  - To go forward the same PWM duty cycle is applied to both the left
//    and right motors.
//  - The length of the delay controls the amount of time the motors are
//    powered.
//  - Direction control for the right motor is provided on PORTB[0]
//    1: forward, 0: reverse
//  - Direction control for the left motor is provided on PORTB[3]
//    1: forward, 0: reverse
/*********************************************************************

void PWM_forward(void)
{

TCCR1A = 0xA1;                        //freq = resonator/510 = 10 MHz/510
                                      //freq = 19.607 kHz
TCCR1B = 0x01;                        //no clock source division

PORTB = PORTB | 0x09;                 //Right and left motor forward
                                      //PORTB[0]=1 and PORTB[3]=1
                                      // 0 0 0 0 _ 1 0 0 1
                                      //            0x09
                                      //Initiate PWM duty cycle variables
                                      //Set PWM for left and right motors
                                      //to 50%
OCR1BH = 0x00;                        //PWM duty cycle CH B left motor
OCR1BL = (unsigned char)(128);
```

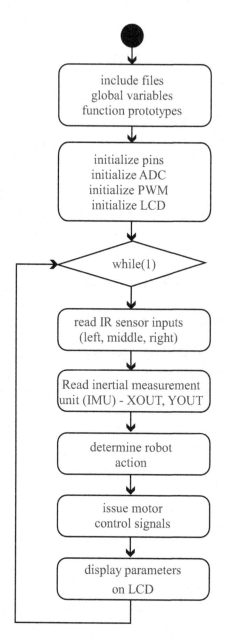

Figure 9.23: Robot UML activity diagram.

```
OCR1AH = 0x00;                          //PWM duty cycle CH B right motor
OCR1AL = (unsigned char)(128);
delay(122);                             //delay 1s
OCR1BL = (unsigned char)(0);            //motors off, no PWM duty cycle
OCR1AL = (unsigned char)(0);
}


//********************************************************************
//void PWM_left(void) //Implementation notes:
//  - The left motor is controlled by PWM channel OC1B
//  - The right motor is controlled by PWM channel OC1A
//  - To go left the left motor is stopped and the right motor is
//    provided a PWM signal.  The robot will pivot about the left motor.
//  - The length of the delay controls the amount of time the motors are
//    powered.
//  - Direction control for the right motor is provided on PORTB[0]
//    1: forward, 0: reverse
//  - Direction control for the left motor is provided on PORTB[3]
//    1: forward, 0: reverse
//********************************************************************

void PWM_left(void)
{
TCCR1A = 0xA1;                          //freq = resonator/510 = 10 MHz/510
                                        //freq = 19.607 kHz
TCCR1B = 0x01;                          //no clock source division

PORTB = PORTB & 0xF7;                   //Left motor reverse:  PORTB[3]=0
PORTB = PORTB | 0x01;                   //Right motor forward: PORTB[0]=1

                                        //Initiate PWM duty cycle variables
                                        //Set PWM for left motor at 0%
                                        //and the right motor to 50%
OCR1BH = 0x00;                          //PWM duty cycle CH B left motor
OCR1BL = (unsigned char)(128);
OCR1AH = 0x00;                          //PWM duty cycle CH B right motor
OCR1AL = (unsigned char)(128);
delay(122);                             //delay 1s
OCR1BL = (unsigned char)(0);            //motors off
```

```
OCR1AL = (unsigned char)(0);
}


//****************************************************************
// void PWM_right(void)
//  - The left motor is controlled by PWM channel OC1B
//  - The right motor is controlled by PWM channel OC1A
//  - To go right the right motor is stopped and the left motor is
//    provided a PWM signal.  The robot will pivot about the right motor.
//  - The length of the delay controls the amount of time the motors are
//    powered.
//  - Direction control for the right motor is provided on PORTB[0]
//    1: forward, 0: reverse
//  - Direction control for the left motor is provided on PORTB[3]
//    1: forward, 0: reverse
//****************************************************************

void PWM_right(void)
{
TCCR1A = 0xA1;                         //freq = resonator/510 = 10 MHz/510
                                       //freq = 19.607 kHz
TCCR1B = 0x01;                         //no clock source division

PORTB = PORTB | 0x08;                  //Left motor forward:  PORTB[3]=1
PORTB = PORTB & 0xfe;                  //Right motor reverse: PORTB[0]=0

                                       //Initiate PWM duty cycle variables
                                       //Set PWM for left motor to 50%
                                       //and right motor to 0%
OCR1BH = 0x00;                         //PWM duty cycle CH B left motor
OCR1BL = (unsigned char)(128);
OCR1AH = 0x00;                         //PWM duty cycle CH B  right motor
OCR1AL = (unsigned char)(128);
delay(122);                            //delay 1s
OCR1BL = (unsigned char)(0);          //motors off
OCR1AL = (unsigned char)(0);
}


//****************************************************************
```

9.9.7   MOUNTAIN MAZE

The mountain maze was constructed from plywood, chicken wire, expandable foam, plaster cloth and Bondo. A rough sketch of the desired maze path was first constructed. Care was taken to insure the pass was wide enough to accommodate the robot. The maze platform was constructed from 3/8 inch plywood on 2 by 4 inch framing material. Maze walls were also constructed from the plywood and supported with steel L brackets.

With the basic structure complete, the maze walls were covered with chicken wire. The chicken wire was secured to the plywood with staples. The chicken wire was then covered with plaster cloth (Creative Mark Artist Products #15006). To provide additional stability, expandable foam was sprayed under the chicken wire (Guardian Energy Technologies, Inc. Foam It Green 12). The mountain scene was then covered with a layer of Bondo for additional structural stability. Bondo is a two-part putty that hardens into a strong resin. Mountain pass construction steps are illustrated in 9.24. The robot is shown in the maze in Figure 9.25

9.9.8   PROJECT EXTENSIONS

- Modify the PWM turning commands such that the PWM duty cycle and the length of time the motors are on are sent in as variables to the function.

- Equip the motor with another IR sensor that looks down toward the maze floor for "land mines." A land mine consists of a paper strip placed in the maze floor that obstructs a portion of the maze. If a land mine is detected, the robot must deactivate the maze by moving slowly back and forth for three seconds and flashing a large LED.

- Develop a function for reversing the robot.

- The current design is a two wheel, front wheel drive system. Modify the design for a two wheel, rear wheel drive system.

- The current design is a two wheel, front wheel drive system. Modify the design for a four wheel drive system.

- Develop a four wheel drive system which includes a tilt sensor. The robot should increase motor RPM (duty cycle) for positive inclines and reduce motor RPM (duty cycle) for negatives inclines.

9.10   EXTENDED EXAMPLE 8: ROBOT WHEEL ODOMETRY

A helpful robot feature is wheel odometry or the ability to measure how far the wheel has traversed. The DF Robotics line of robots may be equipped with wheel encoders for odometry as shown in Figure 9.26.

The encoder consists of a slotted wheel that is connected to the motor axle. As the axle turns, so does the slotted wheel. The slotted wheel is mounted between an optical emitter and detector.

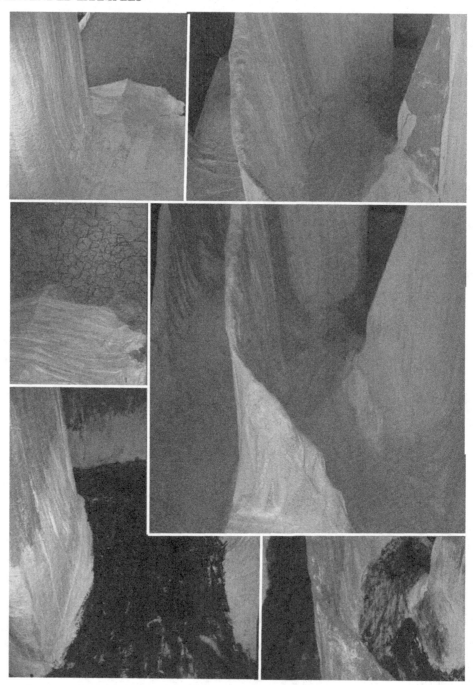

Figure 9.24: Mountain maze.

Figure 9.25: Robot in maze.

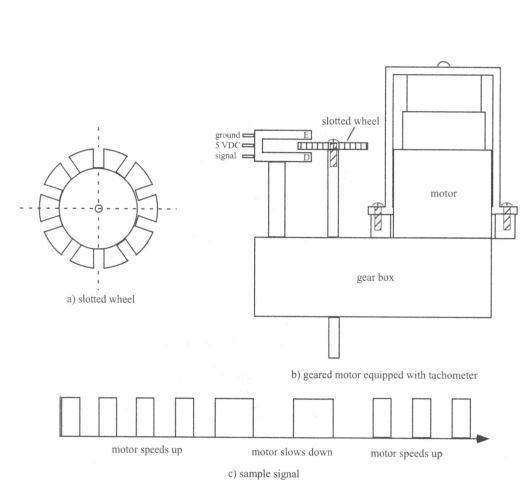

a) slotted wheel

b) geared motor equipped with tachometer

c) sample signal

Figure 9.26: Robot wheel encoder (www.DFRobot.com).

As the slotted wheel turns, a digital signal is provided by the encoder. The slotted wheel has ten vanes or 20 edges per rotation. The edges may be counted to determine distance traversed. Each robot wheel may be equipped with an encoder to monitor turns, distance traveled, slippage, etc. The signals from the encoders may be fed to the external interrupt pins on the Arduino processor board. Sample code to count vane edges is provided on the DF Robotics website (`www.DFRobot.com`).

9.11   SUMMARY

In this chapter, we discussed the voltage and current operating parameters for the Arduino UNO R3 processing board and the Atmel ATmega328 microcontroller. We discussed how this information may be applied to properly design an interface for common input and output circuits. It must be emphasized a properly designed interface allows the microcontroller to operate properly within its parameter envelope. If due to a poor interface design, a microcontroller is used outside its prescribed operating parameter values, spurious and incorrect logic values will result. We provided interface information for a wide range of input and output devices. We also discussed the concept of interfacing a motor to a microcontroller using PWM techniques coupled with high power MOSFET or SSR switching devices. We closed the chapter with a number of extended examples.

9.12   REFERENCES

- Pack D, Barrett S (2002) 68HC12 Microcontroller: Theory and Applications. Prentice–Hall Incorporated, Upper Saddle River, NJ.

- Barrett S, Pack D (2004) Embedded Systems Design with the 68HC12 and HCS12. Prentice––Hall Incorporated, Upper Saddle River, NJ.

- Crydom Corporation, 2320 Paseo de las Americas, Suite 201, San Diego, CA (`www.crydom.com`).

- Sick/Stegmann Incorporated, Dayton, OH, (`www.stegmann.com`).

- Images Company, 39 Seneca Loop, Staten Island, NY 10314.

- *Atmel 8–bit AVR Microcontroller with 16/32/64K Bytes In–System Programmable Flash, ATmega328P/V, ATmega324P/V, 644P/V* data sheet: 8011I–AVR–05/08, Atmel Corporation, 2325 Orchard Parkway, San Jose, CA 95131.

- *Atmel 8–bit AVR Microcontroller with 4/8/16/32K Bytes In–System Programmable Flash, ATmega48PA, 88PA, 168PA, 328P* data sheet: 8171D–AVR–05/11, Atmel Corporation, 2325 Orchard Parkway, San Jose, CA 95131.

- *Atmel 8–bit AVR Microcontroller with 64/128/256K Bytes In–System Programmable Flash, ATmega640/V, ATmega1280/V, 2560/V* data sheet: 2549P–AVR–10/2012, Atmel Corporation, 2325 Orchard Parkway, San Jose, CA 95131.

- Barrett S, Pack D (2006) Microcontrollers Fundamentals for Engineers and Scientists. Morgan and Claypool Publishers. DOI: 10.2200/S00025ED1V01Y200605DCS001

- Barrett S and Pack D (2008) Atmel AVR Microcontroller Primer Programming and Interfacing. Morgan and Claypool Publishers. DOI: 10.2200/S00100ED1V01Y200712DCS015

- Barrett S (2010) Embedded Systems Design with the Atmel AVR Microcontroller. Morgan and Claypool Publishers. DOI: 10.2200/S00225ED1V01Y200910DCS025

- National Semiconductor, *LM34/LM34A/LM34C/LM34CA/LM34D Precision Fahrenheit Temperature Sensor*, 1995.

- Bohm H and Jensen V (1997) Build Your Own Underwater Robot and Other Wet Projects. Westcoast Words.

- seaperch, www.seaperch.org

- SparkFun, www.sparkfun.com

- Pack D and Barrett S (2011) Microcontroller Programming and Interfacing Texas Instruments MSP430. Morgan and Claypool Publishers.

9.13 CHAPTER PROBLEMS

1. What will happen if a microcontroller is used outside of its prescribed operating envelope?

2. Discuss the difference between the terms "sink" and "source" as related to current loading of a microcontroller.

3. Can an LED with a series limiting resistor be directly driven by the Atmel microcontroller? Explain.

4. In your own words, provide a brief description of each of the microcontroller electrical parameters.

5. What is switch bounce? Describe two techniques to minimize switch bounce.

6. Describe a method of debouncing a keypad.

7. What is the difference between an incremental encoder and an absolute encoder? Describe applications for each type.

8. What must be the current rating of the 2N2222 and 2N2907 transistors used in the tri–state LED circuit? Support your answer.

9. Draw the circuit for a six character seven segment display. Fully specify all components. Write a program to display "ATmega328."

10. Repeat the question above for a dot matrix display.

11. Repeat the question above for a LCD display.

12. What is the difference between a unipolar and bipolar stepper motor?

13. What controls the speed of rotation of a stepper motor?

14. A stepper motor provides and angular displacement of 1.8 degrees per step. How can this resolution be improved?

15. Write a function to convert an ASCII numeral representation (0 to 9) to a seven segment display.

16. Why is an interface required between a microcontroller and a stepper motor?

17. For the SeaPerch ROV provide adjustable speed control for each motor. This advanced feature requires several modifications including a proportional joystick and pulse width modulation signals to the motor. The joystick currently in use provides a simple on and off signal. A proportional joystick provides an X and Y output DC signal proportional to the joystick deflection on each axis. These signals may be fed to the analog–to–digital converter aboard the Arduino UNO R3. The X and Y signals captured from the joystick may be used to provide a pulse width modulated signal with a duty cycle proportional to deflection.

18. For the SeaPerch ROV provide a powered dive and surface thruster. To provide for a powered dive and surface capability, the ROV must be equipped with a vertical thruster equipped with an H–bridge to allow for motor forward and reversal.

19. For the SeaPerch ROV provide left and right thruster reverse. Currently the left and right thrusters may only be powered in one direction. To provide additional maneuverability, the left and right thrusters could be equipped with an H–bridge to allow bi–directional motor control.

20. For the SeaPerch ROV provide proportional speed control with bi–directional motor control. Both of these advanced features may be provided by driving the H–bridge circuit with PWM signals.

21. Construct the UML activity diagrams for all functions related to the weather station.

22. It is desired to updated weather parameters every 15 minutes. Write a function to provide a 15 minute delay.

23. Add one of the following sensors to the weather station:

  • anemometer

- barometer

- hygrometer

- rain gauge

- thermocouple

You will need to investigate background information on the selected sensor, develop an interface circuit for the sensor, and modify the weather station code.

24. Modify the weather station software to also employ a 138 x 110 LCD. Display pertinent weather data on the display.

25. Equip the weather station with an MMC/SD flash memory card.

26. Modify the motor speed control circuit interface to provide for bi–directional motor control.

27. Use optical encoder output channels A and B to determine motor direction and speed.

28. Modify the motor speed control algorithm to display motor direction (CW or CCW) and speed in RPM on the LCD.

29. The Blinky 602A robot under microcontroller control abruptly starts and stops when PWM is applied. Modify the algorithm to provide the capability to gradually ramp up (and down) the motor speed.

30. Modify the Blinky 602A circuit and microcontroller code such that the maximum speed of the robot is set with an external potentiometer.

31. Modify the Blinky 602A circuit and microcontroller code such that the IR sensors are only asserted just before a range reading is taken.

32. Apply embedded system design techniques presented throughout the text to a project of your choosing. Follow the design process and provide the following products:

- system description,

- system requirements,

- a structure chart,

- system circuit diagram,

- UML activity diagrams, and the

- microcontroller code.

33. Add the following features to the Blinky 602A platform:

- Line following capability (Hint: Adapt the line following circuitry onboard the Blinky 602A to operate with the Arduino processing board.)

- Two way robot communications (use the IR sensors already aboard)

- LCD display for status and troubleshooting display

- Voice output (Hint: Use SPO–512 chip.)

34. Develop an embedded system controlled submarine (www.seaperch.org).

35. Equip the Arduino UNO R3 with automatic cell phone dialing capability to notify you when a fire is present in your home.

36. Develop an embedded system controlled dirigible/blimp (www.microflight.com, www.rctoys.com).

37. Develop a trip odometer for your bicycle (Hint: use a Hall Effect sensor to detect tire rotation).

38. Develop a timing system for a 4 lane pinewood derby track.

39. Develop a playing board and control system for your favorite game (Yahtzee, Connect Four, Battleship, etc.).

40. You have a very enthusiastic dog that loves to chase balls. Develop a system to launch balls for the dog.

APPENDIX A

ATmega328 Register Set

Address	Name	Bit 7	Bit 6	Bit 5	Bit 4	Bit 3	Bit 2	Bit 1	Bit 0	Page
(0xFF)	Reserved	–	–	–	–	–	–	–	–	
(0xFE)	Reserved	–	–	–	–	–	–	–	–	
(0xFD)	Reserved	–	–	–	–	–	–	–	–	
(0xFC)	Reserved	–	–	–	–	–	–	–	–	
(0xFB)	Reserved	–	–	–	–	–	–	–	–	
(0xFA)	Reserved	–	–	–	–	–	–	–	–	
(0xF9)	Reserved	–	–	–	–	–	–	–	–	
(0xF8)	Reserved	–	–	–	–	–	–	–	–	
(0xF7)	Reserved	–	–	–	–	–	–	–	–	
(0xF6)	Reserved	–	–	–	–	–	–	–	–	
(0xF5)	Reserved	–	–	–	–	–	–	–	–	
(0xF4)	Reserved	–	–	–	–	–	–	–	–	
(0xF3)	Reserved	–	–	–	–	–	–	–	–	
(0xF2)	Reserved	–	–	–	–	–	–	–	–	
(0xF1)	Reserved	–	–	–	–	–	–	–	–	
(0xF0)	Reserved	–	–	–	–	–	–	–	–	
(0xEF)	Reserved	–	–	–	–	–	–	–	–	
(0xEE)	Reserved	–	–	–	–	–	–	–	–	
(0xED)	Reserved	–	–	–	–	–	–	–	–	
(0xEC)	Reserved	–	–	–	–	–	–	–	–	
(0xEB)	Reserved	–	–	–	–	–	–	–	–	
(0xEA)	Reserved	–	–	–	–	–	–	–	–	
(0xE9)	Reserved	–	–	–	–	–	–	–	–	
(0xE8)	Reserved	–	–	–	–	–	–	–	–	
(0xE7)	Reserved	–	–	–	–	–	–	–	–	
(0xE6)	Reserved	–	–	–	–	–	–	–	–	
(0xE5)	Reserved	–	–	–	–	–	–	–	–	
(0xE4)	Reserved	–	–	–	–	–	–	–	–	
(0xE3)	Reserved	–	–	–	–	–	–	–	–	
(0xE2)	Reserved	–	–	–	–	–	–	–	–	
(0xE1)	Reserved	–	–	–	–	–	–	–	–	

Figure A.1: Atmel AVR ATmega328 Register Set. (Figure used with permission of Atmel, Incorporated.)

(0xE0)	Reserved	–	–	–	–	–	–	–	–	
(0xDF)	Reserved	–	–	–	–	–	–	–	–	
(0xDE)	Reserved	–	–	–	–	–	–	–	–	
(0xDD)	Reserved	–	–	–	–	–	–	–	–	
(0xDC)	Reserved	–	–	–	–	–	–	–	–	
(0xDB)	Reserved	–	–	–	–	–	–	–	–	
(0xDA)	Reserved	–	–	–	–	–	–	–	–	
(0xD9)	Reserved	–	–	–	–	–	–	–	–	
(0xD8)	Reserved	–	–	–	–	–	–	–	–	
(0xD7)	Reserved	–	–	–	–	–	–	–	–	
(0xD6)	Reserved	–	–	–	–	–	–	–	–	
(0xD5)	Reserved	–	–	–	–	–	–	–	–	
(0xD4)	Reserved	–	–	–	–	–	–	–	–	
(0xD3)	Reserved	–	–	–	–	–	–	–	–	
(0xD2)	Reserved	–	–	–	–	–	–	–	–	
(0xD1)	Reserved	–	–	–	–	–	–	–	–	
(0xD0)	Reserved	–	–	–	–	–	–	–	–	
(0xCF)	Reserved	–	–	–	–	–	–	–	–	
(0xCE)	Reserved	–	–	–	–	–	–	–	–	
(0xCD)	Reserved	–	–	–	–	–	–	–	–	
(0xCC)	Reserved	–	–	–	–	–	–	–	–	
(0xCB)	Reserved	–	–	–	–	–	–	–	–	
(0xCA)	Reserved	–	–	–	–	–	–	–	–	
(0xC9)	Reserved	–	–	–	–	–	–	–	–	
(0xC8)	Reserved	–	–	–	–	–	–	–	–	
(0xC7)	Reserved	–	–	–	–	–	–	–	–	
(0xC6)	Reserved	USART I/O Data Register								195
(0xC5)	Reserved					USART Baud Rate Register Low				199
(0xC4)	Reserved	USART Baud Rate Register Low								199
(0xC3)	Reserved	–	–	–	–	–	–	–	–	
(0xC2)	Reserved	UMSEL01	UMSEL00	UPM01	UPM00	USB00	UCSZ01/UDORD0	UCSZ00/UCPHA0	UCPOL0	197/212
(0xC1)	Reserved	RXCIE0	TXCIE0	UDRIE0	RXEN0	TXEN0	UCSZ02	RXB80	TXB80	196
(0xC0)	UCSR0A	RXC0	TXC0	UDRE0	FE0	DOR0	UPE0	U2X0	MPCM0	195

Figure A.2: Atmel AVR ATmega328 Register Set. (Figure used with permission of Atmel, Incorporated.)

Address	Name	Bit 7	Bit 6	Bit 5	Bit 4	Bit 3	Bit 2	Bit 1	Bit 0	Page
(0xBF)	Reserved	–	–	–	–	–	–	–	–	
(0xBE)	Reserved	–	–	–	–	–	–	–	–	
(0xBD)	TWAMR	TWAM6	TWAM5	TWAM4	TWAM3	TWAM2	TWAM1	TWAM0	–	244
(0xBC)	TWCR	TWINT	TWEA	TWSTA	TWSTO	TWWC	TWEN	–	TWIE	241
(0xBB)	TWDR	2-wire Serial Interface Data Register								243
(0xBA)	TWAR	TWA6	TWA5	TWA4	TWA3	TWA2	TWA1	TWA0	TWGCE	244
(0xB9)	TWSR	TWS7	TWS6	TWS5	TWS4	TWS3	–	TWPS1	TWPS0	243
(0xB8)	TWBR	2-wire Serial Interface Bit Rate Register								241
(0xB7)	Reserved	–								
(0xB6)	ASSR	–	EXCLK	AS2	TCN2UB	OCR2AUB	OCR2BUB	TCR2AUB	TCR2BUB	164
(0xB5)	Reserved	–	–					–	–	
(0xB4)	OCR2B	Timer/Counter2 Output Compare Register B								162
(0xB3)	OCR2A	Timer/Counter2 Output Compare Register A								162
(0xB2)	TCNT2	Timer/Counter2 (8-bit)								162
(0xB1)	TCCR2B	FOC2A	FOC2B	–	–	WGM22	CS22	CS21	CS20	161
(0xB0)	TCCR2A	COM2A1	COM2A0	COM2B1	COM2B0	–	–	WGM21	WGM20	158
(0xAF)	Reserved	–	–	–	–	–	–	–	–	
(0xAE)	Reserved	–	–	–	–	–	–	–	–	
(0xAD)	Reserved	–	–	–	–	–	–	–	–	
(0xAC)	Reserved	–	–	–	–	–	–	–	–	
(0xAB)	Reserved	–	–	–	–	–	–	–	–	
(0xAA)	Reserved	–	–	–	–	–	–	–	–	
(0xA9)	Reserved	–	–	–	–	–	–	–	–	
(0xA8)	Reserved	–	–	–	–	–	–	–	–	
(0xA7)	Reserved	–	–	–	–	–	–	–	–	
(0xA6)	Reserved	–	–	–	–	–	–	–	–	
(0xA5)	Reserved	–	–	–	–	–	–	–	–	
(0xA4)	Reserved	–	–	–	–	–	–	–	–	
(0xA3)	Reserved	–	–	–	–	–	–	–	–	
(0xA2)	Reserved	–	–	–	–	–	–	–	–	
(0xA1)	Reserved	–	–	–	–	–	–	–	–	
(0xA0)	Reserved	–	–	–	–	–	–	–	–	
(0x9F)	Reserved	–	–	–	–	–	–	–	–	
(0x9E)	Reserved	–	–	–	–	–	–	–	–	
(0x9D)	Reserved	–	–	–	–	–	–	–	–	
(0x9C)	Reserved	–	–	–	–	–	–	–	–	
(0x9B)	Reserved	–	–	–	–	–	–	–	–	
(0x9A)	Reserved	–	–	–	–	–	–	–	–	
(0x99)	Reserved	–	–	–	–	–	–	–	–	
(0x98)	Reserved	–	–	–	–	–	–	–	–	
(0x97)	Reserved	–	–	–	–	–	–	–	–	
(0x96)	Reserved	–	–	–	–	–	–	–	–	
(0x95)	Reserved	–	–	–	–	–	–	–	–	
(0x94)	Reserved	–	–	–	–	–	–	–	–	
(0x93)	Reserved	–	–	–	–	–	–	–	–	
(0x92)	Reserved	–	–	–	–	–	–	–	–	
(0x91)	Reserved	–	–	–	–	–	–	–	–	
(0x90)	Reserved	–	–	–	–	–	–	–	–	
(0x8F)	Reserved	–	–	–	–	–	–	–	–	
(0x8E)	Reserved	–	–	–	–	–	–	–	–	
(0x8D)	Reserved	–	–	–	–	–	–	–	–	
(0x8C)	Reserved	–	–	–	–	–	–	–	–	
(0x8B)	OCR1BH	Timer/Counter1 - Output Compare Register B High Byte								138
(0x8A)	OCR1BL	Timer/Counter1 - Output Compare Register B Low Byte								138
(0x89)	OCR1AH	Timer/Counter1 - Output Compare Register A High Byte								138
(0x88)	OCR1AL	Timer/Counter1 - Output Compare Register A Low Byte								138
(0x87)	ICR1H	Timer/Counter1 - Input Capture Register High Byte								138
(0x86)	ICR1L	Timer/Counter1 - Input Capture Register Low Byte								138
(0x85)	TCNT1H	Timer/Counter1 - Counter Register High Byte								138
(0x84)	TCNT1L	Timer/Counter1 - Counter Register Low Byte								138
(0x83)	Reserved	–	–	–	–	–	–	–	–	
(0x82)	TCCR1C	FOC1A	FOC1B	–	–	–	–	–	–	137
(0x81)	TCCR1B	ICNC1	ICES1	–	WGM13	WGM12	CS12	CS11	CS10	136
(0x80)	TCCR1A	COM1A1	COM1A0	COM1B1	COM1B0	–	–	WGM11	WGM10	134
(0x7F)	DIDR1	–	–	–	–	–	–	AIN1D	AIN0D	249
(0x7E)	DIDR0	–	–	ADC5D	ADC4D	ADC3D	ADC2D	ADC1D	ADC0D	266

Figure A.3: Atmel AVR ATmega328 Register Set. (Figure used with permission of Atmel, Incorporated.)

Address	Name	Bit 7	Bit 6	Bit 5	Bit 4	Bit 3	Bit 2	Bit 1	Bit 0	Page
(0x7D)	Reserved	–	–	–	–	–	–	–	–	
(0x7C)	ADMUX	REFS1	REFS0	ADLAR	–	MUX3	MUX2	MUX1	MUX0	262
(0x7B)	ADCSRB	–	ACME	–	–	–	ADTS2	ADTS1	ADTS0	265
(0x7A)	ADCSRA	ADEN	ADSC	ADATE	ADIF	ADIE	ADPS2	ADPS1	ADPS0	263
(0x79)	ADCH	ADC Data Register High byte								265
(0x78)	ADCL	ADC Data Register Low byte								265
(0x77)	Reserved									
(0x76)	Reserved	–	–	–	–	–	–	–	–	
(0x75)	Reserved	–	–	–	–	–	–	–	–	
(0x74)	Reserved	–	–	–	–	–	–	–	–	
(0x73)	Reserved	–	–	–	–	–	–	–	–	
(0x72)	Reserved	–	–	–	–	–	–	–	–	
(0x71)	Reserved	–	–	–	–	–	–	–	–	
(0x70)	TIMSK2	–	–	–	–	–	OCIE2B	OCIE2A	TOIE2	163
(0x6F)	TIMSK1	–	–	ICIE1	–	–	OCIE1B	OCIE1A	TOIE1	139
(0x6E)	TIMSK0	–	–	–	–	–	OCIE0B	OCIE0A	TOIE0	111
(0x6D)	PCMSK2	PCINT23	PCINT22	PCINT21	PCINT20	PCINT19	PCINT18	PCINT17	PCINT16	74
(0x6C)	PCMSK1	–	PCINT14	PCINT13	PCINT12	PCINT11	PCINT10	PCINT9	PCINT8	74
(0x6B)	PCMSK0	PCINT7	PCINT6	PCINT5	PCINT4	PCINT3	PCINT2	PCINT1	PCINT0	74
(0x6A)	Reserved	–	–	–	–	–	–	–	–	
(0x69)	EICRA	–	–	–	–	ISC11	ISC10	ISC01	ISC00	71
(0x68)	PCICR	–	–	–	–	–	PCIE2	PCIE1	PCIE0	
(0x67)	Reserved	–	–	–	–	–	–	–	–	
(0x66)	OSCCAL	Oscillator Calibration Register								37
(0x65)	Reserved	–	–	–	–	–	–	–	–	
(0x64)	PRR	PRTWI	PRTIM2	PRTIM0	–	PRTIM1	PRSPI	PRUSART0	PRADC	42
(0x63)	Reserved	–	–	–	–	–	–	–	–	
(0x62)	Reserved	–	–	–	–	–	–	–	–	
(0x61)	CLKPR	CLKPCE	–	–	–	CLKPS3	CLKPS2	CLKPS1	CLKPS0	37
(0x60)	WDTCSR	WDIF	WDIE	WDP3	WDCE	WDE	WDP2	WDP1	WDP0	54
0x3F (0x5F)	SREG	I	T	H	S	V	N	Z	C	9
0x3E (0x5E)	SPH	–	–	–	–	–	(SP10)	SP9	SP8	12
0x3D (0x5D)	SPL	SP7	SP6	SP5	SP4	SP3	SP2	SP1	SP0	12
0x3C (0x5C)	Reserved	–	–	–	–	–	–	–	–	
0x3B (0x5B)	Reserved	–	–	–	–	–	–	–	–	
0x3A (0x5A)	Reserved	–	–	–	–	–	–	–	–	
0x39 (0x59)	Reserved	–	–	–	–	–	–	–	–	
0x38 (0x58)	Reserved	–	–	–	–	–	–	–	–	
0x37 (0x57)	SPMCSR	SPMIE	(RWWSB)	–	(RWWSRE)	BLBSET	PGWRT	PGERS	SELFPRGEN	292
0x36 (0x56)	Reserved	–	–	–	–	–	–	–	–	
0x35 (0x55)	MCUCR	–	BODS	BODSE	PUD	–	–	IVSEL	IVCE	44/68/92
0x34 (0x54)	MCUSR	–	–	–	–	WDRF	BORF	EXTRF	PORF	54
0x33 (0x53)	SMCR	–	–	–	–	SM2	SM1	SM0	SE	40
0x32 (0x52)	Reserved	–	–	–	–	–	–	–	–	
0x31 (0x51)	Reserved	–	–	–	–	–	–	–	–	
0x30 (0x50)	ACSR	ACD	ACBG	ACO	ACI	ACIE	ACIC	ACIS1	ACIS0	247
0x2F (0x4F)	Reserved	–	–	–	–	–	–	–	–	
0x2E (0x4E)	SPDR	SPI Data Register								175
0x2D (0x4D)	SPSR	SPIF	WCOL	–	–	–	–	–	SPI2X	174
0x2C (0x4C)	SPCR	SPIE	SPE	DORD	MSTR	CPOL	CPHA	SPR1	SPR0	173
0x2B (0x4B)	GPIOR2	General Purpose I/O Register 2								25
0x2A (0x4A)	GPIOR1	General Purpose I/O Register 1								25
0x29 (0x49)	Reserved	–	–	–	–	–	–	–	–	
0x28 (0x48)	OCR0B	Timer/Counter0 Output Compare Register B								
0x27 (0x47)	OCR0A	Timer/Counter0 Output Compare Register A								
0x26 (0x46)	TCNT0	Timer/Counter0 (8-bit)								
0x25 (0x45)	TCCR0B	FOC0A	FOC0B	–	–	WGM02	CS02	CS01	CS00	
0x24 (0x44)	TCCR0A	COM0A1	COM0A0	COM0B1	COM0B0	–	–	WGM01	WGM00	
0x23 (0x43)	GTCCR	TSM	–	–	–	–	–	PSRASY	PSRSYNC	143/165
0x22 (0x42)	EEARH	(EEPROM Address Register High Byte)								21
0x21 (0x41)	EEARL	EEPROM Address Register Low Byte								21
0x20 (0x40)	EEDR	EEPROM Data Register								21
0x1F (0x3F)	EECR	–	–	EEPM1	EEPM0	EERIE	EEMPE	EEPE	EERE	21
0x1E (0x3E)	GPIOR0	General Purpose I/O Register 0								25
0x1D (0x3D)	EIMSK	–	–	–	–	–	–	INT1	INT0	72
0x1C (0x3C)	EIFR	–	–	–	–	–	–	INTF1	INTF0	72

Figure A.4: Atmel AVR ATmega328 Register Set. (Figure used with permission of Atmel, Incorporated.)

Address	Name	Bit 7	Bit 6	Bit 5	Bit 4	Bit 3	Bit 2	Bit 1	Bit 0	Page
0x1B (0x3B)	PCIFR	–	–	–	–	–	PCIF2	PCIF1	PCIF0	
0x1A (0x3A)	Reserved	–	–	–	–	–	–	–	–	
0x19 (0x39)	Reserved	–	–	–	–	–	–	–	–	
0x18 (0x38)	Reserved	–	–	–	–	–	–	–	–	
0x17 (0x37)	TIFR2	–	–	–	–	–	OCF2B	OCF2A	TOV2	163
0x16 (0x36)	TIFR1	–	–	ICF1	–	–	OCF1B	OCF1A	TOV1	139
0x15 (0x35)	TIFR0	–	–	–	–	–	OCF0B	OCF0A	TOV0	
0x14 (0x34)	Reserved	–	–	–	–	–	–	–	–	
0x13 (0x33)	Reserved	–	–	–	–	–	–	–	–	
0x12 (0x32)	Reserved	–	–	–	–	–	–	–	–	
0x11 (0x31)	Reserved	–	–	–	–	–	–	–	–	
0x10 (0x30)	Reserved	–	–	–	–	–	–	–	–	
0x0F (0x2F)	Reserved	–	–	–	–	–	–	–	–	
0x0E (0x2E)	Reserved	–	–	–	–	–	–	–	–	
0x0D (0x2D)	Reserved	–	–	–	–	–	–	–	–	
0x0C (0x2C)	Reserved	–	–	–	–	–	–	–	–	
0x0B (0x2B)	PORTD	PORTD7	PORTD6	PORTD5	PORTD4	PORTD3	PORTD2	PORTD1	PORTD0	93
0x0A (0x2A)	DDRD	DDD7	DDD6	DDD5	DDD4	DDD3	DDD2	DDD1	DDD0	93
0x09 (0x29)	PIND	PIND7	PIND6	PIND5	PIND4	PIND3	PIND2	PIND1	PIND0	93
0x08 (0x28)	PORTC	–	PORTC6	PORTC5	PORTC4	PORTC3	PORTC2	PORTC1	PORTC0	92
0x07 (0x27)	DDRC	–	DDC6	DDC5	DDC4	DDC3	DDC2	DDC1	DDC0	92
0x06 (0x26)	PINC	–	PINC6	PINC5	PINC4	PINC3	PINC2	PINC1	PINC0	92
0x05 (0x25)	PORTB	PORTB7	PORTB6	PORTB5	PORTB4	PORTB3	PORTB2	PORTB1	PORTB0	92
0x04 (0x24)	DDRB	DDB7	DDB6	DDB5	DDB4	DDB3	DDB2	DDB1	DDB0	92
0x03 (0x23)	PINB	PINB7	PINB6	PINB5	PINB4	PINB3	PINB2	PINB1	PINB0	92
0x02 (0x22)	Reserved	–	–	–	–	–	–	–	–	
0x01 (0x21)	Reserved	–	–	–	–	–	–	–	–	
0x00 (0x20)	Reserved	–	–	–	–	–	–	–	–	

Note: 1. For compatibility with future devices, reserved bits should be written to zero if accessed. Reserved I/O memory addresses should never be written.

2. I/O Registers within the address range 0x00 - 0x1F are directly bit-accessible using the SBI and CBI instructions. In these registers, the value of single bits can be checked by using the SBIS and SBIC instructions.

3. Some of the Status Flags are cleared by writing a logical one to them. Note that, unlike most other AVRs, the CBI and SBI instructions will only operate on the specified bit, and can therefore be used on registers containing such Status Flags. The CBI and SBI instructions work with registers 0x00 to 0x1F only.

4. When using the I/O specific commands IN and OUT, the I/O addresses 0x00 - 0x3F must be used. When addressing I/O Registers as data space using LD and ST instructions, 0x20 must be added to these addresses. The ATmega48PA/88PA/168PA/328P is a complex microcontroller with more peripheral units than can be supported within the 64 location reserved in Opcode for the IN and OUT instructions. For the Extended I/O space from 0x60 - 0xFF in SRAM, only the ST/STS/STD and LD/LDS/LDD instructions can be used.

5. Only valid for ATmega88PA/168PA.

Figure A.5: Atmel AVR ATmega328 Register Set. (Figure used with permission of Atmel, Incorporated.)

APPENDIX B

ATmega328 Header File

During C programming, the contents of a specific register may be referred to by name when an appropriate header file is included within your program. The header file provides the link between the register name used within a program and the hardware location of the register.

Provided below is the ATmega328 header file from the ICC AVR compiler. This header file was provided courtesy of ImageCraft Incorporated.

```c
#ifndef __iom328pv_h
#define __iom328pv_h

/* ATmega328P header file for
 * ImageCraft ICCAVR compiler
 */

/* i/o register addresses
 * >= 0x60 are memory mapped only
 */

/* 2006/10/01 created
 */

/* Port D */
#define PIND (*(volatile unsigned char *)0x29)
#define DDRD (*(volatile unsigned char *)0x2A)
#define PORTD (*(volatile unsigned char *)0x2B)

/* Port C */
#define PINC (*(volatile unsigned char *)0x26)
#define DDRC (*(volatile unsigned char *)0x27)
#define PORTC (*(volatile unsigned char *)0x28)

/* Port B */
#define PINB (*(volatile unsigned char *)0x23)
#define DDRB (*(volatile unsigned char *)0x24)
#define PORTB (*(volatile unsigned char *)0x25)
```

```
/* Port A */
#define PINA (*(volatile unsigned char *)0x20)
#define DDRA (*(volatile unsigned char *)0x21)
#define PORTA (*(volatile unsigned char *)0x22)

/* Timer/Counter Interrupts */
#define TIFR0 (*(volatile unsigned char *)0x35)
#define   OCF0B    2
#define   OCF0A    1
#define   TOV0     0
#define TIMSK0 (*(volatile unsigned char *)0x6E)
#define   OCIE0B   2
#define   OCIE0A   1
#define   TOIE0    0
#define TIFR1 (*(volatile unsigned char *)0x36)
#define   ICF1     5
#define   OCF1B    2
#define   OCF1A    1
#define   TOV1     0
#define TIMSK1 (*(volatile unsigned char *)0x6F)
#define   ICIE1    5
#define   OCIE1B   2
#define   OCIE1A   1
#define   TOIE1    0
#define TIFR2 (*(volatile unsigned char *)0x37)
#define   OCF2B    2
#define   OCF2A    1
#define   TOV2     0
#define TIMSK2 (*(volatile unsigned char *)0x70)
#define   OCIE2B   2
#define   OCIE2A   1
#define   TOIE2    0

/* External Interrupts */
#define EIFR (*(volatile unsigned char *)0x3C)
#define   INTF2    2
#define   INTF1    1
#define   INTF0    0
```

```
#define EIMSK (*(volatile unsigned char *)0x3D)
#define  INT2     2
#define  INT1     1
#define  INT0     0
#define EICRA (*(volatile unsigned char *)0x69)
#define  ISC21    5
#define  ISC20    4
#define  ISC11    3
#define  ISC10    2
#define  ISC01    1
#define  ISC00    0


/* Pin Change Interrupts */
#define PCIFR (*(volatile unsigned char *)0x3B)
#define  PCIF3    3
#define  PCIF2    2
#define  PCIF1    1
#define  PCIF0    0
#define PCICR (*(volatile unsigned char *)0x68)
#define  PCIE3    3
#define  PCIE2    2
#define  PCIE1    1
#define  PCIE0    0
#define PCMSK0 (*(volatile unsigned char *)0x6B)
#define PCMSK1 (*(volatile unsigned char *)0x6C)
#define PCMSK2 (*(volatile unsigned char *)0x6D)
#define PCMSK3 (*(volatile unsigned char *)0x73)


/* GPIOR */
#define GPIOR0 (*(volatile unsigned char *)0x3E)
#define GPIOR1 (*(volatile unsigned char *)0x4A)
#define GPIOR2 (*(volatile unsigned char *)0x4B)


/* EEPROM */
#define EECR (*(volatile unsigned char *)0x3F)
#define  EEPM1    5
#define  EEPM0    4
#define  EERIE    3
#define  EEMPE    2
```

```
#define  EEMWE    2
#define  EEPE     1
#define  EEWE     1
#define  EERE     0
#define EEDR (*(volatile unsigned char *)0x40)
#define EEAR (*(volatile unsigned int *)0x41)
#define EEARL (*(volatile unsigned char *)0x41)
#define EEARH (*(volatile unsigned char *)0x42)


/* GTCCR */
#define GTCCR (*(volatile unsigned char *)0x43)
#define  TSM      7
#define  PSRASY   1
#define  PSR2     1
#define  PSRSYNC  0
#define  PSR10    0


/* Timer/Counter 0 */
#define OCR0B (*(volatile unsigned char *)0x48)
#define OCR0A (*(volatile unsigned char *)0x47)
#define TCNT0 (*(volatile unsigned char *)0x46)
#define TCCR0B (*(volatile unsigned char *)0x45)
#define  FOC0A    7
#define  FOC0B    6
#define  WGM02    3
#define  CS02     2
#define  CS01     1
#define  CS00     0
#define TCCR0A (*(volatile unsigned char *)0x44)
#define  COM0A1   7
#define  COM0A0   6
#define  COM0B1   5
#define  COM0B0   4
#define  WGM01    1
#define  WGM00    0


/* SPI */
#define SPCR (*(volatile unsigned char *)0x4C)
#define  SPIE     7
```

```
#define   SPE       6
#define   DORD      5
#define   MSTR      4
#define   CPOL      3
#define   CPHA      2
#define   SPR1      1
#define   SPR0      0
#define SPSR (*(volatile unsigned char *)0x4D)
#define   SPIF      7
#define   WCOL      6
#define   SPI2X     0
#define SPDR (*(volatile unsigned char *)0x4E)

/* Analog Comparator Control and Status Register */
#define ACSR (*(volatile unsigned char *)0x50)
#define   ACD       7
#define   ACBG      6
#define   ACO       5
#define   ACI       4
#define   ACIE      3
#define   ACIC      2
#define   ACIS1     1
#define   ACIS0     0

/* OCDR */
#define OCDR (*(volatile unsigned char *)0x51)
#define   IDRD      7

/* MCU */
#define MCUSR (*(volatile unsigned char *)0x54)
#define   JTRF      4
#define   WDRF      3
#define   BORF      2
#define   EXTRF     1
#define   PORF      0
#define MCUCR (*(volatile unsigned char *)0x55)
#define   JTD       7
#define   PUD       4
#define   IVSEL     1
```

```
#define  IVCE      0

#define SMCR (*(volatile unsigned char *)0x53)
#define  SM2       3
#define  SM1       2
#define  SM0       1
#define  SE        0

/* SPM Control and Status Register */
#define SPMCSR (*(volatile unsigned char *)0x57)
#define  SPMIE     7
#define  RWWSB     6
#define  SIGRD     5
#define  RWWSRE    4
#define  BLBSET    3
#define  PGWRT     2
#define  PGERS     1
#define  SPMEN     0

/* Stack Pointer */
#define SP   (*(volatile unsigned int *)0x5D)
#define SPL  (*(volatile unsigned char *)0x5D)
#define SPH  (*(volatile unsigned char *)0x5E)

/* Status REGister */
#define SREG (*(volatile unsigned char *)0x5F)

/* Watchdog Timer Control Register */
#define WDTCSR (*(volatile unsigned char *)0x60)
#define WDTCR (*(volatile unsigned char *)0x60)
#define  WDIF      7
#define  WDIE      6
#define  WDP3      5
#define  WDCE      4
#define  WDE       3
#define  WDP2      2
#define  WDP1      1
#define  WDP0      0
```

```
/* clock prescaler control register */
#define CLKPR (*(volatile unsigned char *)0x61)
#define   CLKPCE   7
#define   CLKPS3   3
#define   CLKPS2   2
#define   CLKPS1   1
#define   CLKPS0   0


/* PRR */
#define PRR0 (*(volatile unsigned char *)0x64)
#define   PRTWI      7
#define   PRTIM2     6
#define   PRTIM0     5
#define   PRUSART1   4
#define   PRTIM1     3
#define   PRSPI      2
#define   PRUSART0   1
#define   PRADC      0


/* Oscillator Calibration Register */
#define OSCCAL (*(volatile unsigned char *)0x66)


/* ADC */
#define ADC   (*(volatile unsigned int *)0x78)
#define ADCL (*(volatile unsigned char *)0x78)
#define ADCH (*(volatile unsigned char *)0x79)
#define ADCSRA (*(volatile unsigned char *)0x7A)
#define   ADEN     7
#define   ADSC     6
#define   ADATE    5
#define   ADIF     4
#define   ADIE     3
#define   ADPS2    2
#define   ADPS1    1
#define   ADPS0    0
#define ADCSRB (*(volatile unsigned char *)0x7B)
#define   ACME     6
#define   ADTS2    2
#define   ADTS1    1
```

```
#define  ADTS0    0
#define ADMUX (*(volatile unsigned char *)0x7C)
#define  REFS1    7
#define  REFS0    6
#define  ADLAR    5
#define  MUX4     4
#define  MUX3     3
#define  MUX2     2
#define  MUX1     1
#define  MUX0     0

/* DIDR */
#define DIDR0 (*(volatile unsigned char *)0x7E)
#define  ADC7D    7
#define  ADC6D    6
#define  ADC5D    5
#define  ADC4D    4
#define  ADC3D    3
#define  ADC2D    2
#define  ADC1D    1
#define  ADC0D    0
#define DIDR1 (*(volatile unsigned char *)0x7F)
#define  AIN1D    1
#define  AIN0D    0

/* Timer/Counter1 */
#define ICR1 (*(volatile unsigned int *)0x86)
#define ICR1L (*(volatile unsigned char *)0x86)
#define ICR1H (*(volatile unsigned char *)0x87)
#define OCR1B (*(volatile unsigned int *)0x8A)
#define OCR1BL (*(volatile unsigned char *)0x8A)
#define OCR1BH (*(volatile unsigned char *)0x8B)
#define OCR1A (*(volatile unsigned int *)0x88)
#define OCR1AL (*(volatile unsigned char *)0x88)
#define OCR1AH (*(volatile unsigned char *)0x89)
#define TCNT1 (*(volatile unsigned int *)0x84)
#define TCNT1L (*(volatile unsigned char *)0x84)
#define TCNT1H (*(volatile unsigned char *)0x85)
#define TCCR1C (*(volatile unsigned char *)0x82)
```

```
#define  FOC1A      7
#define  FOC1B      6
#define TCCR1B (*(volatile unsigned char *)0x81)
#define  ICNC1      7
#define  ICES1      6
#define  WGM13      4
#define  WGM12      3
#define  CS12       2
#define  CS11       1
#define  CS10       0
#define TCCR1A (*(volatile unsigned char *)0x80)
#define  COM1A1     7
#define  COM1A0     6
#define  COM1B1     5
#define  COM1B0     4
#define  WGM11      1
#define  WGM10      0

/* Timer/Counter2 */
#define ASSR (*(volatile unsigned char *)0xB6)
#define  EXCLK      6
#define  AS2        5
#define  TCN2UB     4
#define  OCR2AUB    3
#define  OCR2BUB    2
#define  TCR2AUB    1
#define  TCR2BUB    0
#define OCR2B (*(volatile unsigned char *)0xB4)
#define OCR2A (*(volatile unsigned char *)0xB3)
#define TCNT2 (*(volatile unsigned char *)0xB2)
#define TCCR2B (*(volatile unsigned char *)0xB1)
#define  FOC2A      7
#define  FOC2B      6
#define  WGM22      3
#define  CS22       2
#define  CS21       1
#define  CS20       0
#define TCCR2A (*(volatile unsigned char *)0xB0)
#define  COM2A1     7
```

```
#define  COM2A0    6
#define  COM2B1    5
#define  COM2B0    4
#define  WGM21     1
#define  WGM20     0


/* 2-wire SI */
#define TWBR (*(volatile unsigned char *)0xB8)
#define TWSR (*(volatile unsigned char *)0xB9)
#define  TWPS1     1
#define  TWPS0     0
#define TWAR (*(volatile unsigned char *)0xBA)
#define  TWGCE     0
#define TWDR (*(volatile unsigned char *)0xBB)
#define TWCR (*(volatile unsigned char *)0xBC)
#define  TWINT     7
#define  TWEA      6
#define  TWSTA     5
#define  TWSTO     4
#define  TWWC      3
#define  TWEN      2
#define  TWIE      0
#define TWAMR (*(volatile unsigned char *)0xBD)


/* USART0 */
#define UBRR0H (*(volatile unsigned char *)0xC5)
#define UBRR0L (*(volatile unsigned char *)0xC4)
#define UBRR0 (*(volatile unsigned int *)0xC4)
#define UCSR0C (*(volatile unsigned char *)0xC2)
#define  UMSEL01   7
#define  UMSEL00   6
#define  UPM01     5
#define  UPM00     4
#define  USBS0     3
#define  UCSZ01    2
#define  UCSZ00    1
#define  UCPOL0    0
#define UCSR0B (*(volatile unsigned char *)0xC1)
#define  RXCIE0    7
```

```
#define  TXCIE0     6
#define  UDRIE0     5
#define  RXEN0      4
#define  TXEN0      3
#define  UCSZ02     2
#define  RXB80      1
#define  TXB80      0
#define UCSR0A (*(volatile unsigned char *)0xC0)
#define  RXC0       7
#define  TXC0       6
#define  UDRE0      5
#define  FE0        4
#define  DOR0       3
#define  UPE0       2
#define  U2X0       1
#define  MPCM0      0
#define UDR0 (*(volatile unsigned char *)0xC6)


/* USART1 */
#define UBRR1H (*(volatile unsigned char *)0xCD)
#define UBRR1L (*(volatile unsigned char *)0xCC)
#define UBRR1 (*(volatile unsigned int *)0xCC)
#define UCSR1C (*(volatile unsigned char *)0xCA)
#define  UMSEL11    7
#define  UMSEL10    6
#define  UPM11      5
#define  UPM10      4
#define  USBS1      3
#define  UCSZ11     2
#define  UCSZ10     1
#define  UCPOL1     0
#define UCSR1B (*(volatile unsigned char *)0xC9)
#define  RXCIE1     7
#define  TXCIE1     6
#define  UDRIE1     5
#define  RXEN1      4
#define  TXEN1      3
#define  UCSZ12     2
#define  RXB81      1
```

```
#define  TXB81      0
#define UCSR1A (*(volatile unsigned char *)0xC8)
#define  RXC1       7
#define  TXC1       6
#define  UDRE1      5
#define  FE1        4
#define  DOR1       3
#define  UPE1       2
#define  U2X1       1
#define  MPCM1      0
#define UDR1 (*(volatile unsigned char *)0xCE)


/* bits */

/* Port A */
#define  PORTA7     7
#define  PORTA6     6
#define  PORTA5     5
#define  PORTA4     4
#define  PORTA3     3
#define  PORTA2     2
#define  PORTA1     1
#define  PORTA0     0
#define  PA7        7
#define  PA6        6
#define  PA5        5
#define  PA4        4
#define  PA3        3
#define  PA2        2
#define  PA1        1
#define  PA0        0
#define  DDA7       7
#define  DDA6       6
#define  DDA5       5
#define  DDA4       4
#define  DDA3       3
#define  DDA2       2
#define  DDA1       1
```

```
#define   DDA0     0
#define   PINA7    7
#define   PINA6    6
#define   PINA5    5
#define   PINA4    4
#define   PINA3    3
#define   PINA2    2
#define   PINA1    1
#define   PINA0    0

/* Port B */
#define   PORTB7   7
#define   PORTB6   6
#define   PORTB5   5
#define   PORTB4   4
#define   PORTB3   3
#define   PORTB2   2
#define   PORTB1   1
#define   PORTB0   0
#define   PB7      7
#define   PB6      6
#define   PB5      5
#define   PB4      4
#define   PB3      3
#define   PB2      2
#define   PB1      1
#define   PB0      0
#define   DDB7     7
#define   DDB6     6
#define   DDB5     5
#define   DDB4     4
#define   DDB3     3
#define   DDB2     2
#define   DDB1     1
#define   DDB0     0
#define   PINB7    7
#define   PINB6    6
#define   PINB5    5
#define   PINB4    4
```

```
#define  PINB3    3
#define  PINB2    2
#define  PINB1    1
#define  PINB0    0

/* Port C */
#define  PORTC7   7
#define  PORTC6   6
#define  PORTC5   5
#define  PORTC4   4
#define  PORTC3   3
#define  PORTC2   2
#define  PORTC1   1
#define  PORTC0   0
#define  PC7      7
#define  PC6      6
#define  PC5      5
#define  PC4      4
#define  PC3      3
#define  PC2      2
#define  PC1      1
#define  PC0      0
#define  DDC7     7
#define  DDC6     6
#define  DDC5     5
#define  DDC4     4
#define  DDC3     3
#define  DDC2     2
#define  DDC1     1
#define  DDC0     0
#define  PINC7    7
#define  PINC6    6
#define  PINC5    5
#define  PINC4    4
#define  PINC3    3
#define  PINC2    2
#define  PINC1    1
#define  PINC0    0
```

```
/* Port D */
#define   PORTD7    7
#define   PORTD6    6
#define   PORTD5    5
#define   PORTD4    4
#define   PORTD3    3
#define   PORTD2    2
#define   PORTD1    1
#define   PORTD0    0
#define   PD7       7
#define   PD6       6
#define   PD5       5
#define   PD4       4
#define   PD3       3
#define   PD2       2
#define   PD1       1
#define   PD0       0
#define   DDD7      7
#define   DDD6      6
#define   DDD5      5
#define   DDD4      4
#define   DDD3      3
#define   DDD2      2
#define   DDD1      1
#define   DDD0      0
#define   PIND7     7
#define   PIND6     6
#define   PIND5     5
#define   PIND4     4
#define   PIND3     3
#define   PIND2     2
#define   PIND1     1
#define   PIND0     0

/* PCMSK3 */
#define   PCINT31   7
#define   PCINT30   6
#define   PCINT29   5
#define   PCINT28   4
```

```
#define   PCINT27   3
#define   PCINT26   2
#define   PCINT25   1
#define   PCINT24   0
/* PCMSK2 */
#define   PCINT23   7
#define   PCINT22   6
#define   PCINT21   5
#define   PCINT20   4
#define   PCINT19   3
#define   PCINT18   2
#define   PCINT17   1
#define   PCINT16   0
/* PCMSK1 */
#define   PCINT15   7
#define   PCINT14   6
#define   PCINT13   5
#define   PCINT12   4
#define   PCINT11   3
#define   PCINT10   2
#define   PCINT9    1
#define   PCINT8    0
/* PCMSK0 */
#define   PCINT7    7
#define   PCINT6    6
#define   PCINT5    5
#define   PCINT4    4
#define   PCINT3    3
#define   PCINT2    2
#define   PCINT1    1
#define   PCINT0    0


/* Lock and Fuse Bits with LPM/SPM instructions */

/* lock bits */
#define   BLB12     5
#define   BLB11     4
#define   BLB02     3
```

```
#define  BLB01    2
#define  LB2      1
#define  LB1      0

/* fuses low bits */
#define  CKDIV8   7
#define  CKOUT    6
#define  SUT1     5
#define  SUT0     4
#define  CKSEL3   3
#define  CKSEL2   2
#define  CKSEL1   1
#define  CKSEL0   0

/* fuses high bits */
#define  OCDEN    7
#define  JTAGEN   6
#define  SPIEN    5
#define  WDTON    4
#define  EESAVE   3
#define  BOOTSZ1  2
#define  BOOTSZ0  1
#define  BOOTRST  0

/* extended fuses */
#define  BODLEVEL2 2
#define  BODLEVEL1 1
#define  BODLEVEL0 0


/* Interrupt Vector Numbers */

#define iv_RESET          1
#define iv_INT0           2
#define iv_EXT_INT0       2
#define iv_INT1           3
#define iv_EXT_INT1       3
#define iv_INT2           4
#define iv_EXT_INT2       4
```

```
#define iv_PCINT0        5
#define iv_PCINT1        6
#define iv_PCINT2        7
#define iv_PCINT3        8
#define iv_WDT           9
#define iv_TIMER2_COMPA 10
#define iv_TIMER2_COMPB 11
#define iv_TIMER2_OVF   12
#define iv_TIM2_COMPA   10
#define iv_TIM2_COMPB   11
#define iv_TIM2_OVF     12
#define iv_TIMER1_CAPT  13
#define iv_TIMER1_COMPA 14
#define iv_TIMER1_COMPB 15
#define iv_TIMER1_OVF   16
#define iv_TIM1_CAPT    13
#define iv_TIM1_COMPA   14
#define iv_TIM1_COMPB   15
#define iv_TIM1_OVF     16
#define iv_TIMER0_COMPA 17
#define iv_TIMER0_COMPB 18
#define iv_TIMER0_OVF   19
#define iv_TIM0_COMPA   17
#define iv_TIM0_COMPB   18
#define iv_TIM0_OVF     19
#define iv_SPI_STC      20
#define iv_USART0_RX    21
#define iv_USART0_RXC   21
#define iv_USART0_DRE   22
#define iv_USART0_UDRE  22
#define iv_USART0_TX    23
#define iv_USART0_TXC   23
#define iv_ANA_COMP     24
#define iv_ANALOG_COMP  24
#define iv_ADC          25
#define iv_EE_RDY       26
#define iv_EE_READY     26
#define iv_TWI          27
#define iv_TWSI         27
```

```
#define iv_SPM_RDY      28
#define iv_SPM_READY    28
#define iv_USART1_RX    29
#define iv_USART1_RXC   29
#define iv_USART1_DRE   30
#define iv_USART1_UDRE  30
#define iv_USART1_TX    31
#define iv_USART1_TXC   31

/* */

#endif
```

APPENDIX C

ATmega2560 Register Set

Address	Name	Bit 7	Bit 6	Bit 5	Bit 4	Bit 3	Bit 2	Bit 1	Bit 0	Page
(0xBE)	Reserved	-	-	-	-	-	-	-	-	
(0XBD)	TWAMR	TWAM6	TWAM5	TWAM4	TWAM3	TWAM2	TWAM1	TWAM0	-	
(0x13F)	TWCR	TWINT	TWEA	TWSTA	TWST0	TWWC	TWEN	-	TWIE	
(0x13E)	Reserved									
(0x13D)	Reserved									
(0x13C)	Reserved									
(0x13B)	Reserved									
(0x13A)	Reserved									
(0x139)	Reserved									
(0x138)	Reserved									
(0x137)	Reserved									
(0x136)	UDR3				USART3 I/O Data Register					222
(0x135)	UBRR3H	-	-	-	-	USART3 Baud Rate Register High Byte				227
(0x134)	UBRR3L				USART3 Baud Rate Register Low Byte					227
(0x133)	Reserved	-	-	-	-	-	-	-	-	
(0x132)	UCSR3C	UMSEL31	UMSEL30	UPM31	UPM30	USBS3	UCSZ31	UCSZ30	UCPOL3	239
(0x131)	UCSR3B	RXCIE3	TXCIE3	UDRIE3	RXEN3	TXEN3	UCSZ32	RXB83	TXB83	238
(0x130)	UCSR3A	RXC3	TXC3	UDRE3	FE3	DOR3	UPE3	U2X3	MPCM3	238
(0x12F)	Reserved	-	-	-	-	-	-	-	-	
(0x12E)	Reserved	-	-	-	-	-	-	-	-	
(0x12D)	OCR5CH				Timer/Counter5- Output Compare Register C High Byte					165
(0x12C)	OCR5CL				Timer/Counter5- Output Compare Register C Low Byte					165
(0x12B)	OCR5BH				Timer/Counter5- Output Compare Register B High Byte					165
(0x12A)	OCR5BL				Timer/Counter5- Output Compare Register B Low Byte					165
(0x129)	OCR5AH				Timer/Counter5- Output Compare Register A High Byte					164
(0x128)	OCR5AL				Timer/Counter5- Output Compare Register A Low Byte					164
(0x127)	ICR5H				Timer/Counter5- Input Capture Register High Byte					165
(0x126)	ICR5L				Timer/Counter5- Input Capture Register Low Byte					165
(0x125)	TCNT5H				Timer/Counter5- Counter Register High Byte					163
(0x124)	TCNT5L				Timer/Counter5- Counter Register Low Byte					163

Figure C.1: Atmel AVR ATmega2560 Register Set. (Figure used with permission of Atmel, Incorporated.)

Address	Name	Bit 7	Bit 6	Bit 5	Bit 4	Bit 3	Bit 2	Bit 1	Bit 0	Page
(0x123)	Reserved	-	-	-	-	-	-	-	-	
(0x122)	TCCR5C	FOC5A	FOC5B	FOC5C	-	-	-	-	-	162
(0x121)	TCCR5B	ICNC5	ICES5	-	WGM53	WGM52	CS52	CS51	CS50	160
(0x120)	TCCR5A	COM5A1	COM5A0	COM5B1	COM5B0	COM5C1	COM5CO	WGM51	WGM50	158
(0x11F)	Reserved	-	-	-	-	-	-	-	-	
(0x11E)	Reserved	-	-	-	-	-	-	-	-	
(0x11D)	Reserved	-	-	-	-	-	-	-	-	
(0x11C)	Reserved	-	-	-	-	-	-	-	-	
(0x11B)	Reserved	-	-	-	-	-	-	-	-	
(0x11A)	Reserved	-	-	-	-	-	-	-	-	
(0x119)	Reserved	-	-	-	-	-	-	-	-	
(0x118)	Reserved	-	-	-	-	-	-	-	-	
(0x117)	Reserved	-	-	-	-	-	-	-	-	
(0x116)	Reserved	-	-	-	-	-	-	-	-	
(0x115)	Reserved	-	-	-	-	-	-	-	-	
(0x114)	Reserved	-	-	-	-	-	-	-	-	
(0x113)	Reserved	-	-	-	-	-	-	-	-	
(0x112)	Reserved	-	-	-	-	-	-	-	-	
(0x111)	Reserved	-	-	-	-	-	-	-	-	
(0x110)	Reserved	-	-	-	-	-	-	-	-	
(0x10F)	Reserved	-	-	-	-	-	-	-	-	
(0x10E)	Reserved	-	-	-	-	-	-	-	-	
(0x10D)	Reserved	-	-	-	-	-	-	-	-	
(0x10C)	Reserved	-	-	-	-	-	-	-	-	
(0x10B)	PORTL	PORTL7	PORTL6	PORTL5	PORTL4	PORTL3	PORTL2	PORTL1	PORTL0	104
(0x10A)	DDRL	DDL7	DDL6	DDL5	DDL4	DDL3	DDL2	DDL1	DDL0	104
(0x109)	PINL	PINL7	PINL6	PINL5	PINL4	PINL3	PINL2	PINL1	PINL0	104
(0x108)	PORTK	PORTK7	PORTK6	PORTK5	PORTK4	PORTK3	PORTK2	PORTK1	PORTK0	103
(0x107)	DDRK	DDK7	DDK6	DDK5	DDK4	DDK3	DDK2	DDK1	DDR0	103
(0x106)	PINK	PINK7	PINK6	PINK5	PINK4	PINK3	PINK2	PINK1	PINK0	103
(0x105)	PORTJ	PORTJ7	PORTJ6	PORTJ5	PORTJ4	PORTJ3	PORTJ2	PORTJ1	PORTJ0	103
(0x104)	DDRJ	DDJ7	DDJ6	DDJ5	DDJ4	DDJ3	DDJ2	DDJ1	DDRJ0	103
(0x103)	PINJ	PINJ7	PINJ6	PINJ5	PINJ4	PINJ3	PINJ2	PINJ1	PINJ0	103
(0x102)	PORTH	PORTH7	PORTH6	PORTH5	PORTH4	PORTH3	PORTH2	PORTH1	PORTH0	102
(0x101)	DDRH	DDH7	DDH6	DDH5	DDH4	DDH3	DDH2	DDH1	DDH0	103

Figure C.2: Atmel AVR ATmega2560 Register Set. (Figure used with permission of Atmel, Incorporated.)

Address	Name	Bit 7	Bit 6	Bit 5	Bit 4	Bit 3	Bit 2	Bit 1	Bit 0	Page
(0x100)	PINH	PINH7	PINH6	PINH5	PINH4	PINH3	PINH2	PINH1	PINH0	103
(0xFF)	Reserved	-	-	-	-	-	-	-	-	
(0xFE)	Reserved	-	-	-	-	-	-	-	-	
(0xFD)	Reserved	-	-	-	-	-	-	-	-	
(0xFC)	Reserved	-	-	-	-	-	-	-	-	
(0xFB)	Reserved	-	-	-	-	-	-	-	-	
(0xFA)	Reserved	-	-	-	-	-	-	-	-	
(0xF9)	Reserved	-	-	-	-	-	-	-	-	
(0xF8)	Reserved	-	-	-	-	-	-	-	-	
(0xF7)	Reserved	-	-	-	-	-	-	-	-	
(0xF6)	Reserved	-	-	-	-	-	-	-	-	
(0xF5)	Reserved	-	-	-	-	-	-	-	-	
(0xF4)	Reserved	-	-	-	-	-	-	-	-	
(0xF3)	Reserved	-	-	-	-	-	-	-	-	
(0xF2)	Reserved	-	-	-	-	-	-	-	-	
(0xF1)	Reserved	-	-	-	-	-	-	-	-	
(0xF0)	Reserved	-	-	-	-	-	-	-	-	
(0xEF)	Reserved	-	-	-	-	-	-	-	-	
(0xEE)	Reserved	-	-	-	-	-	-	-	-	
(0xED)	Reserved	-	-	-	-	-	-	-	-	
(0xEC)	Reserved	-	-	-	-	-	-	-	-	
(0xEB)	Reserved	-	-	-	-	-	-	-	-	
(0xEA)	Reserved	-	-	-	-	-	-	-	-	
(0xE9)	Reserved	-	-	-	-	-	-	-	-	
(0xE8)	Reserved	-	-	-	-	-	-	-	-	
(0xE7)	Reserved	-	-	-	-	-	-	-	-	
(0xE6)	Reserved	-	-	-	-	-	-	-	-	
(0xE5)	Reserved	-	-	-	-	-	-	-	-	
(0xE4)	Reserved	-	-	-	-	-	-	-	-	
(0xE3)	Reserved	-	-	-	-	-	-	-	-	
(0xE2)	Reserved	-	-	-	-	-	-	-	-	
(0xE1)	Reserved	-	-	-	-	-	-	-	-	
(0xE0)	Reserved	-	-	-	-	-	-	-	-	
(0xDF)	Reserved	-	-	-	-	-	-	-	-	
(0xDE)	Reserved	-	-	-	-	-	-	-	-	
(0xDD)	Reserved	-	-	-	-	-	-	-	-	
(0xDC)	Reserved	-	-	-	-	-	-	-	-	
(0xDB)	Reserved	-	-	-	-	-	-	-	-	
(0xDA)	Reserved	-	-	-	-	-	-	-	-	
(0xD9)	Reserved	-	-	-	-	-	-	-	-	
(0xD8)	Reserved	-	-	-	-	-	-	-	-	
(0xD7)	Reserved	-	-	-	-	-	-	-	-	
(0xD6)	UDR2	USART2 I/O Data Register								222
(0xD5)	UBRR2H					USART2 Baud Rate Register High Byte				227
(0xD4)	UBRR2L	USART2 Baud Rate Register Low Byte								227
(0xD3)	Reserved	-	-	-	-	-	-	-	-	
(0xD2)	UCSR2C	UMSEL21	UMSEL20	UPM21	UPM20	USBS2	UCSZ21	UCSZ20	UCPOL2	239
(0xD1)	UCSR2B	RXCIE2	TXCIE2	UDRIE2	RXEN2	TXEN2	UCSZ22	RXB82	TXB82	238
(0xD0)	UCSR2A	RXC2	TXC2	UDRIE2	FE2	DOR2	UPE2	U2X2	MPCM2	238
(0xCF)	Reserved	-	-	-	-	-	-	-	-	
(0xCE)	UDR1	USART1 I/O Data Register								222
(0xCD)	UBRR1H	-	-	-	-	USART1 Baud Rate Register High Byte				227
(0xCC)	UBRR1L	USART2 Baud Rate Register Low Byte								227
(0xCB)	Reserved	-	-	-	-	-	-	-	-	
(0xCA)	UCSR1C	UMSEL11	UMSEL10	UPM11	UPM10	USBS1	UCSZ11	UCSZ10	UCPOL1	239
(0xC9)	UCSR1B	RXCIE1	TXCIE1	UDRIE1	RXEN1	TXEN1	UCSZ12	RXB81	TXB81	238
(0xC8)	UCSR1A	RXC1	TXC1	UDRE1	FE1	DOR1	UPE1	U2X1	MPCM1	238
(0xC7)	Reserved	-	-	-	-	-	-	-	-	
(0xC6)	UDR0	USART0 I/O Data Register								222
(0xC5)	UBRR0H	-	-	-	-	USART0 Baud Rate Register High Byte				227
(0xC4)	UBRR0L	USART0 Baud Rate Register Low Byte								227
(0xC3)	Reserved	-	-	-	-	-	-	-	-	
(0xC2)	UCSR0C	UMSEL01	UMSEL00	UPM01	UPM00	USBS0	UCSZ01	UCSZ00	UCPOL0	239
(0xC1)	UCSR0B	RXCIE0	TXCIE0	UDRIE0	RXEN0	TXEN0	UCSZ02	RXB80	TXB80	238
(0xC0)	UCSR0A	RXC0	TXC0	UDRE0	FE0	DOR0	UPE0	U2X0	MPCM0	238
(0xBF)	Reserved	-	-	-	-	-	-	-	-	

Figure C.3: Atmel AVR ATmega2560 Register Set. (Figure used with permission of Atmel, Incorporated.)

Address	Name	Bit 7	Bit 6	Bit 5	Bit 4	Bit 3	Bit 2	Bit 1	Bit 0	Page
(0xBE)	Reserved	-	-	-	-	-	-	-	-	
(0xBD)	TWAMR	TWAM6	TWAM5	TWAM4	TWAM3	TWAM2	TWAM1	TWAM0	-	269
(0xBC)	TWCR	TWINT	TWEA	TWSTA	TWST0	TWWC	TWEN	-	TWIE	266
(0xBB)	TWDR	2-wire Serial Interface Data Register								268
(0xBA)	TWAR	TWA6	TWA5	TWA4	TWA3	TWA2	TWA1	TWA0	TWGCE	269
(0xB9)	TWSR	TWS7	TWS6	TWS5	TWS4	TWS3	-	TWPS1	TWPS0	268
(0xB8)	TWBR	2-wire Serial Interface Bit Rate Register								266
(0xB7)	Reserved	-	-	-	-	-	-	-	-	
(0xB6)	ASSR	-	EXCLK	AS2	TCN2UB	OCR2AUB	OCR2BUB	TCR2AUB	TCR2BUB	184
(0xB5)	Reserved	-	-	-	-	-	-	-	-	
(0xB4)	OCR2B	Timer/Counter2 Output Compare Register B								191
(0xB3)	OCR2A	Timer/Counter2 Output Compare Register A								191
(0xB2)	TCNT2	Timer/Counter2 (8Bit)								191
(0xB1)	TCCR2B	FOC2A	FOC2B	-	-	WGM22	CS22	CS21	CS20	190
(0xB0)	TCCR2A	COM2A1	COM2A0	COM2B1	COM2B0	-	-	WGM21	WGM20	191
(0xAF)	Reserved	-	-	-	-	-	-	-	-	
(0xAE)	Reserved	-	-	-	-	-	-	-	-	
(0xAD)	OCR4CH	Timer/Counter4- Output Compare Register C High Byte								164
(0xAC)	OCR4CL	Timer/Counter4- Output Compare Register C Low Byte								164
(0xAB)	OCR4BH	Timer/Counter4- Output Compare Register B High Byte								164
(0xAA)	OCR4BL	Timer/Counter4- Output Compare Register B Low Byte								164
(0xA9)	OCR4AH	Timer/Counter4- Output Compare Register A High Byte								164
(0xA8)	OCR4AL	Timer/Counter4- Output Compare Register A Low Byte								164
(0xA7)	ICR4H	Timer/Counter4- Input Capture Register High Byte								165
(0xA6)	ICR4L	Timer/Counter4- Input Capture Register Low Byte								165
(0xA5)	TCNT4H	Timer/Counter4- Counter Register High Byte								163
(0xA4)	TCNT4L	Timer/Counter4- Counter Register Low Byte								163
(0xA3)	Reserved	-	-	-	-	-	-	-	-	
(0xA2)	TCCR4C	FOC4A	FOC4B	FOC4C	-	-	-	-	-	162
(0xA1)	TCCR4B	ICNC4	ICES4	-	WGM43	WGM42	CS42	CS41	CS40	160
(0xA0)	TCCR4A	COM4A1	COM4A0	COM4B1	COM4B0	COM4C1	COM4CO	WGM41	WGM40	158
(0x9F)	Reserved	-	-	-	-	-	-	-	-	
(0x9E)	Reserved	-	-	-	-	-	-	-	-	
(0x9D)	OCR3CH	Timer/Counter3- Output Compare Register C High Byte								163
(0x9C)	OCR3CL	Timer/Counter3- Output Compare Register C Low Byte								163
(0x9B)	OCR3BH	Timer/Counter3- Output Compare Register B High Byte								163
(0x9A)	OCR3BL	Timer/Counter3- Output Compare Register B Low Byte								163
(0x99)	OCR3AH	Timer/Counter3- Output Compare Register A High Byte								163
(0x98)	OCR3AL	Timer/Counter3- Output Compare Register A Low Byte								163
(0x97)	ICR3H	Timer/Counter3- Input Capture Register High Byte								165
(0x96)	ICR3L	Timer/Counter3- Input Capture Register Low Byte								165
(0x95)	TCNT3H	Timer/Counter3- Counter Register High Byte								162
(0x94)	TCNT3L	Timer/Counter3- Counter Register Low Byte								162
(0x93)	Reserved	-	-	-	-	-	-	-	-	
(0x92)	TCCR3C	FOC3A	FOC3B	FOC3C	-	-	-	-	-	162
(0x91)	TCCR3B	ICNC3	ICES3	-	WGM33	WGM32	CS32	CS31	CS30	160
(0x90)	TCCR3A	COM3A1	COM3A0	COM3B1	COM3B0	COM3C1	COM3B1	WGM31	WGM30	158
(0x8F)	Reserved	-	-	-	-	-	-	-	-	
(0x8E)	Reserved	-	-	-	-	-	-	-	-	
(0x8D)	OCR1CH	Timer/Counter1- Output Compare Register C High Byte								163
(0x8C)	OCR3CL	Timer/Counter1- Output Compare Register C Low Byte								163
(0x8B)	OCR1BH	Timer/Counter1- Output Compare Register B High Byte								163
(0x8A)	OCR1BL	Timer/Counter1- Output Compare Register B Low Byte								163
(0x89)	OCR1AH	Timer/Counter1- Output Compare Register A High Byte								163
(0x88)	OCR1AL	Timer/Counter1- Output Compare Register A Low Byte								163
(0x87)	ICR1H	Timer/Counter1- Input Capture Register High Byte								165
(0x86)	ICR1L	Timer/Counter1- Input Capture Register Low Byte								165
(0x85)	TCNT1H	Timer/Counter1- Counter Register High Byte								162
(0x84)	TCNT1L	Timer/Counter1- Counter Register Low Byte								162
(0x83)	Reserved	-	-	-	-	-	-	-	-	
(0x82)	TCCR1C	FOC1A	FOC1B	FOC1C	-	-	-	-	-	161
(0x81)	TCCR1B	ICNC1	ICES1	-	WGM13	WGM12	CS12	CS11	CS10	160
(0x80)	TCCR1A	COM1A1	COM1A0	COM1B1	COM1B0	COM1C1	COM1C0	WGM11	WGM10	158
(0x7F)	DIDR1	-	-	-	-	-	-	AIN1D	AIN0D	274
(0x7E)	DIDR0	ADC7D	ADC6D	ADC5D	ADC4D	ADC3D	ADC2D	ADC1D	ADC0D	295
(0x7D)	DIDR2	ADC15D	ADC14D	ADC13D	ADC12D	ADC11D	ADC10D	ADC9D	ADC8D	295

Figure C.4: Atmel AVR ATmega2560 Register Set. (Figure used with permission of Atmel, Incorporated.)

Address	Name	Bit 7	Bit 6	Bit 5	Bit 4	Bit 3	Bit 2	Bit 1	Bit 0	Page
(0x7C)	ADMUX	REFS1	REFS0	ADLAR	MUX4	MUX3	MUX2	MUX1	MUX0	289
(0x7B)	ADCSR	-	ACME			MUX5	ADTS2	ADTS1	ADTS0	272, 290, 294
(0x7A)	ADCSR	ADEN	ADSC	ADATE	ADIF	ADIE	ADPS2	ADPS1	ADPS0	292
(0x79)	ADCH	ADC Data Register High byte								294
(0x78)	ADCL	ADC Data Register Low byte								294
(0x77)	Reserved	-	-	-	-	-	-	-	-	
(0x76)	Reserved	-	-	-	-	-	-	-	-	
(0x75)	XMCRB	XMBK	-			-	XMM2	XMM1	XMM0	38
(0x74)	XMCRA	SRE	SRL2	SRL1	SRL0	SRW11	SRW10	SRW01	SRW00	37
(0x73)	TIMSK5	-	-	ICIE5	-	OCIE5C	OCIE5B	OCIE5A	TOIE5	166
(0x72)	TIMSK4	-	-	ICIE4	-	OCIE4C	OCIE4B	OCIE4A	TOIE4	166
(0x71)	TIMSK3	-	-	ICIE3	-	OCIE3C	OCIE3B	OCIE3A	TOIE3	166
(0x70)	TIMSK2	-	-	-	-	-	OCIE2B	OCIE2A	TOIE2	193
(0x6F)	TIMSK1	-	-	ICIE1	-	OCIE1C	OCIE1B	OCIE1A	TOIE1	166
(0x6E)	TIMSK0	-	-	-	-	-	OCIE0B	OCIE0A	TOIE0	134
(0x6D)	PCMSK2	PCINT 23	PCINT22	PCINT21	PCINT20	PCINT19	PCINT18	PCINT17	PCINT16	116
(0x6C)	PCMSK1	PCINT 15	PCINT14	PCINT13	PCINT12	PCINT11	PCINT10	PCINT9	PCINT8	116
(0x6B)	PCMSK0	PCINT 7	PCINT16	PCINT5	PCINT4	PCINT3	PCINT2	PCINT1	PCINT0	117
(0x6A)	EICRB	ISC71	ISC70	ISC61	ISC60	ISC51	ISC50	ISC41	ISC40	114
(0x69)	EICRA	ISC31	ISC30	ISC21	ISC20	ISC11	ISC10	ISC01	ISC00	113
(0x68)	PCICR	-	-	-	-	-	PCIE2	PCIE1	PCIE0	115
(0x67)	Reserved	-	-	-	-	-	-	-	-	
(0x66)	OSCCAL	Oscillator Calibration Register								50
(0x65)	PRR1	-	-	PRTIM5	PRTIM4	PRTIM3	PRUSART3	PRUSART2	PRUSART1	57
(0x64)	PRR0	ISC31	PRTIM2	PRTIM0	-	PRTIM1	PRSPI	PRUSART0	PRADC	56
(0x63)	Reserved	-	-	-	-	-	-	-	-	
(0x62)	Reserved	-	-	-	-	-	-	-	-	
(0x61)	CLKPR	CLKPCE	-	-	-	CLKPS3	CLKPS2	CLKPS1	CLKPS0	50
(0x60)	WDTCSR	WDIF	WDIE	WDP3	WDCE	WDE	WDP2	WDP1	WDP0	67
0x3F (0x5F)	SREG	I	T	H	S	V	N	Z	C	14
0x3E (0x5E)	SPH	SP15	SP14	SP13	SP12	SP11	SP10	SP9	SP8	16
0x3D (0x5D)	SPL	SP7	SP6	SP5	SP4	SP3	SP2	SP1	SP0	16
0x3C (0x5C)	EIND	-	-	-	-	-	-	-	EIND0	17
0x3B (0x5B)	RAMPZ	-	-	-	-	-	-	RAMPZ1	RAMPZ0	17
0x3A (0x5A)	Reserved	-	-	-	-	-	-	-	-	
0x39 (0x59)	Reserved	-	-	-	-	-	-	-	-	
0x38 (0x58)	Reserved	-	-	-	-	-	-	-	-	
0x37 (0x57)	SPMCSR	SPMIE	RWWSB	SIGRD	RWWSRE	BLBSET	PGWRT	PGERS	SPMEN	332
0x36 (0x56)	Reserved	-	-	-	-	-	-	-	-	
0x35 (0x55)	MCUCR	JTD	-	-	PUD	-	-	IVSEL	IVCE	67, 110, 100, 308
0x34 (0x54)	MCUSR	-	-	-	JTRF	WDRF	BORF	EXTRF	PORF	308
0x33 (0x53)	SMCR	-	-	-	-	SM2	SM1	SM0	SE	52
0x32 (0x52)	Reserved	-	-	-	-	-	-	-	-	
0x31 (0x51)	OCDR	OCDR7	OCDR6	OCDR5	OCDR4	OCDR3	OCDR2	OCDR1	OCDR0	301
0x30 (0x50)	ACSR	ACD	ACBG	ACO	ACI	ACIE	ACIC	ACIS1	ACIS0	272
0x2F (0x4F)	Reserved	-	-	-	-	-	-	-	-	
0x2E (0x4E)	SPDR	SPI Data Register								204
0x2D (0x4D)	SPSR	SPIF	WCOL	-	-	-	-	-	SPI2X	203
0x2C (0x4C)	SPCR	SPIE	SPE	DORD	MSTR	CPOL	CPHA	SPR1	SPR0	202
0x2B (0x4B)	GPIOR2	General Purpose I/O Register 2								37
0x2A (0x4A)	GPIOR1	General Purpose I/O Register 1								37
0x29 (0x49)	Reserved	-	-	-	-	-	-	-	-	
0x28 (0x48)	OCR0B	Timer/Counter0 Output Compare Register B								133
0x27 (0x47)	OCR0A	Timer/Counter0 Output Compare Register A								133
0x26 (0x46)	TCNT0	Timer/Counter0 (8 Bit)								133
0x25 (0x45)	TCCR0B	FOC0A	FOC0B	-	-	WGM02	CS02	CS01	CS00	132
0x24 (0x44)	TCCR0A	COM0A1	COM0A0	COM0B1	COM0B0	-	-	WGM01	WGM00	129
0x23 (0x43)	GTCCR	TSM	-	-	-	-	-	PSRASY	PSRSYNC	170, 194
0x22 (0x42)	EEARH	-	-	-	-	EEPROM Address Register High Byte				35
0x21 (0x41)	EEARL	EEPROM Address Register Low Byte								35
0x20 (0x40)	EEDT	EEPROM Data Register								35
0x1F (0x3F)	EECR	-	-	EEPM1	EEPM0	EERIE	EEMPE	EEPE	EERE	35
0x1E (0x3E)	GPIOR0	General Purpose I/O Register 0								37
0x1D (0x3D)	EIMSK	INT7	INT6	INT5	INT4	INT3	INT2	INT1	INT0	115
0x1C (0x3C)	EIFR	INTF7	INTF6	INTF5	INTF4	INTF3	INTF2	INTF1	INTF0	115
0x1B (0x3B)	PCIFR	-	-	-	-	-	PCIF2	PCIF1	PCIF0	116

Figure C.5: Atmel AVR ATmega2560 Register Set. (Figure used with permission of Atmel, Incorporated.)

Address	Name	Bit 7	Bit 6	Bit 5	Bit 4	Bit 3	Bit 2	Bit 1	Bit 0	Page
0x1A (0x3A)	TIFR5	-	-	ICF5	-	OCF5C	OCF5B	OCF5A	TOV5	166
0x19 (0x39)	TIFR4	-	-	ICF4	-	OCF4C	OCF4B	OCF4A	TOV4	167
0x18 (0x38)	TIFR3	-	-	ICF3	-	OCF3C	OCF3B	OCF3A	TOV3	167
0x17 (0x37)	TIFR2	-	-	-	-	-	OCF2B	OCF2A	TOV2	193
0x16 (0x36)	TIFR1	-	-	ICF1	-	OCF1C	OCF1B	OCF1A	TOV1	167
0x15 (0x35)	TIFR0	-	-	-	-	-	OCF0B	OCF0A	TOV0	134
0x14 (0x34)	PORTG	-	-	PORTG5	PORTG4	PORTG3	PORTG2	PORTG1	PORTG0	102
0x13 (0x33)	DDRG	-	-	DDG5	DDG4	DDG3	DDG2	DDG1	DDG0	102
0x12 (0x32)	PING	-	-	PING5	PING4	PING3	PING2	PING1	PING0	102
0x11 (0x31)	PORTF	PORTF7	PORTF6	PORTF5	PORTF4	PORTF3	PORTF2	PORTF1	PORTF0	101
0x10 (0x30)	DDRF	DDF7	DDF6	DDF5	DDF4	DDF3	DDF2	DDF1	DDF0	102
0x0F (0x2F)	PINF	PINF7	PINF6	PINF5	PINF4	PINF3	PINF2	PINF1	PINF0	102
0x0E (0x2E)	PORTE	PORTE7	PORTE6	PORTE5	PORTE4	PORTE3	PORTE2	PORTE1	PORTE0	101
0x0D (0x2D)	DDRE	DDRE7	DDRE6	DDRE5	DDRE4	DDRE3	DDRE2	DDRE1	DDRE0	101
0x0C (0x2C)	PINE	PINE7	PINE6	PINE5	PINE4	PINE3	PINE2	PINE1	PINE0	102
0x0B (0x2B)	PORTD	PORTD7	PORTD6	PORTD5	PORTD4	PORTD3	PORTD2	PORTD1	PORTD0	101
0x0A (0x2A)	DDRD	DDRD7	DDRD6	DDRD5	DDRD4	DDRD3	DDRD2	DDRD1	DDRD0	101
0x09 (0x29)	PIND	PIND7	PIND6	PIND5	PIND4	PIND3	PIND2	PIND1	PIND0	101
0x08(0x28)	PORTC	PORTC7	PORTC6	PORTC5	PORTC4	PORTC3	PORTC2	PORTC1	PORTC0	101
0x07 (0x27)	DDRC	DDRC7	DDRC6	DDRC5	DDRC4	DDRC3	DDRC2	DDRC1	DDRC0	101
0x06 (0x26)	PINC	PINC7	PINC6	PINC5	PINC4	PINC3	PINC2	PINC1	PINC0	101
0x05 (0x25)	PORTB	PORTB7	PORTB6	PORTB5	PORTB4	PORTB3	PORTB2	PORTB1	PORTB0	100
0x04 (0x24)	DDRB	DDRB7	DDRB6	DDRB5	DDRB4	DDRB3	DDRB2	DDRB1	DDRB0	100
0x03 (0x23)	PINB	PINB7	PINB6	PINB5	PINB4	PINB3	PINB2	PINB1	PINB0	100
0x02 (0x22)	PORTA	PORTA7	PORTA6	PORTA5	PORTA4	PORTA3	PORTA2	PORTA1	PORTA0	100
0x01 (0x21)	DDRA	DDRA7	DDRA6	DDRA5	DDRA4	DDRA3	DDRA2	DDRA1	DDRA0	100
0x00 (0x20)	PINA	PINA7	PINA6	PINA5	PINA4	PINA3	PINA2	PINA1	PINA0	100

Figure C.6: Atmel AVR ATmega2560 Register Set. (Figure used with permission of Atmel, Incorporated.)

APPENDIX D

ATmega2560 Header File

During C programming, the contents of a specific register may be referred to by name when an appropriate header file is included within your program. The header file provides the link between the register name used within a program and the hardware location of the register.

Provided below is the ATmega2560 header file from the ICC AVR compiler. This header file was provided courtesy of ImageCraft Incorporated.

```
//**********************************************************
#ifndef __iom2560v_h
#define __iom2560v_h

/* ATmega2560 header file for
 * ImageCraft ICCAVR compiler
 */

/* 2008/04/22 created
 */

#include <_iom640to2561v.h>

/* */
//**********************************************************

#ifndef ___iom640to2561v_h
#define ___iom640to2561v_h

/* ATmega640..2561 io header file for
 * ImageCraft ICCAVR compiler
 */

/* 2004/12/19 created as iom2560v.h
   2005/03/10 fixed SPI bits
   2005/05/12 fixed EECR "P" bits
   2005/05/16 added GPIOR0..2
   2006/02/11 fixed device #define and CKOUT
```

```
   2008/04/21 added USART bits for SPI mode
 */


/* Port L */
#define PINL (*(volatile unsigned char *)0x109) /* m/m */
#define DDRL (*(volatile unsigned char *)0x10A) /* m/m */
#define PORTL (*(volatile unsigned char *)0x10B) /* m/m */


/* Port K */
#define PINK (*(volatile unsigned char *)0x106) /* m/m */
#define DDRK (*(volatile unsigned char *)0x107) /* m/m */
#define PORTK (*(volatile unsigned char *)0x108) /* m/m */


/* Port J */
#define PINJ (*(volatile unsigned char *)0x103) /* m/m */
#define DDRJ (*(volatile unsigned char *)0x104) /* m/m */
#define PORTJ (*(volatile unsigned char *)0x105) /* m/m */


/* Port H */
#define PINH (*(volatile unsigned char *)0x100) /* m/m */
#define DDRH (*(volatile unsigned char *)0x101) /* m/m */
#define PORTH (*(volatile unsigned char *)0x102) /* m/m */


/* Port G */
#define PING (*(volatile unsigned char *)0x32)
#define DDRG (*(volatile unsigned char *)0x33)
#define PORTG (*(volatile unsigned char *)0x34)


/* Port F */
#define PINF (*(volatile unsigned char *)0x2F)
#define DDRF (*(volatile unsigned char *)0x30)
#define PORTF (*(volatile unsigned char *)0x31)


/* Port E */
#define PINE (*(volatile unsigned char *)0x2C)
#define DDRE (*(volatile unsigned char *)0x2D)
#define PORTE (*(volatile unsigned char *)0x2E)


/* Port D */
```

```
#define PIND (*(volatile unsigned char *)0x29)
#define DDRD (*(volatile unsigned char *)0x2A)
#define PORTD (*(volatile unsigned char *)0x2B)

/* Port C */
#define PINC (*(volatile unsigned char *)0x26)
#define DDRC (*(volatile unsigned char *)0x27)
#define PORTC (*(volatile unsigned char *)0x28)

/* Port B */
#define PINB (*(volatile unsigned char *)0x23)
#define DDRB (*(volatile unsigned char *)0x24)
#define PORTB (*(volatile unsigned char *)0x25)

/* Port A */
#define PINA (*(volatile unsigned char *)0x20)
#define DDRA (*(volatile unsigned char *)0x21)
#define PORTA (*(volatile unsigned char *)0x22)

/* ADC */
#define ADC  (*(volatile unsigned int *)0x78)
#define ADCL (*(volatile unsigned char *)0x78)
#define ADCH (*(volatile unsigned char *)0x79)
#define ADCSRA (*(volatile unsigned char *)0x7A)
#define   ADEN     7
#define   ADSC     6
#define   ADATE    5
#define   ADIF     4
#define   ADIE     3
#define   ADPS2    2
#define   ADPS1    1
#define   ADPS0    0
#define ADCSRB (*(volatile unsigned char *)0x7B)
#define   ACME     6
#define   MUX5     3
#define   ADTS2    2
#define   ADTS1    1
#define   ADTS0    0
#define ADMUX (*(volatile unsigned char *)0x7C)
```

```
#define   REFS1    7
#define   REFS0    6
#define   ADLAR    5
#define   MUX4     4
#define   MUX3     3
#define   MUX2     2
#define   MUX1     1
#define   MUX0     0

/* Analog Comparator Control and Status Register */
#define ACSR (*(volatile unsigned char *)0x50)
#define   ACD      7
#define   ACBG     6
#define   ACO      5
#define   ACI      4
#define   ACIE     3
#define   ACIC     2
#define   ACIS1    1
#define   ACIS0    0

/* DIDR */
#define DIDR2 (*(volatile unsigned char *)0x7D) /* m/m */
#define   ADC15D   7
#define   ADC14D   6
#define   ADC13D   5
#define   ADC12D   4
#define   ADC11D   3
#define   ADC10D   2
#define   ADC9D    1
#define   ADC8D    0
#define DIDR0 (*(volatile unsigned char *)0x7E) /* m/m */
#define   ADC7D    7
#define   ADC6D    6
#define   ADC5D    5
#define   ADC4D    4
#define   ADC3D    3
#define   ADC2D    2
#define   ADC1D    1
#define   ADC0D    0
```

```
#define DIDR1 (*(volatile unsigned char *)0x7F) /* m/m */
#define   AIN1D     1
#define   AIN0D     0


/* USART0 */
#define UBRR0H (*(volatile unsigned char *)0xC5) /* m/m */
#define UBRR0L (*(volatile unsigned char *)0xC4) /* m/m */
#define UBRR0 (*(volatile unsigned int *)0xC4) /* m/m */
#define UCSR0C (*(volatile unsigned char *)0xC2) /* m/m */
#define   UMSEL01   7
#define   UMSEL00   6
#define   UPM01     5
#define   UPM00     4
#define   USBS0     3
#define   UCSZ01    2
#define   UCSZ00    1
#define   UCPOL0    0
/* in SPI mode */
#define   UDORD0    2
#define   UCPHA0    1
#define UCSR0B (*(volatile unsigned char *)0xC1) /* m/m */
#define   RXCIE0    7
#define   TXCIE0    6
#define   UDRIE0    5
#define   RXEN0     4
#define   TXEN0     3
#define   UCSZ02    2
#define   RXB80     1
#define   TXB80     0
#define UCSR0A (*(volatile unsigned char *)0xC0) /* m/m */
#define   RXC0      7
#define   TXC0      6
#define   UDRE0     5
#define   FE0       4
#define   DOR0      3
#define   UPE0      2
#define   U2X0      1
#define   MPCM0     0
#define UDR0 (*(volatile unsigned char *)0xC6) /* m/m */
```

```
/* USART1 */
#define UBRR1H (*(volatile unsigned char *)0xCD) /* m/m */
#define UBRR1L (*(volatile unsigned char *)0xCC) /* m/m */
#define UBRR1 (*(volatile unsigned int *)0xCC) /* m/m */
#define UCSR1C (*(volatile unsigned char *)0xCA) /* m/m */
#define  UMSEL11  7
#define  UMSEL10  6
#define  UPM11    5
#define  UPM10    4
#define  USBS1    3
#define  UCSZ11   2
#define  UCSZ10   1
#define  UCPOL1   0
/* in SPI mode */
#define  UDORD1   2
#define  UCPHA1   1
#define UCSR1B (*(volatile unsigned char *)0xC9) /* m/m */
#define  RXCIE1   7
#define  TXCIE1   6
#define  UDRIE1   5
#define  RXEN1    4
#define  TXEN1    3
#define  UCSZ12   2
#define  RXB81    1
#define  TXB81    0
#define UCSR1A (*(volatile unsigned char *)0xC8) /* m/m */
#define  RXC1     7
#define  TXC1     6
#define  UDRE1    5
#define  FE1      4
#define  DOR1     3
#define  UPE1     2
#define  U2X1     1
#define  MPCM1    0
#define UDR1 (*(volatile unsigned char *)0xCE) /* m/m */

/* USART2 */
#define UBRR2H (*(volatile unsigned char *)0xD5) /* m/m */
```

```
#define UBRR2L (*(volatile unsigned char *)0xD4) /* m/m */
#define UBRR2 (*(volatile unsigned int *)0xD4) /* m/m */
#define UCSR2C (*(volatile unsigned char *)0xD2) /* m/m */
#define  UMSEL21  7
#define  UMSEL20  6
#define  UPM21    5
#define  UPM20    4
#define  USBS2    3
#define  UCSZ21   2
#define  UCSZ20   1
#define  UCPOL2   0
/* in SPI mode */
#define  UDORD2   2
#define  UCPHA2   1
#define UCSR2B (*(volatile unsigned char *)0xD1) /* m/m */
#define  RXCIE2   7
#define  TXCIE2   6
#define  UDRIE2   5
#define  RXEN2    4
#define  TXEN2    3
#define  UCSZ22   2
#define  RXB82    1
#define  TXB82    0
#define UCSR2A (*(volatile unsigned char *)0xD0) /* m/m */
#define  RXC2     7
#define  TXC2     6
#define  UDRE2    5
#define  FE2      4
#define  DOR2     3
#define  UPE2     2
#define  U2X2     1
#define  MPCM2    0
#define UDR2 (*(volatile unsigned char *)0xD6) /* m/m */

/* USART3 */
#define UBRR3H (*(volatile unsigned char *)0x135) /* m/m */
#define UBRR3L (*(volatile unsigned char *)0x134) /* m/m */
#define UBRR3 (*(volatile unsigned int *)0x134) /* m/m */
#define UCSR3C (*(volatile unsigned char *)0x132) /* m/m */
```

```
#define  UMSEL31   7
#define  UMSEL30   6
#define  UPM31     5
#define  UPM30     4
#define  USBS3     3
#define  UCSZ31    2
#define  UCSZ30    1
#define  UCPOL3    0
/* in SPI mode */
#define  UDORD3    2
#define  UCPHA3    1
#define UCSR3B (*(volatile unsigned char *)0x131) /* m/m */
#define  RXCIE3    7
#define  TXCIE3    6
#define  UDRIE3    5
#define  RXEN3     4
#define  TXEN3     3
#define  UCSZ32    2
#define  RXB83     1
#define  TXB83     0
#define UCSR3A (*(volatile unsigned char *)0x130) /* m/m */
#define  RXC3      7
#define  TXC3      6
#define  UDRE3     5
#define  FE3       4
#define  DOR3      3
#define  UPE3      2
#define  U2X3      1
#define  MPCM3     0
#define UDR3 (*(volatile unsigned char *)0x136) /* m/m */


/* 2-wire SI */
#define TWBR (*(volatile unsigned char *)0xB8) /* m/m */
#define TWSR (*(volatile unsigned char *)0xB9) /* m/m */
#define  TWPS1     1
#define  TWPS0     0
#define TWAR (*(volatile unsigned char *)0xBA) /* m/m */
#define  TWGCE     0
#define TWDR (*(volatile unsigned char *)0xBB) /* m/m */
```

```
#define TWCR (*(volatile unsigned char *)0xBC) /* m/m */
#define  TWINT    7
#define  TWEA     6
#define  TWSTA    5
#define  TWSTO    4
#define  TWWC     3
#define  TWEN     2
#define  TWIE     0
#define TWAMR (*(volatile unsigned char *)0xBD) /* m/m */


/* SPI */
#define SPCR (*(volatile unsigned char *)0x4C)
#define  SPIE     7
#define  SPE      6
#define  DORD     5
#define  MSTR     4
#define  CPOL     3
#define  CPHA     2
#define  SPR1     1
#define  SPR0     0
#define SPSR (*(volatile unsigned char *)0x4D)
#define  SPIF     7
#define  WCOL     6
#define  SPI2X    0
#define SPDR (*(volatile unsigned char *)0x4E)


/* EEPROM */
#define EECR (*(volatile unsigned char *)0x3F)
#define  EEPM1    5
#define  EEPM0    4
#define  EERIE    3
#define  EEMPE    2
#define  EEMWE    2
#define  EEPE     1
#define  EEWE     1
#define  EERE     0
#define EEDR (*(volatile unsigned char *)0x40)
#define EEAR (*(volatile unsigned int *)0x41)
#define EEARL (*(volatile unsigned char *)0x41)
```

```
#define EEARH (*(volatile unsigned char *)0x42)

/* GTCCR */
#define GTCCR (*(volatile unsigned char *)0x43)
#define   TSM       7
#define   PSRASY    1
#define   PSRSYNC   0

/* Watchdog Timer Control Register */
#define WDTCSR (*(volatile unsigned char *)0x60) /* m/m */
#define WDTCR (*(volatile unsigned char *)0x60) /* m/m */
#define   WDIF      7
#define   WDIE      6
#define   WDP3      5
#define   WDCE      4
#define   WDE       3
#define   WDP2      2
#define   WDP1      1
#define   WDP0      0

/* OCDR */
#define OCDR (*(volatile unsigned char *)0x51)
#define MONDR (*(volatile unsigned char *)0x51)

/* Timer/Counter 0 */
#define OCR0B (*(volatile unsigned char *)0x48)
#define OCR0A (*(volatile unsigned char *)0x47)
#define TCNT0 (*(volatile unsigned char *)0x46)
#define TCCR0B (*(volatile unsigned char *)0x45)
#define   FOC0      7
#define   FOC0A     7
#define   FOC0B     6
#define   WGM02     3
#define   CS02      2
#define   CS01      1
#define   CS00      0
#define TCCR0A (*(volatile unsigned char *)0x44)
#define   COM0A1    7
#define   COM0A0    6
```

```
#define  COM0B1   5
#define  COM0B0   4
#define  WGM01    1
#define  WGM00    0


/* Timer/Counter1 */
#define ICR1 (*(volatile unsigned int *)0x86) /* m/m */
#define ICR1L (*(volatile unsigned char *)0x86) /* m/m */
#define ICR1H (*(volatile unsigned char *)0x87) /* m/m */
#define OCR1C (*(volatile unsigned int *)0x8C) /* m/m */
#define OCR1CL (*(volatile unsigned char *)0x8C) /* m/m */
#define OCR1CH (*(volatile unsigned char *)0x8D) /* m/m */
#define OCR1B (*(volatile unsigned int *)0x8A) /* m/m */
#define OCR1BL (*(volatile unsigned char *)0x8A) /* m/m */
#define OCR1BH (*(volatile unsigned char *)0x8B) /* m/m */
#define OCR1A (*(volatile unsigned int *)0x88) /* m/m */
#define OCR1AL (*(volatile unsigned char *)0x88) /* m/m */
#define OCR1AH (*(volatile unsigned char *)0x89) /* m/m */
#define TCNT1 (*(volatile unsigned int *)0x84) /* m/m */
#define TCNT1L (*(volatile unsigned char *)0x84) /* m/m */
#define TCNT1H (*(volatile unsigned char *)0x85) /* m/m */
#define TCCR1C (*(volatile unsigned char *)0x82) /* m/m */
#define  FOC1A    7
#define  FOC1B    6
#define  FOC1C    5
#define TCCR1B (*(volatile unsigned char *)0x81) /* m/m */
#define  ICNC1    7
#define  ICES1    6
#define  WGM13    4
#define  WGM12    3
#define  CS12     2
#define  CS11     1
#define  CS10     0
#define TCCR1A (*(volatile unsigned char *)0x80) /* m/m */
#define  COM1A1   7
#define  COM1A0   6
#define  COM1B1   5
#define  COM1B0   4
#define  COM1C1   3
```

```
#define   COM1C0    2
#define   WGM11     1
#define   WGM10     0


/* Timer/Counter2 */
#define ASSR (*(volatile unsigned char *)0xB6) /* m/m */
#define   EXCLK     6
#define   AS2       5
#define   TCN2UB    4
#define   OCR2AUB   3
#define   OCR2BUB   2
#define   TCR2AUB   1
#define   TCR2BUB   0
#define OCR2B (*(volatile unsigned char *)0xB4) /* m/m */
#define OCR2A (*(volatile unsigned char *)0xB3) /* m/m */
#define TCNT2 (*(volatile unsigned char *)0xB2) /* m/m */
#define TCCR2B (*(volatile unsigned char *)0xB1) /* m/m */
#define   FOC2A     7
#define   FOC2B     6
#define   WGM22     3
#define   CS22      2
#define   CS21      1
#define   CS20      0
#define TCCR2A (*(volatile unsigned char *)0xB0) /* m/m */
#define   COM2A1    7
#define   COM2A0    6
#define   COM2B1    5
#define   COM2B0    4
#define   WGM21     1
#define   WGM20     0


/* Timer/Counter3 */
#define OCR3CH (*(volatile unsigned char *)0x9D) /* m/m */
#define OCR3CL (*(volatile unsigned char *)0x9C) /* m/m */
#define OCR3C (*(volatile unsigned int *)0x9C) /* m/m */
#define OCR3BH (*(volatile unsigned char *)0x9B) /* m/m */
#define OCR3BL (*(volatile unsigned char *)0x9A) /* m/m */
#define OCR3B (*(volatile unsigned int *)0x9A) /* m/m */
#define OCR3AH (*(volatile unsigned char *)0x99) /* m/m */
```

```
#define OCR3AL (*(volatile unsigned char *)0x98) /* m/m */
#define OCR3A (*(volatile unsigned int *)0x98) /* m/m */
#define ICR3H (*(volatile unsigned char *)0x97) /* m/m */
#define ICR3L (*(volatile unsigned char *)0x96) /* m/m */
#define ICR3 (*(volatile unsigned int *)0x96) /* m/m */
#define TCNT3H (*(volatile unsigned char *)0x95) /* m/m */
#define TCNT3L (*(volatile unsigned char *)0x94) /* m/m */
#define TCNT3 (*(volatile unsigned int *)0x94) /* m/m */
#define TCCR3C (*(volatile unsigned char *)0x92) /* m/m */
#define   FOC3A     7
#define   FOC3B     6
#define   FOC3C     5
#define TCCR3B (*(volatile unsigned char *)0x91) /* m/m */
#define   ICNC3     7
#define   ICES3     6
#define   WGM33     4
#define   WGM32     3
#define   CS32      2
#define   CS31      1
#define   CS30      0
#define TCCR3A (*(volatile unsigned char *)0x90) /* m/m */
#define   COM3A1    7
#define   COM3A0    6
#define   COM3B1    5
#define   COM3B0    4
#define   COM3C1    3
#define   COM3C0    2
#define   WGM31     1
#define   WGM30     0


/* Timer/Counter4 */
#define OCR4CH (*(volatile unsigned char *)0xAD) /* m/m */
#define OCR4CL (*(volatile unsigned char *)0xAC) /* m/m */
#define OCR4C (*(volatile unsigned int *)0xAC) /* m/m */
#define OCR4BH (*(volatile unsigned char *)0xAB) /* m/m */
#define OCR4BL (*(volatile unsigned char *)0xAA) /* m/m */
#define OCR4B (*(volatile unsigned int *)0xAA) /* m/m */
#define OCR4AH (*(volatile unsigned char *)0xA9) /* m/m */
#define OCR4AL (*(volatile unsigned char *)0xA8) /* m/m */
```

```c
#define OCR4A (*(volatile unsigned int *)0xA8) /* m/m */
#define ICR4H (*(volatile unsigned char *)0xA7) /* m/m */
#define ICR4L (*(volatile unsigned char *)0xA6) /* m/m */
#define ICR4 (*(volatile unsigned int *)0xA6) /* m/m */
#define TCNT4H (*(volatile unsigned char *)0xA5) /* m/m */
#define TCNT4L (*(volatile unsigned char *)0xA4) /* m/m */
#define TCNT4 (*(volatile unsigned int *)0xA4) /* m/m */
#define TCCR4C (*(volatile unsigned char *)0xA2) /* m/m */
#define   FOC4A    7
#define   FOC4B    6
#define   FOC4C    5
#define TCCR4B (*(volatile unsigned char *)0xA1) /* m/m */
#define   ICNC4    7
#define   ICES4    6
#define   WGM43    4
#define   WGM42    3
#define   CS42     2
#define   CS41     1
#define   CS40     0
#define TCCR4A (*(volatile unsigned char *)0xA0) /* m/m */
#define   COM4A1   7
#define   COM4A0   6
#define   COM4B1   5
#define   COM4B0   4
#define   COM4C1   3
#define   COM4C0   2
#define   WGM41    1
#define   WGM40    0


/* Timer/Counter5 */
#define OCR5CH (*(volatile unsigned char *)0x12D) /* m/m */
#define OCR5CL (*(volatile unsigned char *)0x12C) /* m/m */
#define OCR5C (*(volatile unsigned int *)0x12C) /* m/m */
#define OCR5BH (*(volatile unsigned char *)0x12B) /* m/m */
#define OCR5BL (*(volatile unsigned char *)0x12A) /* m/m */
#define OCR5B (*(volatile unsigned int *)0x12A) /* m/m */
#define OCR5AH (*(volatile unsigned char *)0x129) /* m/m */
#define OCR5AL (*(volatile unsigned char *)0x128) /* m/m */
#define OCR5A (*(volatile unsigned int *)0x128) /* m/m */
```

```
#define ICR5H (*(volatile unsigned char *)0x127) /* m/m */
#define ICR5L (*(volatile unsigned char *)0x126) /* m/m */
#define ICR5 (*(volatile unsigned int *)0x126) /* m/m */
#define TCNT5H (*(volatile unsigned char *)0x125) /* m/m */
#define TCNT5L (*(volatile unsigned char *)0x124) /* m/m */
#define TCNT5 (*(volatile unsigned int *)0x124) /* m/m */
#define TCCR5C (*(volatile unsigned char *)0x122) /* m/m */
#define   FOC5A     7
#define   FOC5B     6
#define   FOC5C     5
#define TCCR5B (*(volatile unsigned char *)0x121) /* m/m */
#define   ICNC5     7
#define   ICES5     6
#define   WGM53     4
#define   WGM52     3
#define   CS52      2
#define   CS51      1
#define   CS50      0
#define TCCR5A (*(volatile unsigned char *)0x120) /* m/m */
#define   COM5A1    7
#define   COM5A0    6
#define   COM5B1    5
#define   COM5B0    4
#define   COM5C1    3
#define   COM5C0    2
#define   WGM51     1
#define   WGM50     0


/* Oscillator Calibration Register */
#define OSCCAL (*(volatile unsigned char *)0x66) /* m/m */


/* clock prescaler control register */
#define CLKPR (*(volatile unsigned char *)0x61) /* m/m */
#define   CLKPCE    7
#define   CLKPS3    3
#define   CLKPS2    2
#define   CLKPS1    1
#define   CLKPS0    0
```

```
/* PRR */
#define PRR0 (*(volatile unsigned char *)0x64) /* m/m */
#define   PRTWI     7
#define   PRTIM2    6
#define   PRTIM0    5
#define   PRTIM1    3
#define   PRSPI     2
#define   PRUSART0 1
#define   PRADC     0
#define PRR1 (*(volatile unsigned char *)0x65) /* m/m */
#define   PRTIM5    5
#define   PRTIM4    4
#define   PRTIM3    3
#define   PRUSART3 2
#define   PRUSART2 1
#define   PRUSART1 0

/* MCU */
#define MCUSR (*(volatile unsigned char *)0x54)
#define   JTRF      4
#define   WDRF      3
#define   BORF      2
#define   EXTRF     1
#define   PORF      0
#define MCUCR (*(volatile unsigned char *)0x55)
#define   JTD       7
#define   PUD       4
#define   IVSEL     1
#define   IVCE      0

#define SMCR (*(volatile unsigned char *)0x53)
#define   SM2       3
#define   SM1       2
#define   SM0       1
#define   SE        0

/* SPM Control and Status Register */
#define SPMCSR (*(volatile unsigned char *)0x57)
#define   SPMIE     7
```

```
#define  RWWSB      6
#define  RWWSRE     4
#define  BLBSET     3
#define  PGWRT      2
#define  PGERS      1
#define  SPMEN      0


/* Timer/Counter Interrupts */
#define TIFR0 (*(volatile unsigned char *)0x35)
#define  OCF0B      2
#define  OCF0A      1
#define  TOV0       0
#define TIMSK0 (*(volatile unsigned char *)0x6E) /* m/m */
#define  OCIE0B     2
#define  OCIE0A     1
#define  TOIE0      0
#define TIFR1 (*(volatile unsigned char *)0x36)
#define  ICF1       5
#define  OCF1C      3
#define  OCF1B      2
#define  OCF1A      1
#define  TOV1       0
#define TIMSK1 (*(volatile unsigned char *)0x6F) /* m/m */
#define  ICIE1      5
#define  OCIE1C     3
#define  OCIE1B     2
#define  OCIE1A     1
#define  TOIE1      0
#define TIFR2 (*(volatile unsigned char *)0x37)
#define  OCF2B      2
#define  OCF2A      1
#define  TOV2       0
#define TIMSK2 (*(volatile unsigned char *)0x70) /* m/m */
#define  OCIE2B     2
#define  OCIE2A     1
#define  TOIE2      0
#define TIFR3 (*(volatile unsigned char *)0x38)
#define  ICF3       5
#define  OCF3C      3
```

```
#define   OCF3B      2
#define   OCF3A      1
#define   TOV3       0
#define TIMSK3 (*(volatile unsigned char *)0x71) /* m/m */
#define   ICIE3      5
#define   OCIE3C     3
#define   OCIE3B     2
#define   OCIE3A     1
#define   TOIE3      0
#define TIFR4 (*(volatile unsigned char *)0x39)
#define   ICF4       5
#define   OCF4C      3
#define   OCF4B      2
#define   OCF4A      1
#define   TOV4       0
#define TIMSK4 (*(volatile unsigned char *)0x72) /* m/m */
#define   ICIE4      5
#define   OCIE4C     3
#define   OCIE4B     2
#define   OCIE4A     1
#define   TOIE4      0
#define TIFR5 (*(volatile unsigned char *)0x3A)
#define   ICF5       5
#define   OCF5C      3
#define   OCF5B      2
#define   OCF5A      1
#define   TOV5       0
#define TIMSK5 (*(volatile unsigned char *)0x73) /* m/m */
#define   ICIE5      5
#define   OCIE5C     3
#define   OCIE5B     2
#define   OCIE5A     1
#define   TOIE5      0

/* External Interrupts */
#define EIFR (*(volatile unsigned char *)0x3C)
#define   INTF7      7
#define   INTF6      6
#define   INTF5      5
```

```
#define   INTF4      4
#define   INTF3      3
#define   INTF2      2
#define   INTF1      1
#define   INTF0      0
#define EIMSK (*(volatile unsigned char *)0x3D)
#define   INT7       7
#define   INT6       6
#define   INT5       5
#define   INT4       4
#define   INT3       3
#define   INT2       2
#define   INT1       1
#define   INT0       0
#define EICRB (*(volatile unsigned char *)0x6A) /* m/m */
#define   ISC71      7
#define   ISC70      6
#define   ISC61      5
#define   ISC60      4
#define   ISC51      3
#define   ISC50      2
#define   ISC41      1
#define   ISC40      0
#define EICRA (*(volatile unsigned char *)0x69) /* m/m */
#define   ISC31      7
#define   ISC30      6
#define   ISC21      5
#define   ISC20      4
#define   ISC11      3
#define   ISC10      2
#define   ISC01      1
#define   ISC00      0

/* Pin Change Interrupts */
#define PCIFR (*(volatile unsigned char *)0x3B)
#define   PCIF2      2
#define   PCIF1      1
#define   PCIF0      0
#define PCICR (*(volatile unsigned char *)0x68) /* m/m */
```

```
#define  PCIE2     2
#define  PCIE1     1
#define  PCIE0     0
#define PCMSK0 (*(volatile unsigned char *)0x6B) /* m/m */
#define PCMSK1 (*(volatile unsigned char *)0x6C) /* m/m */
#define PCMSK2 (*(volatile unsigned char *)0x6D) /* m/m */


/* eXternal Memory Control Register */
#define XMCRB (*(volatile unsigned char *)0x75) /* m/m */
#define  XMBK      7
#define  XMM2      2
#define  XMM1      1
#define  XMM0      0
#define XMCRA (*(volatile unsigned char *)0x74) /* m/m */
#define  SRE       7
#define  SRL2      6
#define  SRL1      5
#define  SRL0      4
#define  SRW11     3
#define  SRW10     2
#define  SRW01     1
#define  SRW00     0


/* GPIO */
#define GPIOR0 (*(volatile unsigned char *)0x3E)
#define GPIOR1 (*(volatile unsigned char *)0x4A)
#define GPIOR2 (*(volatile unsigned char *)0x4B)


/* RAM page Z-pointer */
#define RAMPZ (*(volatile unsigned char *)0x5B)


/* EIND */
#define EIND (*(volatile unsigned char *)0x5C)


/* Stack Pointer */
#define SP   (*(volatile unsigned int *)0x5D)
#define SPL  (*(volatile unsigned char *)0x5D)
#define SPH  (*(volatile unsigned char *)0x5E)
```

```
/* Status REGister */
#define SREG (*(volatile unsigned char *)0x5F)



/* bits */

/* Port A */
#define  PORTA7    7
#define  PORTA6    6
#define  PORTA5    5
#define  PORTA4    4
#define  PORTA3    3
#define  PORTA2    2
#define  PORTA1    1
#define  PORTA0    0
#define  PA7       7
#define  PA6       6
#define  PA5       5
#define  PA4       4
#define  PA3       3
#define  PA2       2
#define  PA1       1
#define  PA0       0
#define  DDA7      7
#define  DDA6      6
#define  DDA5      5
#define  DDA4      4
#define  DDA3      3
#define  DDA2      2
#define  DDA1      1
#define  DDA0      0
#define  PINA7     7
#define  PINA6     6
#define  PINA5     5
#define  PINA4     4
#define  PINA3     3
#define  PINA2     2
#define  PINA1     1
#define  PINA0     0
```

```
/* Port B */
#define   PORTB7   7
#define   PORTB6   6
#define   PORTB5   5
#define   PORTB4   4
#define   PORTB3   3
#define   PORTB2   2
#define   PORTB1   1
#define   PORTB0   0
#define   PB7      7
#define   PB6      6
#define   PB5      5
#define   PB4      4
#define   PB3      3
#define   PB2      2
#define   PB1      1
#define   PB0      0
#define   DDB7     7
#define   DDB6     6
#define   DDB5     5
#define   DDB4     4
#define   DDB3     3
#define   DDB2     2
#define   DDB1     1
#define   DDB0     0
#define   PINB7    7
#define   PINB6    6
#define   PINB5    5
#define   PINB4    4
#define   PINB3    3
#define   PINB2    2
#define   PINB1    1
#define   PINB0    0

/* Port C */
#define   PORTC7   7
#define   PORTC6   6
#define   PORTC5   5
```

```
#define  PORTC4   4
#define  PORTC3   3
#define  PORTC2   2
#define  PORTC1   1
#define  PORTC0   0
#define  PC7      7
#define  PC6      6
#define  PC5      5
#define  PC4      4
#define  PC3      3
#define  PC2      2
#define  PC1      1
#define  PC0      0
#define  DDC7     7
#define  DDC6     6
#define  DDC5     5
#define  DDC4     4
#define  DDC3     3
#define  DDC2     2
#define  DDC1     1
#define  DDC0     0
#define  PINC7    7
#define  PINC6    6
#define  PINC5    5
#define  PINC4    4
#define  PINC3    3
#define  PINC2    2
#define  PINC1    1
#define  PINC0    0

/* Port D */
#define  PORTD7   7
#define  PORTD6   6
#define  PORTD5   5
#define  PORTD4   4
#define  PORTD3   3
#define  PORTD2   2
#define  PORTD1   1
#define  PORTD0   0
```

```
#define   PD7        7
#define   PD6        6
#define   PD5        5
#define   PD4        4
#define   PD3        3
#define   PD2        2
#define   PD1        1
#define   PD0        0
#define   DDD7       7
#define   DDD6       6
#define   DDD5       5
#define   DDD4       4
#define   DDD3       3
#define   DDD2       2
#define   DDD1       1
#define   DDD0       0
#define   PIND7      7
#define   PIND6      6
#define   PIND5      5
#define   PIND4      4
#define   PIND3      3
#define   PIND2      2
#define   PIND1      1
#define   PIND0      0

/* Port E */
#define   PORTE7     7
#define   PORTE6     6
#define   PORTE5     5
#define   PORTE4     4
#define   PORTE3     3
#define   PORTE2     2
#define   PORTE1     1
#define   PORTE0     0
#define   PE7        7
#define   PE6        6
#define   PE5        5
#define   PE4        4
#define   PE3        3
```

```
#define   PE2      2
#define   PE1      1
#define   PE0      0
#define   DDE7     7
#define   DDE6     6
#define   DDE5     5
#define   DDE4     4
#define   DDE3     3
#define   DDE2     2
#define   DDE1     1
#define   DDE0     0
#define   PINE7    7
#define   PINE6    6
#define   PINE5    5
#define   PINE4    4
#define   PINE3    3
#define   PINE2    2
#define   PINE1    1
#define   PINE0    0

/* Port F */
#define   PORTF7   7
#define   PORTF6   6
#define   PORTF5   5
#define   PORTF4   4
#define   PORTF3   3
#define   PORTF2   2
#define   PORTF1   1
#define   PORTF0   0
#define   PF7      7
#define   PF6      6
#define   PF5      5
#define   PF4      4
#define   PF3      3
#define   PF2      2
#define   PF1      1
#define   PF0      0
#define   DDF7     7
#define   DDF6     6
```

```
#define   DDF5       5
#define   DDF4       4
#define   DDF3       3
#define   DDF2       2
#define   DDF1       1
#define   DDF0       0
#define   PINF7      7
#define   PINF6      6
#define   PINF5      5
#define   PINF4      4
#define   PINF3      3
#define   PINF2      2
#define   PINF1      1
#define   PINF0      0

/* Port G */
#define   PORTG5     5
#define   PORTG4     4
#define   PORTG3     3
#define   PORTG2     2
#define   PORTG1     1
#define   PORTG0     0
#define   PG5        5
#define   PG4        4
#define   PG3        3
#define   PG2        2
#define   PG1        1
#define   PG0        0
#define   DDG5       5
#define   DDG4       4
#define   DDG3       3
#define   DDG2       2
#define   DDG1       1
#define   DDG0       0
#define   PING5      5
#define   PING4      4
#define   PING3      3
#define   PING2      2
#define   PING1      1
```

```c
#define  PING0     0

/* Port H */
#define  PORTH7    7
#define  PORTH6    6
#define  PORTH5    5
#define  PORTH4    4
#define  PORTH3    3
#define  PORTH2    2
#define  PORTH1    1
#define  PORTH0    0
#define  PH7       7
#define  PH6       6
#define  PH5       5
#define  PH4       4
#define  PH3       3
#define  PH2       2
#define  PH1       1
#define  PH0       0
#define  DDH7      7
#define  DDH6      6
#define  DDH5      5
#define  DDH4      4
#define  DDH3      3
#define  DDH2      2
#define  DDH1      1
#define  DDH0      0
#define  PINH7     7
#define  PINH6     6
#define  PINH5     5
#define  PINH4     4
#define  PINH3     3
#define  PINH2     2
#define  PINH1     1
#define  PINH0     0

/* Port J */
#define  PORTJ7    7
#define  PORTJ6    6
```

483

```
                             5
                             4
            .3               3
           .TJ2             2
        ORTJ1              1
        PORTJ0             0
    .e  PJ7               7
   ine  PJ6               6
 .efine PJ5               5
#define PJ4               4
#define PJ3               3
#define PJ2               2
#define PJ1               1
#define PJ0               0
#define DDJ7              7
#define DDJ6              6
#define DDJ5              5
#define DDJ4              4
#define DDJ3              3
#define DDJ2              2
#define DDJ1              1
#define DDJ0              0
#define PINJ7             7
#define PINJ6             6
#define PINJ5             5
#define PINJ4             4
#define PINJ3             3
#define PINJ2             2
#define PINJ1             1
#define PINJ0             0

/* Port K */
#define PORTK7            7
#define PORTK6            6
#define PORTK5            5
#define PORTK4            4
#define PORTK3            3
#define PORTK2            2
#define PORTK1            1
```

```
#define   PORTK0    0
#define   PK7       7
#define   PK6       6
#define   PK5       5
#define   PK4       4
#define   PK3       3
#define   PK2       2
#define   PK1       1
#define   PK0       0
#define   DDK7      7
#define   DDK6      6
#define   DDK5      5
#define   DDK4      4
#define   DDK3      3
#define   DDK2      2
#define   DDK1      1
#define   DDK0      0
#define   PINK7     7
#define   PINK6     6
#define   PINK5     5
#define   PINK4     4
#define   PINK3     3
#define   PINK2     2
#define   PINK1     1
#define   PINK0     0

/* Port L */
#define   PORTL7    7
#define   PORTL6    6
#define   PORTL5    5
#define   PORTL4    4
#define   PORTL3    3
#define   PORTL2    2
#define   PORTL1    1
#define   PORTL0    0
#define   PL7       7
#define   PL6       6
#define   PL5       5
#define   PL4       4
```

```
485                            3
                               2
                               1
             ᴗ                 0
           ᴗDL7                7
           DDL6                6
       ᴗe  DDL5               5
     ᴗine  DDL4                4
   ᴗefine  DDL3                3
#define   DDL2                2
#define   DDL1                1
#define   DDL0                0
#define   PINL7               7
#define   PINL6               6
#define   PINL5               5
#define   PINL4               4
#define   PINL3               3
#define   PINL2               2
#define   PINL1               1
#define   PINL0               0

/* PCMSK2 */
#define   PCINT23             7
#define   PCINT22             6
#define   PCINT21             5
#define   PCINT20             4
#define   PCINT19             3
#define   PCINT18             2
#define   PCINT17             1
#define   PCINT16             0
/* PCMSK1 */
#define   PCINT15             7
#define   PCINT14             6
#define   PCINT13             5
#define   PCINT12             4
#define   PCINT11             3
#define   PCINT10             2
#define   PCINT9              1
#define   PCINT8              0
```

```c
/* PCMSK0 */
#define  PCINT7    7
#define  PCINT6    6
#define  PCINT5    5
#define  PCINT4    4
#define  PCINT3    3
#define  PCINT2    2
#define  PCINT1    1
#define  PCINT0    0


/* Lock and Fuse Bits with LPM/SPM instructions */

/* lock bits */
#define  BLB12     5
#define  BLB11     4
#define  BLB02     3
#define  BLB01     2
#define  LB2       1
#define  LB1       0

/* fuses low bits */
#define  CKDIV8    7
#define  CKOUT     6
#define  SUT1      5
#define  SUT0      4
#define  CKSEL3    3
#define  CKSEL2    2
#define  CKSEL1    1
#define  CKSEL0    0

/* fuses high bits */
#define  OCDEN     7
#define  JTAGEN    6
#define  SPIEN     5
#define  WDTON     4
#define  EESAVE    3
#define  BOOTSZ1   2
#define  BOOTSZ0   1
```

es */
LEVEL2 2
DLEVEL1 1
BODLEVEL0 0

Interrupt Vector Numbers */

```
#define iv_RESET          1
#define iv_INT0           2
#define iv_EXT_INT0       2
#define iv_INT1           3
#define iv_EXT_INT1       3
#define iv_INT2           4
#define iv_EXT_INT2       4
#define iv_INT3           5
#define iv_EXT_INT3       5
#define iv_INT4           6
#define iv_EXT_INT4       6
#define iv_INT5           7
#define iv_EXT_INT5       7
#define iv_INT6           8
#define iv_EXT_INT6       8
#define iv_INT7           9
#define iv_EXT_INT7       9
#define iv_PCINT0         10
#define iv_PCINT1         11
#define iv_PCINT2         12
#define iv_WDT            13
#define iv_TIMER2_COMPA   14
#define iv_TIMER2_COMPB   15
#define iv_TIMER2_OVF     16
#define iv_TIM2_COMPA     14
#define iv_TIM2_COMPB     15
#define iv_TIM2_OVF       16
#define iv_TIMER1_CAPT    17
#define iv_TIMER1_COMPA   18
```

```c
#define iv_TIMER1_COMPB 19
#define iv_TIMER1_COMPC 20
#define iv_TIMER1_OVF   21
#define iv_TIM1_CAPT    17
#define iv_TIM1_COMPA   18
#define iv_TIM1_COMPB   19
#define iv_TIM1_COMPC   20
#define iv_TIM1_OVF     21
#define iv_TIMER0_COMPA 22
#define iv_TIMER0_COMPB 23
#define iv_TIMER0_OVF   24
#define iv_TIM0_COMPA   22
#define iv_TIM0_COMPB   23
#define iv_TIM0_OVF     24
#define iv_SPI_STC      25
#define iv_USART0_RX    26
#define iv_USART0_RXC   26
#define iv_USART0_DRE   27
#define iv_USART0_UDRE  27
#define iv_USART0_TX    28
#define iv_USART0_TXC   28
#define iv_ANA_COMP     29
#define iv_ANALOG_COMP  29
#define iv_ADC          30
#define iv_EE_RDY       31
#define iv_EE_READY     31
#define iv_TIMER3_CAPT  32
#define iv_TIMER3_COMPA 33
#define iv_TIMER3_COMPB 34
#define iv_TIMER3_COMPC 35
#define iv_TIMER3_OVF   36
#define iv_TIM3_CAPT    32
#define iv_TIM3_COMPA   33
#define iv_TIM3_COMPB   34
#define iv_TIM3_COMPC   35
#define iv_TIM3_OVF     36
#define iv_USART1_RX    37
#define iv_USART1_RXC   37
#define iv_USART1_DRE   38
```

```
                      JRE      38
               _TX             39
            T1_TXC             39
          T                    40
        TWSI                   40
      .v_SPM_RDY               41
    iv_SPM_READY               41
ne  iv_TIMER4_CAPT            42
fine  iv_TIMER4_COMPA         43
define  iv_TIMER4_COMPB       44
#define  iv_TIMER4_COMPC       45
#define  iv_TIMER4_OVF         46
#define  iv_TIM4_CAPT          42
#define  iv_TIM4_COMPA         43
#define  iv_TIM4_COMPB         44
#define  iv_TIM4_COMPC         45
#define  iv_TIM4_OVF           46
#define  iv_TIMER5_CAPT        47
#define  iv_TIMER5_COMPA       48
#define  iv_TIMER5_COMPB       49
#define  iv_TIMER5_COMPC       50
#define  iv_TIMER5_OVF         51
#define  iv_TIM5_CAPT          47
#define  iv_TIM5_COMPA         48
#define  iv_TIM5_COMPB         49
#define  iv_TIM5_COMPC         50
#define  iv_TIM5_OVF           51
#define  iv_USART2_RX          52
#define  iv_USART2_RXC         52
#define  iv_USART2_DRE         53
#define  iv_USART2_UDRE        53
#define  iv_USART2_TX          54
#define  iv_USART2_TXC         54
#define  iv_USART3_RX          55
#define  iv_USART3_RXC         55
#define  iv_USART3_DRE         56
#define  iv_USART3_UDRE        56
#define  iv_USART3_TX          57
#define  iv_USART3_TXC         57
```

```
/*ja*/

#endif


#endif

//***************************************************
```

491

Author's Biography

STEVEN F. BARRETT

Steven F. Barrett, Ph.D., P.E., received the BS Electronic Engineering Technology from the University of Nebraska at Omaha in 1979, the M.E.E.E. from the University of Idaho at Moscow in 1986, and the Ph.D. from The University of Texas at Austin in 1993. He was formally an active duty faculty member at the United States Air Force Academy, Colorado and is now the Associate Dean of Academic Programs at the University of Wyoming. He is a member of IEEE (senior) and Tau Beta Pi (chief faculty advisor). His research interests include digital and analog image processing, computer-assisted laser surgery, and embedded controller systems. He is a registered Professional Engineer in Wyoming and Colorado. He Co–wrote with Dr. Daniel Pack several textbooks on microcontrollers and embedded systems. In 2004, Barrett was named "Wyoming Professor of the Year" by the Carnegie Foundation for the Advancement of Teaching and in 2008 was the recipient of the National Society of Professional Engineers (NSPE) Professional Engineers in Higher Education, Engineering Education Excellence Award.

493